MOLECULAR MECHANISMS OF TRANSPORT

DEVELOPMENTS IN BIOCHEMISTRY

Volume 1 – Mechanisms of Oxidizing Enzymes, edited by Thomas P. Singer and Raul N. Ondarza, 1978
Volume 2 – Electrophoresis '78, edited by Nicholas Catsimpoolas, 1978
Volume 3 – Physical Aspects of Protein Interactions, edited by Nicholas Catsimpoolas, 1978
Volume 4 – Chemistry and Biology of Pteridines, edited by Roy L. Kisliuk and Gene M. Brown, 1989
Volume 5 – Cytochrome Oxidase, edited by Tsoo E. King, Yutaka Orii, Britton Chance and Kazuo Okunuki, 1979
Volume 6 – Drug Action and Design. Mechanism-Based Enzyme Inhibitors, edited by Thomas I. Kalman, 1979
Volume 7 – Electrofocus/78, edited by Herman Haglund, John G. Westerfeld and Jack T. Ball, Jr., 1979
Volume 8 – The Regulation of Coagulation, edited by Kenneth G. Mann and Fletcher B. Taylor, Jr., 1980
Volume 9 – Red Blood Cell and Lens Metabolism, edited by Satish K. Srivastava, 1980
Volume 10 – Frontiers in Protein Chemistry, edited by Teh-Yung Liu, Gunji Mamiya and Kerry T. Yasunobu, 1980
Volume 11 – Chemical and Biochemical Aspects of Superoxide and Superoxide Dismutase, edited by J.V. Bannister and H.A.O. Hill, 1980
Volume 12 – Biological and Clinical Aspects of Superoxide and Superoxide Dismutase, edited by W.H. Bannister and J.V. Bannister, 1980
Volume 13 – Biochemistry, Biophysics and Regulation of Cytochrome P-450, edited by Jan-Åke Gustafsson, Jan Carlstedt-Duke, Agnela Mode and Joseph Rafter, 1980
Volume 14 – Calcium-Binding Proteins: Structure and Function, edited by Frank L. Siegel, Ernesto Carafoli, Robert H. Kretsinger, David H. MacLennan and Robert H. Wasserman, 1980
Volume 15 – Gene Families of Collagen and Other Proteins, edited by Darwin J. Prockop and Pamela C. Champe, 1980
Volume 16 – Biochemical and Medical Aspects of Trytophan Metabolism, edited by Osamu Hayaishi, Yuzuru Ishimura and Ryo Kido, 1980
Volume 17 – Chemical Synthesis and Sequencing of Peptides and Proteins, edited by Teh-Yung Liu, Alan N. Schechter, Robert L. Heinrickson and Peter G. Condliffe, 1981
Volume 18 – Metabolism and Clinical Implications of Branched Chain Amino Acids, edited by Mackenzie Walser and John Williamson, 1981
Volume 19 – Molecular Basis of Drug Action, edited by Thomas P. Singer and Raul N. Ondarza, 1981
Volume 20 – Energy Coupling in Photosynthesis, edited by Bruce R. Selman and Suzanne Selman-Reimer, 1981
Volume 21 – Flavins and Flavoproteins, edited by Vincent Massey and Charles H. Williams, 1982
Volume 22 – The Chemistry and Biology of Mineralized Connective Tissues, edited by Artur Veis, 1982
Volume 23 – Cytochrome P-450, Biochemistry, Biophysics and Environmental Implications, edited by Eino Hietanen, Matti Laitinen and Osmo Hänninen, 1982
Volume 24 – Interaction of Translational and Transcriptional Controls in the Regulation of Gene Expression, edited by Marianne Grunberg-Manago and Brian Safer, 1982
Volume 25 – Calcium-Binding Proteins, 1983, edited by B. de Bernard, G.L. Sottocasa, G. Sandri, E. Carafoli, A.N. Taylor, T.C. Vanaman and R.J.P. Williams, 1983
Volume 26 – The Biology and Chemistry of Active Oxygen, edited by J.V. Bannister and W.H. Bannister, 1984
Volume 27 – Cytochrome P-450, Biochemistry, Biophysics and Induction, edited by L. Vereczkey and K. Magyar, 1985
Volume 28 – Structure, Function and Biogenesis of Energy Transfer Systems, edited by E. Quagliariello, S. Papa, F. Palmieri and C. Saccone, 1990
Volume 29 – Molecular mechanisms of transport, edited by E. Quagliariello and F. Palmieri, 1992

MOLECULAR MECHANISMS OF TRANSPORT

Proceedings of the 22nd Bari Meeting on Bioenergetics, International Symposium on Molecular Mechanisms of Transport, Bari, Italy, 29 September – 1 October 1991

Editors:

ERNESTO QUAGLIARIELLO
Department of Biochemistry and Molecular Biology
University of Bari
Bari, Italy

FERDINANDO PALMIERI
Department of Pharmaco-Biology
University of Bari
Bari, Italy

1992

ELSEVIER SCIENCE PUBLISHERS
AMSTERDAM · NEW YORK · LONDON · TOKYO

WILLIAM MADISON RANDALL LIBRARY UNC
AT WILMINGTON

© 1992 Elsevier Science Publishers B.V. All rights reserved.

No part of this publication may be reproduced, stored in a retrieval system or transmitted in any form or by any means, electronic, mechanical, photocopying, recording or otherwise without the prior written permission of the publisher, Elsevier Science Publishers B.V., Permissions Department, P.O. Box 521, 1000 AM Amsterdam, The Netherlands.

No responsibility is assumed by the Publisher for any injury and/or damage to persons or property as a matter of products liability, negligence or otherwise, or from any use or operation of any methods, products, instructions or ideas contained in the material herein. Because of rapid advances in the medical sciences, the Publisher recommends that independent verification of diagnoses and drug dosages should be made.

Special regulations for readers in the USA – This publication has been registered with the Copyright Clearance Center Inc. (CCC), 27 Congress Street, Salem, MA 01970, USA. Information can be obtained from the CCC about conditions under which photocopies of parts of this publication may be made in the USA. All other copyright questions, including photocopying outside the USA, should be referred to the copyright owner, Elsevier Science Publishers B.V., unless otherwise specified.

ISSN 0165-1714
ISBN 0 444 89408 X (Vol. 29)

This book is printed on acid-free paper.

Published by:
Elsevier Science Publishers B.V.
(Biomedical Division)
P.O. Box 211
1000 AE Amsterdam
The Netherlands

Sole distributors for the USA and Canada:
Elsevier Science Publishing Company Inc.
655 Avenue of the Americas
New York, NY 10010
USA

Library of Congress Cataloging in Publication Data:

Printed in The Netherlands

QH
509
.I 635
1991

PREFACE

This volume comprises the proceedings of the International Symposium on 'Molecular Mechanisms of Transport' which was held at Selva di Fasano near Bari, September 29 – October 1, 1991. This symposium is the 22nd of the 'Bari Meetings' that began in Bari in 1965 with the symposium 'Regulation of Mitochondrial Processes in Mitochondria'. The organizers feel that the old-style 'Bari-meetings' are still very stimulating scientific events with, not too broad topics, limited numbers of participants, and with emphasis on the presentation of papers by young members of the biochemical community. This volume contains the lectures presented at the symposium.

The topic of this meeting represents a particularly timely and, at present, exciting area. The vast field of biomembrane transport has, in recent years, made great strides towards a better understanding of its mechanisms by the molecular approach. More transport systems are being defined in terms of their catalysts, pumping systems and carriers. Also the description of transport has advanced to a better understanding of its regulation, control and the mechanism by which energy drives transport.

It is felt that the present proceedings represents a comprehensive and up-to-date record of the biochemical aspects of biomembrane transport. Major achievements as well as new openings in the field have been stressed in many of the contributions to this book. Thus, it represents a valuable source and reference book, comprising the most recent results in this area. Since transport systems in a variety of membranes such as cell membranes, mitochondria and bacteria are discussed, it should attract the attention of scientists from various fields, who are interested in biological transport.

The volume has been published as quickly as possible thanks to the efforts of the publishers whose co-operation is gratefully acknowledged.

The Editors

CONTENTS

METHODS OF STRUCTURAL ANALYSIS

© 1992 Elsevier Science Publishers B.V. All rights reserved.
Molecular mechanisms of transport. E. Quagliariello, F. Palmieri, eds.

SPECTROSCOPIC STUDIES OF MEMBRANE PROTEINS AND ASSOCIATED POLYPEPTIDES

Parvez I. Haris and Dennis Chapman

Department of Protein and Molecular Biology
Royal Free Hospital School of Medicine
University of London
Rowland Hill Street
London, NW3 2PF, United Kingdom

INTRODUCTION

Elucidation of the structure of membrane proteins remains an important challenge in biochemistry. Many important biological processes such as the transport of ions and molecules are mediated by membrane spanning proteins. Rapid advances in molecular biology are providing a wealth of information about the amino acid sequence of these proteins. However, as detailed structural data on these proteins are scarce, the molecular mechanism of such biological processes are as yet little understood. Application of X-ray diffraction and nuclear magnetic resonance (NMR) spectroscopy to the structural analysis of membrane proteins are being made but at a slow pace. We are exploring the technique of Fourier Transform Infrared (FT-IR) spectroscopy for studying proteins and polypeptides in a membrane environment.

Infrared spectroscopy is one of the earliest techniques to be applied to the study of protein structure. The potential of the technique was first demonstrated by the pioneering work of Elliott and Ambrose [1]. Initially, the problems encountered in the study of biomolecules included the low sensitivity of infrared spectrometers, absorption of liquid water over much of the infrared spectrum as well as difficulties in extracting information from the broad overlapping infrared bands. The advent of computerised FT-IR instrumentation has largely overcome most of these problems. It is now possible to obtain high signal-to-noise ratio spectra of dilute (1mM) protein solutions in H_2O. The background H_2O absorption can be digitally subtracted from the spectrum of the protein solution. Furthermore, the broad overlapping bands can be examined in detail by the use of several mathematical procedures including deconvolution and second-derivative methods [2,3,4].

MEMBRANE PROTEINS

FT-IR spectroscopic studies of a large number of membrane proteins have been performed. These include, bacteriorhodopsin, rhodopsin, Ca^{2+}-ATPase, Na^+/K^+-ATPase, H^+/K^+-ATPase, photosynthetic reaction centres, cytochrome c oxidase, acetylcholine receptor and porin [3,4,5]. A particular feature of these membrane systems is the remarkable consistency in the frequency of the main amide I band associated with the α-helical structure present, except for the membrane proteins bacteriorhodopsin and porin. The latter is known to differ from other membrane proteins in having a ß-barrel structure [2,3,5]. Bacteriorhodpsin is an integral membrane protein found in the purple membrane of *Halobacterium halobium*. It is a light driven proton pump consisting of 248 amino acid residues with an *all-trans* retinal chromophore. FT-IR spectroscopic studies conducted in our laboratory as well as other laboratories have shown that the amide I frequency of this membrane protein is unusually high compared with what is expected for normal α-helical structure [6,7,8]. This high frequency band for bacteriorhodopsin has been attempted to be explained in terms of α_{II}-helices [7] or the presence of short 3_{10}-helical regions in addition to normal α-helices [8].

Hydrogen-deuterium exchange of membrane proteins studied using FT-IR spectroscopy has also been valuable for understanding membrane protein structure. A large number of proteins have been examined [2,3]. These include bacteriorhodopsin, rhodopsin, glucose transporter etc. The glucose transporter protein is found to have a remarkably high rate of hydrogen-deuterium exchange [9]. Almost 90% of its amide protons are exchanged within one hour. This was explained in terms of an aqueous pore present in this protein [9], an interpretation supported by other techniques [10].

Studies have also been performed to investigate the effect of ligands, lipid composition, pH and temperature as well as drugs on the structure of membrane proteins [2,3,4]. In a recent study for example we demonstrated dramatic changes in the infrared spectra of photosystem II reaction centre of higher plants as a function of light and temperature [11].

Information obtained using derivative and deconvolution methods applied to FT-IR spectra of proteins have been mainly of a qualitative nature. Quantitative studies have been made by curve-fitting analysis of the amide I band [2,3,4]. However, this approach has some difficulties and recent studies in our laboratory have led to the development of a more satisfactory procedure [12]. This method involves factor analysis of the infrared spectra of 17 proteins whose crystal structures are known from X-ray studies. A good correlation is observed between the infrared estimates and those calculated from X-ray data. This method gives standard errors of prediction 3.9% for α-helix, 8.3 % for ß-sheet, and 6.6% for turns. No preliminary treatment of the bands such as deconvolution is required. More recently we have applied the partial least squares method [PLS, see ref. 13] to our calibration set of 17 proteins and also extended the method to the FT-IR spectra of membrane proteins (see Table I). The accuracies of the FT-IR method compares very well with other spectroscopic

TABLE 1 Secondary Structure Prediction for Membrane Proteins using PLS

Membrane Protein	**% α-helix**	**% ß-sheet**	**% turn**
Ca^{+2}-ATPase	65.1	15.7	6.8
H^{+}/K^{+}-ATPase	47.6	33.0	21.5
Photosystem II Reaction Centre	60.0	17.6	22.2
Bacteriorhodopsin	71.5	5.8	-11.2

techniques such as Circular Dichroism (CD) and Raman spectroscopy. The results obtained for the membrane proteins (Table I) are consistent with previous estimations based on predictions from sequence data, electron diffraction and CD and Raman spectroscopic data.

MEMBRANE ASSOCIATED POLYPEPTIDES

The large size and complexity of membrane proteins makes it difficult to apply many of the techniques normally used in protein structural studies. Thus there is a need to find ways to simplify these systems so they can be made amenable for structural studies. Using solid-phase methods, polypeptide segments of large membrane proteins can be synthesised. We have began to use this approach to study polypeptide segments of a number of membrane proteins.

Mitochondrial transit peptide: Many proteins are synthesised in the cytoplasm of eukaryotic cells and then are translocated to their final destinations in subcellular organelles, for example the mitochondria, crossing one or more membranes in the process [14]. The information which a protein needs to select one intracellular membrane from several is contained in the initially-translated protein. The preprotein consists of the mature protein plus an N-terminal extension,referred to as a signal or transit sequence, of between 17 and 60 residues in length. In recent studies we have been using a combination of FT-IR, CD and NMR spectroscopy to understand the structural features of some of these systems. We have used these techniques to determine the structure of a mitochondrial transit peptide in trifluroethanol. The peptide MLSALARPVGAALRRSFSTSAQNNAK, is the N-terminal sequence of the precursor of rat malate dehydrogenase and two residues of the mature protein.

The primary structure of mitochondrial transit peptides are not conserved; however, they do have secondary and tertiary structural features in common. They are relatively rich in positively-charged (basic) amino acids (mainly Arg), they lack acidic residues, have a high content of hydroxylated residues and small groups of adjacent hydrophobic residues. Hydropathicity analyses have suggested that these peptides may form amphiphilic helices, with a hydrophobic face and a positively charged face, and also the possibility of a two domain structure, consisting of helical segments connected by a turn or bend.

Our FT-IR and CD results (see Figure 1) indicate that the mitochondrial transit peptide (sequence shown above) adopts a predominantly α-helical conformation in trifluroethanol solution. This observation is supported by 2D NMR results which shows nuclear Overhauser enhancement (NOE) patterns characteristic of

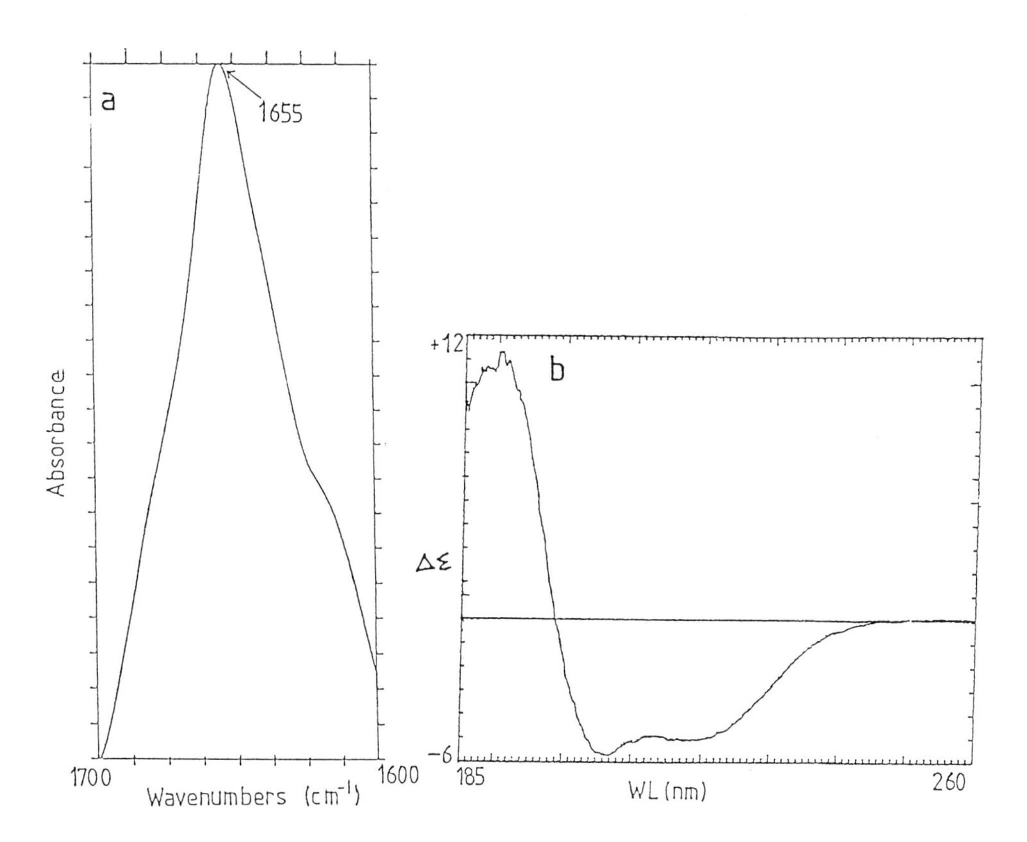

Fig. 1: a) FT-IR spectrum of the mitochondrial transit peptide. The amide I maximum at 1655cm^{-1} is consistent with a predominantly α-helical peptide conformation.
b) CD spectrum of the mitochondrial transit peptide. The minimum at 207nm and shoulder around 220nm are indicative of the presence of α-helical structure in the peptide.

α-helical secondary structure (i.e. strong NH-NH NOE's and weaker NH-Hα NOE's, and also by small values of ^{3J}NH-Hα). A semi-quantitative structure for the peptide has been constructed using distance constraints derived from the observed backbone NOE's. The derived structure is characterised by two predominantly α-helical stretches, between Arg7 and Ser16, and between Thr19 and Gln22, and a pronounced amphiphilic nature. Our FTIR studies also confirm that this polypeptide is α-helical in cardiolipin and dimyristoylphosphatidylglycerol lipid water system.

K^+-Ion Channel Peptides: We have started studies to attempt to determine the secondary and tertiary structure of ion channel proteins. These proteins have the ability to distinguish between various ions such as sodium, potassium and calcium. The structural basis for this selectivity has been the subject of research for some time. However, only recently have workers identified [15,16,17] the amino acid sequence that constitutes the ion selective pore in the voltage gated K^+-channels. It has been speculated that this stretch of amino acids may form two anti-parallel ß-strands traversing the membrane. The pore is not what might have been expected. It contains no charged residues and furthermore the pore forming sequence is not particularly hydrophilic. The amino acid sequence of this pore is:

DAFWWAVVTMTTVGYGDMT

We have chemically synthesised the sequence shown and are presently carrying out spectroscopic studies of this polypeptide in phospholipids using CD, FT-IR and 2D NMR spectroscopy. Preliminary FTIR and CD studies of this pore forming polypeptide shows that it possesses an α-helical structure in phospholipid systems rather than the proposed ß-sheet structure. We also have studies in progress to determine the conformation of other membrane spanning segments of the K^+-channel protein eg. the S2, S4 and S5 polypeptides within lipid bilayer structures.

ACKNOWLEDGEMENTS

We wish to thank the Wellcome Trust for support for this work.

REFERENCES

1 Elliott A and Ambrose EJ. Nature 1950;165:921-922.
2 Surewicz WK and Mantsch HH. In: Narang SA, ed. Protein Engineering: Approaches to the Manipulation of Protein Folding. Butterworths, 1990;131-157.
3 Jackson M, Haris PI and Chapman D. J Mol Struct 1989;214:329-355.
4 Susi H and Byler DM. Methods Enzymol 1986;130:290-311.
5 Haris PI and Chapman D. Biochem Soc Trans 1989;17:161-162.
6 Rothschild KJ and Clark NA. Biophys J 1979;25:473-488.
7 Krimm S and Dwivedi AM. Science 1982;216:407-408.
8 Haris PI and Chapman D. Biochem Biophys Acta 1988;943:375-380.
9 Alvarez J, Lee DC, Baldwin SA and Chapman D. J Biol Chem 1987;262:3502-3509.
10 Fischbarg J, Kuang K, Vera JC, Arant S, *et al.* Proc Natl Acad Sci 1990;87:3244-3247.
11 He WZ, Newell WR, Haris PI, Chapman D and Barber J. Biochemistry 1991;30:4552-4559.
12 Lee DC, Haris PI, Chapman D and Mitchell RC. Biochemistry 1990;29:9185-9193.
13 Lee DC, Haris PI, Chapman D and Mitchell RC. In: Hester RE and Girling RB, eds. Spectroscopy of Biological Molecules. Royal Society of Chemistry 1991;7-10.
14 Gierasch LM. Biochemistry 1989;28:923-930.
15 Yool AJ and Schwarz TL. Nature 1991;349:700-704.
16 Yellen G, Jurman ME, Abramson T and MacKinnon, R. Science 1991;251: 939-942.
17 Hartmann HA, Kirsch GE, Drewe JA, Taglialatela M, et al. Science 1991;251:942-944.

© 1992 Elsevier Science Publishers B.V. All rights reserved.
Molecular mechanisms of transport. E. Quagliariello, F. Palmieri, eds.

The organisation of the α-helices within the B800-850 light-harvesting complex from *Rhodopseudomonas acidophila* strain 10050

Richard J. Cogdell[a] Anna M. Hawthornthwaite[a] J. Gordon Lindsay[b] Miroslav Z. Papiz[c] and Dan Donnelly[d]

[a]Departments of Botany, University of Glasgow, Glasgow, G12 8QQ, UK

[b]Department of Biochemistry, University of Glasgow, Glasgow, G12 8QQ, UK

[c]SERC Daresbury Laboratory, Warrington, Cheshire, WA4 4AD, UK

[d]Department of Crystallography, Birkbeck College, University of London, London, WE1 7HX, UK

INTRODUCTION

The initial reaction of purple bacterial photosynthesis involves the absorption of light by the light-harvesting or antenna pigments, usually bacteriochlorophyll a and carotenoids, followed by the efficient transfer of that absorbed energy to the reaction centres. In the reaction centres the light-energy is 'trapped' and used to drive the primary oxidation-reduction reactions, which then initiate cyclic electron transport, ATP synthesis and CO_2 fixation. The light-harvesting pigments are organised into discrete well-defined pigment-protein complexes, where the pigments are non-covalently attached to small, hydrophobic apoproteins (1)

The molecular details of how the photosynthetic reaction centre works are known in great detail, especially now since the structure of the reaction centre has been determined to high-resolution (2). Unfortunately, a similar level of understanding of the light-harvesting process is lacking, mainly due to the absence of detailed structural information. In this paper we shall describe our crystallographic studies designed to try and determine a high-resolution structure of a bacterial antenna complex. This work is still in progress but we have been able to use a Patterson analysis of the native data set to get a rather good model of how the α-helices are arranged within the native complex.

CRYSTALLISATION

When this project was initiated 10-12 different antenna complexes isolated from a range of species of purple bacteria were screened for their suitability for crystal formation. This initial trial produced two successful cases, the B800-850-complexes from *Rps. acidophila* and from *Rhodobacter sphaeroides*. Of these two, the complex from *Rps. acidophila* has proved the easiest to proceed with. Our initial crystallisations used the basic method pioneered by Michel (3) and Garavito (4) using the addition of small amphiphiles to prevent phase separation prior to crystallisation. Using ammonium sulphate as the precipitant and heptane-triol as the amphiphile long thin needle-like crystals were produced. These were, however, unsuitable for X-ray diffraction studies. Now, finally, after extensive trials with different detergents, different amphiphiles, different precipitants and different pH's we have discovered conditions which yield large single crystals that are suitable for X-ray studies. An example of one of these is shown in Fig. 1. This crystal was grown using potassium phosphate as the precipitant and benzamidine hydrochloride as the amphiphile (5).

Figure 1. A crystal of the B800-850-complex from *Rps. acidophila* strain 10050. The crystal is ~400μ in length.

X-RAY CHARACTERISATION OF THE CRYSTALS OF THE B800-850-COMPLEX FROM *RPS. ACIDOPHILA*

Two crystal forms were examined by X-ray diffraction, using the synchrotron radiation source at Daresbury. The first form was grown with complexes solubilised in the detergent LDAO (lauryldimethylamine-N-oxide). These crystals had the P_4 space group and unit cell dimensions of a = b = 75.8 Å, c = 97.5 Å. However, they only diffracted to a resolution of 12 Å. The second crystal form was grown with the complex exchanged into β-octylglucoside. These crystals diffract X-rays to a much higher resolution and we have collected a full native data set to 3.2 Å, with an R_{merge} of ~9%. These crystals have the rhombohedral space group R32 and unit cell dimensions of a = 121.1 Å, α = 60°. A heavy atom search is currently underway and we have so far obtained a mercury derivative which is isomorphous.

ORGANISATION OF THE α-HELICES WITHIN THE B800-850-COMPLEX

The antenna apoproteins from a range of purple bacteria have been isolated, purified and sequenced (1). They form a very homologous group of proteins. Typically they are short proteins, usually between 50 and 60 amino acids in length, and they come in two types, the α- and β-apoproteins. The α-apoproteins contain a single conserved histidine residue, while the β-apoproteins contain two conserved histidines. They all show a marked three domain structure. Their N- and C-termini are polar, while internally they contain a hydrophobic region of between 21-24 amino acids. Naturally this domain structure has led to the suggestion that they should lie across the photosynthetic membrane with a single membrane-spanning α-helix. Topological studies have confirmed this picture (6).

In the absence of a high-resolution structure we have used the native crystallographic data to gain interim insight into the structure of the antenna complex. Fig. 2 shows a slice through Patterson space calculated from the full native data set to a resolution of 3.2 Å. In this case the slice is in the a/b plane. In simple terms this type of calculation takes each atom in turn, puts it at the origin of the projection and calculates the vector between it and every other atom in the molecule. Clearly regular structures, i.e. α-helices will build up density in this type of calculation. The pattern seen in Fig. 2 can be readily understood by a consideration of what type of α-helical packing might be present in the complex. The density at the origin goes up and down the 'c' axis (at right angles to the projection) for 22-23 Å. If we make the reasonable assumption that this comes from the α-helices, then they must be lined up in the crystal almost parallel to the 'c' axis and be 44-46 Å in length. There are rings of density, going out from the origin, at 10.5 Å. 14.2 Å and 24 Å. These can be readily explained by reference to Fig.3. If two α-helical rods lie next to and parallel to each other, then the distance of the centre of gravity of one to the centre of gravity of the next is 10.5 Å. If four, parallel α-helical rods are arranged in a square, then the distance across the diagonal, from the centre of gravity of one to the centre of gravity of the next, is 14.2 Å. At 24 Å in an R32 space group six peaks would be expected just from the symmetry of this space group. However, 18 peaks are seen. This means that there is an extra, non-crystallographic threefold axis of symmetry. All of these features can be accounted for by assuming that the intact antenna complex contains three bundles of 4 α-helices, arranged as in Fig.3. Each bundle of 4 would represent an $\alpha_2\beta_2$ unit, giving an intact structure of $\alpha_6\beta_6$, which corresponds exactly to the molecular weight of the native complex of 84,000.

Figure 2. A slice through the Patterson space calculated from the native data set collected from crystals of the B800-850-complex of *Rps. acidophila* to a resolution of 3.2 Å. The slice is cut through the a/b plane.

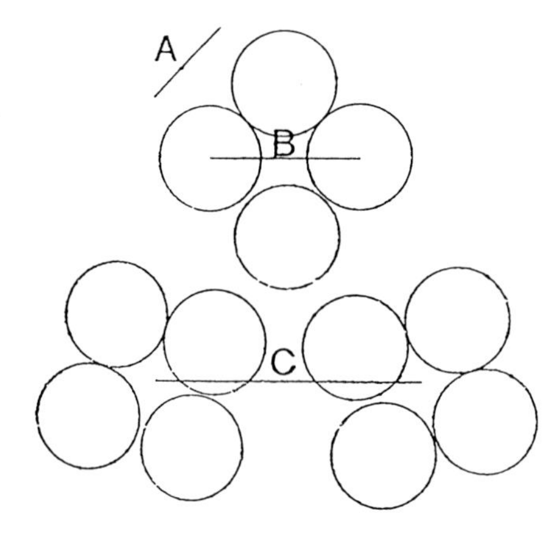

A is 10.5 Å distance between helicies
B is 14.2 Å diagonal packing
C is 24.0 Å distance of monomers

Figure 3. A model of the arrangement of the α-helices of the B800-850-complex from *Rps. acidophila* based upon an interpretation of the Patterson shown in Fig. 2.

ACKNOWLEDGMENTS
This work was supported by a grant from the SERC.

REFERENCES
1 Zuber H, Brunisholz RA In: Scheer H, ed. Chlorophylls. Boca Raton, CRC Press, 1991: 627-703.
2 Deisenhofer J, Epp O, Miki K, Huber R, Michel H. Nature 1985; 318: 19-26.
3 Michel H, Oesterhelt D. Proc. Nat. Acad. Sci. USA, 1980: 77; 1283-1285.
4 Garavito RM, Rosenbusch JP. J. Cell Biol. 1980: 86; 327-329.
5 Papiz M, Hawthornthwaite AM, Cogdell RJ, Woolley KJ, Wightman PA, Ferguson LA, Lindsay JG. J. Mol. Biol. 1989: 209; 833-835.
6 Brunisholz RA, Zuber H, Valentine J, Lindsay JG, Woolley KJ, Cogdell RJ. Biochim. Biophys. Acta. 1986: 849; 295-303.

© 1992 Elsevier Science Publishers B.V. All rights reserved.
Molecular mechanisms of transport. E. Quagliariello, F. Palmieri, eds.

A strategy for determination of the sequences of the major proteins in the inner mitochondrial membrane

J. E. Walker, J. M. Arizmendi, I. M. Fearnley, S. M. Medd, S. J. Pilkington, M. J. Runswick and J. M. Skehel

The Medical Research Council Laboratory of Molecular Biology, Hills Road, Cambridge CB2 2QH, U. K.

Introduction

ATP synthesis in mitochondria depends upon a series of complex enzymes in the inner membrane to generate an electrochemical proton potential gradient to promote phosphorylation of ADP. Supply of substrates is provided by two transport proteins, ADP/ATP translocase and the phosphate carrier, and some of the essential metabolites for mitochondrial function are brought into the organelle by other carrier proteins. Altogether these functions account for more than 90 proteins in the inner membrane of the organelle (see Table 1). Thirteen of them are encoded in the mitochondrial genome and their sequences are known; the remainder, the products of nuclear genes, are imported into the organelle from the cytoplasm.

Table 1
The major proteins of the inner membrane of bovine heart mitochondria.

Enzyme	Number of different proteins	
	Nuclear coded	Coded in mt-DNA
Complex I	>30 (28)	7
Complex II	4 (2)	0
Complex III	10 (10)	1
Complex IV	10 (10)	3
ATP synthase	12 (12)	2
Transhydrogenase	1 (1)	0
Transport proteins	>13 (3)	0
Total	>80 (66)	13

Figures in parentheses are the numbers of nuclear proteins that have been sequenced.

The determination of the sequences of the nuclear encoded subunits is an essential part in unravelling the structures and mechanisms of the enzymes of oxidative phosphorylation and of the transport proteins. The sequences of the subunits of the three respiratory enzymes ATP synthase, cytochrome c oxidase and cytochrome c reductase, and of 3 carrier proteins (ADP/ATP translocase, the phosphate carrier and the oxoglutarate/malate carrier) have already been established. In contrast, the sequence analysis of subunits of NADH:ubiquinone oxidoreductase (complex I) has been almost entirely ignored until recently. This is probably a consequence of its complexity; for example, the enzyme in bovine heart mitochondria contains at least 30 different nuclear coded subunits, in addition to the 7 hydrophobic subunits of known sequence that are encoded in mitochondrial DNA. As described below, the sequences of 28 nuclear components have now been determined, mostly by use of a novel strategy, which employs the polymerase chain reaction to generate cDNAs encoding specific subunits [1-6]. This strategy is suitable for the determination of the sequences of all of the major proteins in an organelle, and is discussed with reference to the inner membrane of bovine heart mitochondria.

Protein complexity of the inner membrane of bovine heart mitochondria

At least 70 major components can be detected in 2-dimensional gels of bovine heart sub-mitochondrial particles (inner membrane vesicles; see Figure 1).

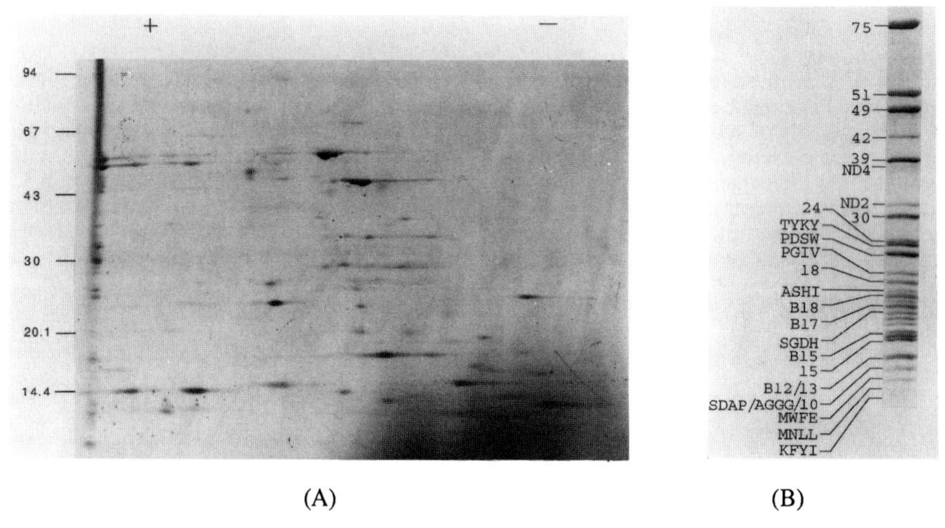

Figure 1. Analysis of the proteins present in (A) bovine heart submitochondrial proteins by 2-dimensional gel electrophoresis and (B) of the subunits of complex I in a 1-dimensional gel.

This technique underestimates the number of major proteins that are present in the membrane, since, for example, none of the 13 hydrophobic subunits encoded in mt-DNA enters the isoelectric focussing gel, and they are not represented in the 2-dimensional gel. Moreover, these and other hydrophobic proteins (for example, the c subunit of ATP synthase), stain poorly with Coomassie Blue dye, and, even if present in substantial quantity, would be difficult to detect. At present, with the possible exception of the abundant α and β subunits of ATP synthase, the identities of the spots are not known. Therefore, in order to avoid duplication of earlier work, it is necessary to determine which proteins in this map have been sequenced and which have not. The extent of the latter category is uncertain, but it is likely to include several metabolite carriers, and possibly proteins involved in import of proteins into the organelle. Where the proteins have a free unmodified N-terminus, discrimination of known from unknown proteins can be accomplished easily by N-terminal sequence analysis, after first transferring the proteins from the 2-dimensional gel to a poly(vinylidene difluoride) (pvdf) membrane. Proteins with modified α-amino groups cannot be identified by this means, and other methods such as use of specific antibodies to identify proteins by Western blotting and mass spectrometric analysis (see below) will have to be employed. For the purpose of the following discussion, it is assumed that the discrimination of known from unknown proteins can be accomplished. The rest of this paper will be concerned with the strategy for the sequence analysis of the individual proteins, illustrated with examples from a substantial subset of the proteins of the inner mitochondrial membrane, namely the nuclear coded components of complex I.

Outline of the protein sequencing strategy

In its simplest form, the strategy has two parts, A and B (see Figure 2).

Part A. A protein sequence of about 18 consecutive amino acids is determined in the protein of interest. Where the N-terminus of the intact protein is modified, or if for other reasons the N-terminal region of the protein proves to be difficult to sequence, peptides are sequenced from digests of the protein. Then two oligonucleotide mixtures, "the primers", are synthesised; one is sense, the other anti-sense. They are 17 bases in length (plus appropriate linker sequences) and their sequences are based on the flanking hexapeptide sequences. These primers are used in polymerase chain reactions with total unfractionated poly A^+ double stranded cDNA as template in order to amplify a short cDNA representing the protein sequence. The products of this reaction are separated by agarose gel electrophoresis and are subjected to Southern blotting, using as hybridisation probe a third mixed 17 base synthetic oligonucleotide corresponding to a hexapeptide segment between those employed to make the primers. Hybridising bands are recovered and cloned into an appropriate M13 vector. Clones containing the cDNA of interest are identified by plaque hybridization with the same oligonucleotide probe, and are sequenced. This short cDNA encodes the segment or segments of known protein sequence, but because its 5' and 3' extremities arise from mixed primers, only the central part of its sequence between the primers is accurate and unique.

Part B. The partial unique DNA sequence derived in part A is extended to the 3' and 5' extremities of the complete cDNA, again by use of the polymerase chain reaction. Firstly, with a unique synthetic oligonucleotide primer based on the known DNA sequence as forward primer and oligo (dT) (plus linker) as reverse primer, a cDNA is made to extend from the established unique sequence to the 3' end of the complete cDNA. Secondly, a clone extending from the established unique sequence to the 5' extremity is made using as primers the complement of the unique primer above and oligo (dT) (or oligo (dC)), employing single stranded template cDNA that has been modified by addition of a 5' homopolymer (dA) (or (dG)) tail with terminal deoxynucleotide transferase. The desired products in these reactions are recognised by hybridisation with mixed oligonucleotides based upon the the flanking hexapeptide sequences of the original 18 amino acid sequence.

By means of this strategy, in many cases it is possible using minimal protein sequence information to generate in three polymerase chain reactions overlapping cDNAs which represent the complete cDNA for a protein. In other cases, related but somewhat more elaborate strategies are required. These strategies, and other important experimental parameters, are discussed in detail elsewhere [1]. The use of this strategy has permitted us to determine the sequences of 28 different proteins from the inner membranes of bovine heart mitochondria over a period of about 2 years. These are 2 subunits of ATP synthase [7, 8], the oxoglutarate / malate carrier protein [9] and 25 subunits of complex I [1-6].

Application of the strategy to subunits of complex I from bovine heart mitochondria

Subunits of complex I were separated in 1-dimensional (see Figure 1B) and 2-dimensional gels (not shown), transferred to pvdf membranes and N-terminal sequences were determined on 20 of its subunits. Other subunits proved to have modified N-terminals, and in these cases, the proteins were isolated by chromatography, digested and internal sequences

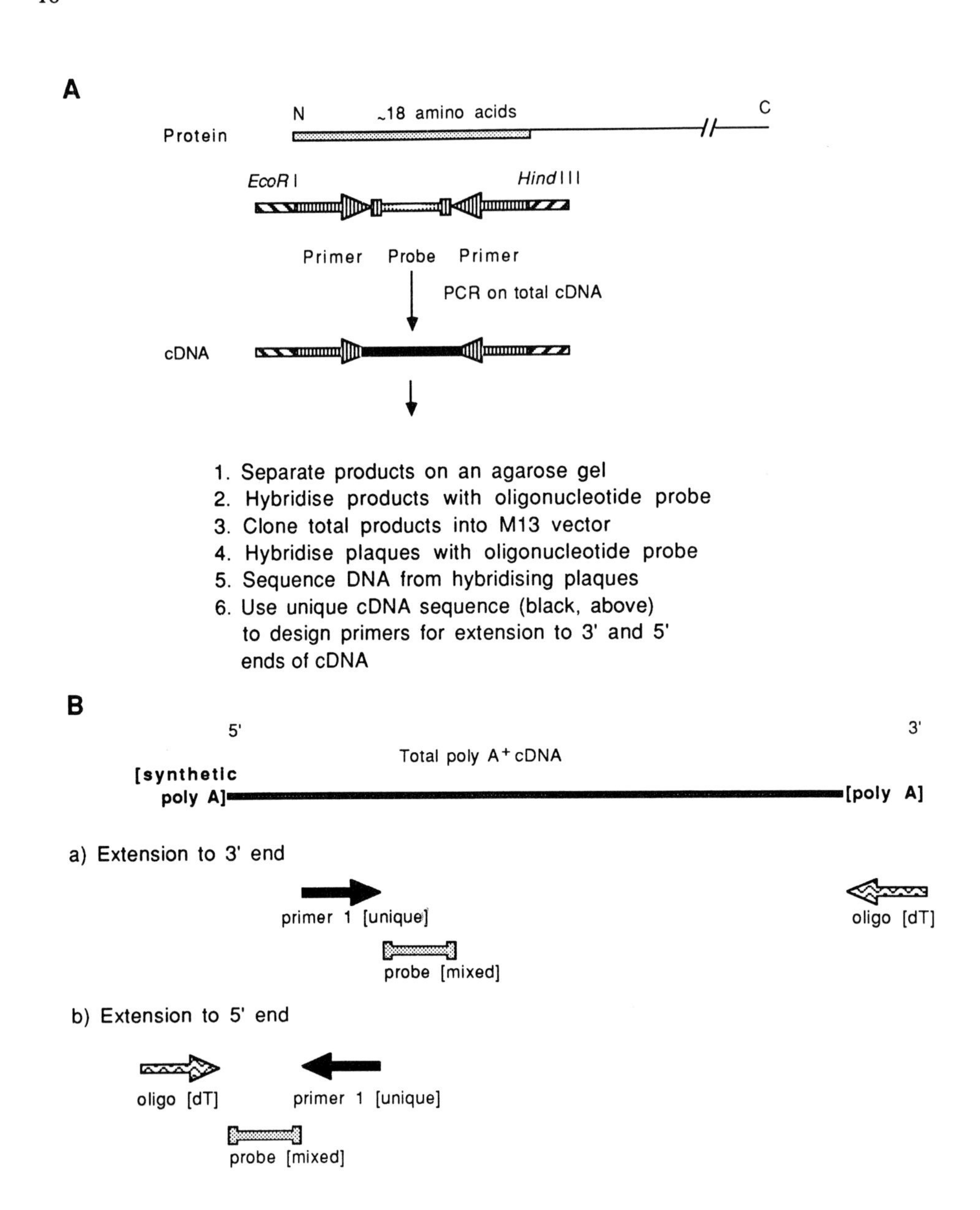

Figure 2. A strategy for sequence analysis of proteins (see text for further details).

were determined on peptides. Four of the subunits with unmodified N-terminals (the 75, 49, 30 and 24 kDa subunits) were cloned by conventional means [10-13]; in 24 other subunits the directly determined protein sequences provided the information for design of syntheses of mixtures of oligonucleotides, which were then employed as primers and probes in polymerase chain reactions as described above, and the sequence of the 30 kDa subunit was also verified by this means using unique primers. The most complex of these oligonucleotide mixtures contained over 3,000 different sequences (see ref. [1]).

As discussed elsewhere [1], the sequences of 28 of the nuclear coded subunits of bovine complex I that have now been completed, provide many important clues to their function, to the location of substrate and cofactor binding sites and of iron-sulphur centres. Together they contain 5,039 amino acids, and with the known post-translational modifications (not including iron-sulphur centres, of which the number, type and location all are uncertain), contribute a molecular mass of 574,757. The 7 additional subunits of complex I that are encoded in mitochondrial DNA have a combined molecular mass of 238,462, and so, if unit stoichiometries for all the subunits are assumed, the molecular mass of bovine complex I is greater than 813,219. At least 3 more subunits, with estimated molecular weights of 22, 15 and 10 kDa, remain to be sequenced.

Verification of protein sequences by electrospray mass spectrometry

As an independent check of the protein sequences deduced from the cDNAs, the molecular masses of the subunits of complex I have been determined by electrospray mass spectrometry. This is a technique which allows the molecular weights of proteins to be determined with an accuracy of 0.01% [14]. Although mass measurements have been made on proteins over 100 kDa, our success with components of complex I is restricted to subunits with masses below 30 kDa. The reason why spectra have not been observed on the larger subunits is not known. With one exception, the values obtained for the proteins with unmodified N-terminals with masses below 30 kDa agree with those calculated from the sequences (Table 2). The exception is subunit SDAP which is an acyl carrier protein, and carries a pantetheine-4'-phosphate moiety, probably on serine-44 [15].

In the cases of subunits B17, B13, B14 and B8, close agreement between the measured values and those calculated from their sequences is obtained if the initiator methionines predicted from the cDNA sequences are removed and the new N-terminal residues are N-acetylated. Similarly, the measured and calculated values of subunit B18 can be reconciled if the initiator methionine has been lost and the new N-terminal glycine has been myristylated. The measured values of molecular mass for subunits B12 and B9 cannot be reconciled with the calculated values, and they appear to be modified in an as yet unknown way. In order to remove the possibility that their cDNA sequences are incorrect, because, for example, of introduction of errors during polymerase chain reactions, complete cDNAs for these subunits were amplified from total bovine heart cDNA in polymerase chain reactions that were independent of those first carried out, and the sequences were completely re-determined. In both instances, the sequences agreed with the original ones, and no errors were found.

Conclusions and perspectives

As part of an effort to understand the mechanisms of the enzymes of oxidative phosphorylation by structural analysis, the sequences of about 67 nuclear coded proteins present in the inner membranes of bovine heart mitochondria have now been established. The strategy that has been developed for the analysis of the subunits of complex I also offers the possibility to sequence the other major subunits in the inner membrane that have not been sequenced so far. These include 3 or more further subunits of complex I, 2 membrane subunits of complex II, possibly 10 or more transport proteins and perhaps other hitherto unknown proteins. The group of transport proteins is of particular current interest, and directly relevant to the topic of this Meeting.

Table 2.
Molecular weights of some subunits of NADH:ubiquinone oxidoreductase from bovine heart mitochondria. Values were estimated by gel electrophoresis, calculated from sequences and measured by electrospray mass spectrometry of the proteins.

Subunit	Molecular weight		
	by gel (kDa)[1]	from sequence	by ms (standard deviation)
30 kD (IP)	30[a]	26,431.8	26,435 (0.6)
PDSW	22	20,833.6	20,831 (2.2)
ASHI	19	18,737.0	18,742 (4.8)
SGDH	16	16,726.3	16,728 (2.4)
B18	18	16,476.8[2]	16,477 (2.5)
B17	16.5	15,435.9[3]	15,438 (0.6)
18 kD (IP)	18[a]	15,337.2	15,341 (2.9)
B14	14[a]	14,964.2[3]	14,965 (1.0)
B13	13	13,226.4[3]	13,227 (0.75)
15 kD (IP)	15[a]; 13	12,536.4	12,534 (1.7)
B12	12	11,009.4[4]	11,038 (0.5)
B8	8[a]	10,990.6[3]	10,990 (0.55)
13 kD (IP)	13	10,535.7	10,539 (2.4)
SDAP	8	10,109.5	10,752 (0.2)
B9	9[a]	9,217.7[4]	9,298 (1.3)
AGGG	7.9	8,493.3	8,495 (1.12)
10 kD (FP)	10	8,437.3	8,438 (0.26)
MWFE	7.5	8,135.4	nd
MNLL	7	6,966.1	6,967 (0.91)
KFYI	6	5,828.7	5,829 (0.65)

[1] Molecular weights were estimated using either the gel system of Schägger and von Jagow [16] or [a] of Laemmli [17]; [2] calculated without initiator methionine and with $N\alpha$-myristylation of new N-terminal glycine; [3] assumes removal of N-terminal methionine and acetylation of new N-terminal alanine or serine; [4] excluding initiator methionine; nd, not determined.

Many of the transport proteins appear to have similar molecular weights, and it is likely that they will belong to the same multi-gene family as the ADP/ATP translocase, the phosphate carrier, the oxoglutarate/malate carrier and the uncoupling protein from brown adipose tissue [9]. Further progress with the analysis of this family depends on the continuation of present efforts to isolate and reconstitute individual carriers, but might also be aided by an independent sequence analysis of all of the unknown proteins of the inner membrane, which might reveal new members of the carrier multi-gene family.

In principle, the strategy described above could be applied to the determination of the sequences of the major proteins of any organelle or sub-organellar population of proteins; the mitochondrial ribosome, the mitochondrial matrix proteins and various compartments of the chloroplast are possible applications with direct interest to bioenergetics.

A major advantage of the strategy is its speed in comparison with conventional methods of cDNA cloning, where either antibodies or oligonucleotides are employed to screen libraries of clones. This advantage is derived in part from the polymerase chain reaction, which permits a wide range of experimental conditions to be explored rapidly. The requirement for a minimum of 18 amino acids (possibly in three separate hexapeptides, but preferably in the same stretch of sequence) can be met easily, but when the N-terminus of the protein is modified this often is a difficult obstacle that can be surmounted by isolation of the protein, and sequence analysis of peptides derived from it. Alleviation of this problem may come from new mass spectrometric methods of protein analysis, which take protein ions *in vacuo*, such as those produced by the electro-spray method, fragment the protein ions and analyse the fragments to produce sequence information [18]. Providing that this can be accomplished on the quantities of protein that can be recovered from a 2-dimensional gel, this may provide a convenient means of obtaining the requisite partial protein sequence data for the strategy described above.

REFERENCES

1 Walker JE, Arizmendi JM, Dupuis A, Fearnley IM et al. J Mol Biol 1991; submitted for publication.
2 Pilkington SJ, Skehel JM, Gennis RB, Walker JE. Biochemistry 1991; 30: 2166-2175.
3 Dupuis A, Skehel JM, Walker JE. Biochem J 1991; 277: 11-15.
4 Dupuis A, Skehel JM, Walker JE. Biochemistry 1991; 30: 2954-2960.
5 Fearnley IM, Finel M, Skehel JM, Walker JE. Biochem J 1991; 278: 821-829.
6 Skehel JM, Pilkington SJ, Runswick MJ, Fearnley IM et al. FEBS Lett 1991; 282: 135-138.
7 Runswick MJ, Medd SM, Walker JE. Biochem J 1990; 266: 421-426.
8 Walker JE, Lutter R, Dupuis A, Runswick MJ. Biochemistry 1991; 30: 5369-5378.
9 Runswick MJ, Walker JE, Bisaccia F, Iacobazzi V et al. Biochemistry 1990; 29: 11033-11040.
10 Runswick MJ, Gennis RB, Fearnley IM, Walker JE. Biochemistry 1989; 28: 9452-9459.
11 Fearnley IM, Runswick MJ, Walker JE. EMBO J 1989; 8: 665-672.
12 Pilkington SJ, Skehel JM, Walker JE. Biochemistry 1991; 30: 1901-1908.
13 Pilkington SJ, Walker JE. Biochemistry 1989; 28: 3257-3264.
14 Fenn JB, Mann M, Meng CK, Wong SF et al. Science 1989; 246: 64-71.
15 Runswick MJ, Fearnley IM, Skehel JM, Walker JE. FEBS Lett 1991; 286: 121-124.
16 Schägger H, von Jagow G. Anal Biochem 1987; 166: 368-379.
17 Laemmli UK. Nature 1970; 227: 680-685.
18 Loo JA, Edmonds CG, Smith RD. Science 1990; 248: 201-204.

ATPASES

© 1992 Elsevier Science Publishers B.V. All rights reserved.
Molecular mechanisms of transport. E. Quagliariello, F. Palmieri, eds.

PRIMARY Na^+ TRANSPORT SYSTEMS OF BACTERIA

P. Dimroth

Mikrobiologisches Institut, Eidgenössische Technische Hochschule, ETH-Zentrum, Schmelzbergstr. 7, CH-8092 Zürich, Switzerland

INTRODUCTION

Most bacteria employ a cycling of protons over their cytoplasmic membrane to couple exergonic and endergonic membrane reactions. However, protons are not the only coupling ions used for bioenergetic energy conversions. In recent years, it has been detected that several bacterial species employ a Na^+ cycle for energy coupling [1-4]. Some examples are shown in Table 1. Most common are the Na^+ symport systems, which are widely distributed among bacteria. Organisms living at high Na^+ concentrations (halophilic and marine bacteria) or at high pH values (alkaliphilic bacteria) use these Na^+ coupled transport systems primarily, while the Na^+-solute symporters are less frequent in neutrophilic fresh water bacteria. Another endergonic reaction that may be driven by $\Delta\tilde{\mu}Na^+$ is the flagellar motor, as shown for marine Vibrios [19] or alkaliphilic bacteria [21].

Table 1
Bacteria with Na^+ bioenergetics

Organism	Primary Na^+ pump	Na^+/H^+ antiporter present	Endergonic reaction linked to Na^+	Ref.
Klebsiella pneumoniae	Oxaloacetate decarboxylase	+	Citrate transport NADH reduction (?)	5-8
Veillonella alcalescens	Methylmalonyl-CoA decarboxylase	?	?	9,10
Acidaminococcus fermentans	Glutaconyl-CoA decarboxylase	?	?	11
Propionigenium modestum	Methylmalonyl-CoA decarboxylase	?	ATP synthesis	12-15
Vibrio alginolyticus	NADH:ubiquinone oxidoreductase	+	Transport of aminoiso-butyric acid and sucrose, flagellar motion	16-19
Bacillus alcalophilus	----	+	Transport of amino acids, malate, flagellar motion	20,21
Escherichia coli	----	+	Transport of glutamate, proline, serine, threonine, melibiose	22
Methanosarcina barkeri	Reduction leading to methyl-CoM	+	Reduction of CO_2	23
Vitreoscilla	Cytochrome O			24

Most bacteria use the Na^+/H^+ antiporter for $\Delta\tilde{\mu}Na^+$ formation, which therefore is a secondary event that is dependent on the $\Delta\tilde{\mu}H^+$ established by a primary proton pump. However, a few bacterial species are capable of $\Delta\tilde{\mu}Na^+$ formation by a primary Na^+ pump. The first primary Na^+ pump detected in a bacterium was oxaloacetate decarboxylase of *Klebsiella pneumoniae* fermenting citrate [5]. Subsequently, the related Na^+ pumps methylmalonyl-CoA decarboxylase [9,10] and glutaconyl-CoA decarboxylase [11] were detected in other anaerobic bacteria. *Vibrio alginolyticus* was shown to pump Na^+ ions out of the cell by an NADH:ubiquinone oxidoreductase [16-18]. The most prominent example of a bacterium using Na^+ bioenergetics is *Propionigenium modestum*. The entire ATP synthesis in this organism depends on a Na^+ circuit between the Na^+-pumping methylmalonyl-CoA decarboxylase and the Na^+-translocating F_1F_0 ATPase [12-15].

Bacteria thus generate an electrochemical gradient of Na^+ ions either by the Na^+/H^+ antiporter at the expense of a $\Delta\tilde{\mu}H^+$ or with a primary Na^+ pump and use the $\Delta\mu Na^+$ as driving force for active transport, flagellar motion or ATP synthesis.

Na^+ AS COUPLING ION IN *KLEBSIELLA PNEUMONIAE* FERMENTING CITRATE

The fermentation of citrate in *K.pneumoniae* involves the decarboxylation of oxaloacetate by the membrane-bound and biotin-containing oxaloacetate decarboxylase that acts as a Na^+ pump [1-4]. The Na^+ ion gradient thus established is used to drive the uptake of the growth substrate citrate into the cell [5,6]. Citrate is degraded anaerobically in these cells to two acetate, CO_2 and formate, which are probably secreted from the cell in symport with protons. A ΔpH would counteract such end product extrusion. The resulting accumulation of these acids in the cytoplasm could lead to a self poisoning of the cells. To overcome these problems, the cells generate a $\Delta\tilde{\mu}Na^+$ (not $\Delta\tilde{\mu}H^+$) with oxaloacetate decarboxylase and use the Na^+ gradient for citrate uptake.

It is interesting in this context that *Klebsiella pneumoniae* degrades citrate aerobically via the tricarboxylic acid cycle and does not express an oxaloacetate decarboxylase Na^+ pump. The citrate carrier of these cells is a H^+ symporter that uses the $\Delta\tilde{\mu}H^+$ created by the respiratory chain for citrate accumulation [5]. As aerobic cells oxidize citrate completely to CO_2, they do not have to excrete large amounts of acidic end products, and the development of a ΔpH is therefore not a disadvantage.

Citrate fermenting *K.pneumoniae* cells have still another Na^+-translocating system within their membrane. This is an NADH:quinone oxidoreductase functioning as an Na^+ pump [8]. The physiological role of this enzyme in an anaerobic bacterium is unknown. Possibly, reversed electron transfer from quinol to NAD^+ is driven by an Na^+ gradient and accounts for NADH biosynthesis in these cells. It should be noted that NADH is not formed during citrate fermentation and can neither be generated by the tricarboxylic acid cycle, because this is not functioning under anaerobic conditions.

STRUCTURE AND MECHANISM OF OXALOACETATE DECARBOXYLASE

The molecular properties and the catalytic mechanism of oxaloacetate decarboxylase have been studied in considerable detail and are summarized in Fig. 1. The enzyme consists of three different subunits, α, β, γ with M_r of 63,600, 34,600 and 8,900, respectively [25]. The genes for the three subunits are located on the chromosome in an operon-like structure in the order γ-α-β [26-28]. An evaluation of the sequences indicates that the γ-chain is located in the membrane with one putative transmembrane helix and that the β-chain is incorporated into the membrane with five or six membrane-spanning helices [28]. The α-subunit is hydrophilic and is bound to the membrane via protein/protein interaction with the β- and γ-subunits [25,27]. The α-subunit is readily cleaved by limited proteolysis with e.g.

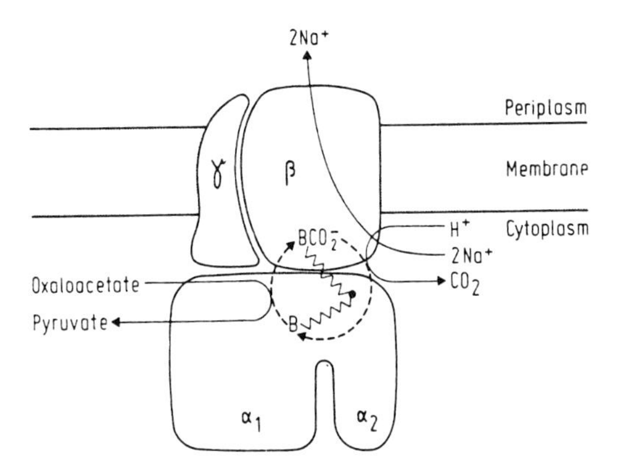

Figure 1. A hypothetical model linking structure and function of the oxaloacetate decarboxylase subunits. B, biotin.

trypsin into a larger N-terminal domain and a smaller C-terminal domain containing the biotin. The N-terminal domain harbours the catalytic site of the carboxyltransferase; its amino acid sequence is strikingly homologous to that of the 5S subunit of transcarboxylase from *Propionibacterium shermanii*, which catalyzes exactly the same carboxyltransfer reaction. The biotin domain is homologous to biotin peptides of other biotin enzymes. A remarkable sequence between these domains is a strech of amino acids consisting mainly of alanine and proline residues which is believed to be a hinge, giving the prostethic biotin group the flexibility required to move between the carboxyltransferase catalytic site (α_1) and the active center of the decarboxylase that is located on the β-subunit [27].

The mechanism of coupling between Na^+ transport and the decarboxylation reaction has been studied with the related Na^+ pump methylmalonyl-CoA decarboxylase. The reconstituted enzyme catalyzed the transport of $2Na^+$ ions into the proteoliposomes per molecule of malonyl-CoA decarboxylated [29]. Only in the initial phase leads this influx of Na^+ to net accumulation. Interestingly, the influx of Na^+ and the decarboxylation of malonyl-CoA do not stop in the steady state, where net accumulation has ceased, but remain to be coupled to each other at the stoichiometry of $2Na^+$ ions per decarboxylation reaction [30]. Obviously, then the influx of Na^+ must be accompanied by Na^+ efflux. The rate of this efflux is zero in the absence of a Na^+ gradient and increases continuously to $2Na^+$ per reaction in the steady state, where the rates of Na^+ influx and efflux must be equal. These results suggest a pump cycle with two potential steps for Na^+ translocation. We suggested a mechanism in which these Na^+ translocating steps are associated with conformational transitions of the β-subunit. Sodium ions bound to the β-subunit, associated with carboxybiotin, from the outside of the proteoliposomes, would become exposed to the inside of these particles upon decarboxylation. This step causes Na^+ influx at a defined stoichiometry ($2Na^+$ per reaction). The Na^+ binding site on the free β-subunit returns from the inside to the outside upon rebinding of carboxybiotin. This could cause Na^+ efflux, because depending on the magnitude of the Na^+ gradient, this step may proceed with or without Na^+ translocation [30].

These details of the coupling mechanism of the methylmalonyl-CoA decarboxylase Na^+ pump were elucidated by following the fluxes of Na^+ ions using the isotope $^{22}Na^+$ [30]. This method cannot be applied to H^+-coupled transport systems and the mechanistic details of these systems are therefore more difficult to decipher. We do not know yet, whether the features of the coupling mechanism determined for methylmalonyl-CoA decarboxylase are unique to this enzyme and perhaps the family of Na^+-translocating decarboxylases or may more generally apply to primary transport systems which convert chemical energy into an electrochemical ion gradient.

Na^+-COUPLED ATP SYNTHESIS IN *PROPIONIGENIUM MODESTUM*

P.modestum is the most prominent example of a bacterium depending on an Na^+ cycle for energy transduction. This strict anaerobe grows by fermenting succinate to propionate and CO_2. The pathway of energy metabolism is shown in Fig. 2 [12]. The only step in this pathway that is sufficiently exergonic to be used for energy conservation is the decarboxylation of methylmalonyl-CoA. The corresponding decarboxylase is a membrane-bound biotin-containing sodium pump related to the sodium ion transport decarboxylases discussed above [12].

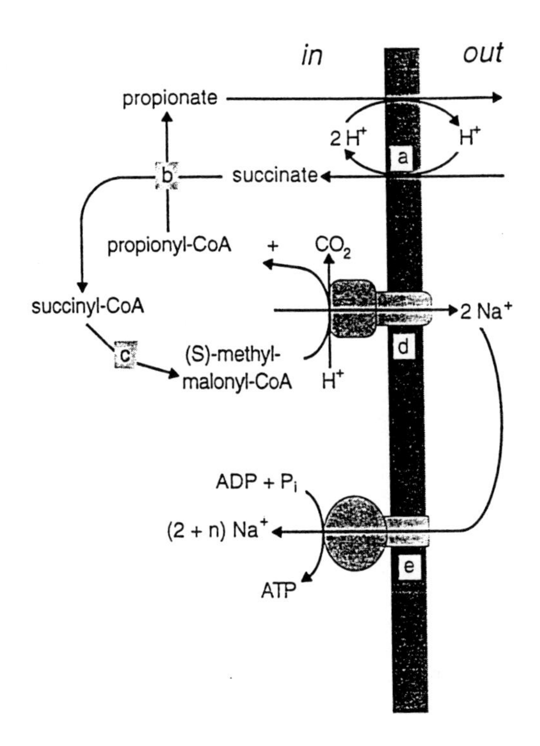

Figure 2. Energy metabolism of *Propionigenium modestum* with a Na^+ cycle coupling the exergonic decarboxylation of (*S*)-methylmalonyl-CoA to endergonic ATP synthesis. A hypothetical proton circuit could couple succinate uptake with the extrusion of propionate and CO_2. a succinate uptake system; b succinate propionyl-CoA:CoA transferase; c methylmalonyl-CoA mutase and methylmalonyl-CoA epimerase; d methylmalonyl-CoA decarboxylase; e ATPase.

P.modestum had to solve the problem of energizing ATP synthesis by the electrochemical Na^+ gradient generated by methylmalonyl-CoA decarboxylase. For this purpose, these bacteria contain a unique ATPase that couples ATP synthesis/hydrolysis to Na^+ translocation [13-15]. However, in spite of this unusual feature, the *P.modestum* ATPase has a typical F_1F_0 structure. The F_1 portion, which contains the ATP hydrolyzing activity, has been dissociated from the membrane-bound F_0 moiety and purified to homogeneity. It consists of five different subunit α, β, γ, δ, ε, with similar molecular masses as the corresponding five F_1 ATPase subunits of *E.coli*. The purified F_1 ATPase was not stimulated by Na^+ ions in contrast to the membrane-bound holoenzyme (F_1F_0). Upon reconstruction of the enzyme complex from purified F_1 and F_1-depleted membranes, the activation by Na^+ ions was recovered, thus indicating that the Na^+ binding site is located on the F_0 moiety, not on F_1 [13].

The F_1F_0 ATPase complex contains, in addition to the F_1 subunits, the subunits a, b and c, that correspond in molecular mass to the three related subunits of the *E.coli* ATPase [14]. The ATPase activity of F_1F_0 was stimulated about 10-fold by Na^+ ions (5 mM) and was severely inhibited by the typical inhibitors of prokaryotic F_1F_0 ATPases dicyclohexylcarbodiimide (DCCD), venturicidin, tributyltin chloride and azide. The DCCD was specifically bound to subunit c of the enzyme, which forms unusually stable aggregates of probably six polypeptide chains, which resist dissociation by SDS even at 100°C; 121° is required to dissociate the complex [14].

The genes of the *P.modestum* ATPase seem to be organized in an operon-like structure like in *E.coli*. The amino acid sequences of subunits β were 69% identical between *P.modestum* and *E.coli* [31], while the identities between the corresponding subunits a, b and c were 18%, 11%, and 17%, respectively [32,33]. Site directed mutagenesis studies with the *E.coli* ATPase have shown that the conserved amino acid residues on the a subunit arg-210, glu-219 and his-245 are essential for the function of H^+ translocation [34,35]. The conserved arg-210 residue is found in all known ATPase sequences including that of *P.modestum*. Position 219 contains either aspartate or histidine. ATPases from mitochondria, which have histidine at position 219, have glutamate at position 245, while those with glu-219 may have histidine, glycine, proline or serine at position 245. Results from site directed mutagenesis studies with the *E.coli* enzyme indicated partial retainment of H^+ translsocation activity by replacing glu-219 by asp or his, but no activity in the mutant with a leu substitution. A (a his245 → glu) mutation produced a severe defect in the F_0 mediated proton translocation, but interestingly, the double mutant (a glu219 → his, his245 → glu) yielded an ATPase complex with improved proton translocation as compared to the single mutants [35]. These results may indicate a close interaction of residues 219 and 245 of the a subunit and a requirement for an acidic residue in either of these positions in order to catalyze proton translocation. The *P.modestum* ATPase has methionine at position 219 and aspartate at position 245 [33]. This enzyme is therefore related to the other ATPases by the retainment of an acidic amino acid in one of these positions. A methionine at position 219, however, has never been found before in an ATPase sequence. Whether this substitution is essential for the change in ion specificity is unknown.

Subunit c contains a conserved aspartate or glutamate residue within the membrane (asp-61 in *E.coli*) that is specifically modified with low concentrations of DCCD, which abolishes the ATPase activity and the translocation of protons (Na^+ ions in the case of the *P.modestum* ATPase) through F_0 [36,37]. Asp-61 is therefore a good candidate to participate in the pathway of ion translocation. Subunit c of the *P.modestum* ATPase contains this conserved aspartate residue [32]. The highly conserved residues in the polar loop region between the two membrane spanning helices ala-40, arg-41, glu-41, pro-43, and asp-44 are also found in the *P.modestum* ATPase [32]. This region may be important for the interaction with the F_1 subunits, which would explain its resistance against mutagenesis.

In summary, these results clearly indicate the phylogenetic relationship of the Na^+-translocating ATPase of *P.modestum* with the H^+-translocating ATPases from other organisms.

The function of the *P.modestum* ATPase as a primary Na^+ pump was firmly secured by reconstitution experiments. Sodium ions were pumped into proteoliposomes containing the

ATPase in response to ATP hydrolysis [14]. The transport and the chemical reaction were simultaneously inhibited by DCCD. The transport rate increased about fourfold after dissipation of the rate-determining membrane potential with either valinomycin or the uncoupler carbonylcyanide-*m*-chloro-phenylhydrazone (CCCP). Stimulation rather than inhibition of Na^+ transport by the uncoupler firmly substantiates the functioning of the ATPase as a primary Na^+ pump and excludes the possibility of an Na^+ transport by the combined action of an H^+-translocating ATPase and an Na^+/H^+ antiporter [14].

Another interesting feature of the *P.modestum* ATPase is a switch from Na^+ to H^+ translocation at unphysiologically low (< 1mM) Na^+ concentrations [15]. Similar Na^+ concentrations (0.2 - 0.8mM) yielded half maximal rates of Na^+ transport and ATP hydrolysis, and reduced the rate of proton transport by one half. The ATPase, therefore, appears to switch between Na^+ and H^+ translocation in response to the Na^+ concentration applied. The results also suggest that the mechanism for the translocation of these two different cations is the same.

These results are further supported by recent results obtained with hybrid ATPases. A reconstituted proteoliposomal system consisting of F_0 from *P.modestum* and F_1 from *E.coli* or the thermophilic bacterium PS3 was capable of either H^+ or Na^+ pumping, and Na^+ prevented H^+ pumping at about the same concentration (> 1mM) as in the reconstituted system with the homologous *P.modestum* ATPase [38]. The ATPase of *E.coli* was on the other hand unable to catalyze Na^+ translocation. The F_0 part thus clearly defines the specificity for the coupling cation, and the ion translocation through F_0 apparently triggers ATP synthesis within the F_1 moiety independent of whether the translocated cation is Na^+ or H^+. The change in cation specificity in F_0 could be caused by relatively minor alterations at the cation binding site, not significantly affecting the recognition site for F_1, as suggested from the formation of functional hybrids described above.

Proteoliposomes containing only F_0 of *P.modestum* performed Na^+ or H^+ translocation (in the absence of Na^+) in response to a potassium diffusion potential induced by valinomycin (unpublished results). The proton translocation was impaired in the presence of Na^+ ions, and both Na^+ or H^+ translocation were abolished by DCCD treatment or after reconstitution of the holoenzyme by F_1 addition. The maximal rate of Na^+ translocation by F_0 was similar to that of ATP-dependent Na^+ uptake catalyzed by F_1F_0. Unless much of our enzyme was denatured during the preparation of the F_0-containing proteoliposomes, these results argue against a highly active Na^+ (H^+) channel for *P.modestum* F_0. The observed rates are more compatible with a transporter-type mechanism for F_0. It is suggested that the F_0 moiety contains a cation binding site that in a distinct conformation is accessible from one side of the membrane only and that the translocation involves a conformational change, by which the cations bound from one side of the membrane get access to the other surface, where they dissociate. In isolated F_0, a membrane potential may trigger the conformational transition, because very low Na^+ conductivity was observed in its absence. In contrast, the Na^+ carrier monensin catalyzed a very rapid efflux of Na^+ ions from liposomes even in the absence of a membrane potential.

These results have important implications for the ATP synthesis/hydrolysis mechanism in general. Thus, all models in which protons have a specific role that cannot be performed by another cation as well, are unlikely to be correct. Our experiments clearly demonstrate that the cation binding site is situated in the F_0 sector of the molecule, not in F_1. We suppose that, in agreement with the model of Boyer [39], the binding of the cations to these sites on F_0, from the side of high electrochemical potential, triggers a conformational change which exposes the cations to the other surface of the membrane and brings about ATP synthesis on the F_1 moiety of the enzyme complex. After dissociation of the cations, the enzyme returns to its original conformation. It is also apparent that cation conduction through F_0 cannot proceed via a network of a hydrogen-bonded chain ("proton wire") [40], but more likely requires specific binding sites. As shown for crown ethers, such sites could bind either Na^+ or H_3O^+, indicating that H_3O^+ rather than H^+ may be the transported species [41].

REFERENCES

1 Dimroth P. Microbiol Rev 1987; 51: 320-340.
2 Dimroth P. Trans R Soc Lond B 1190; 326: 465-477.
3 Dimroth P. In: Hauska G, Thauer R, eds. 41. Colloquium Mosbach: The Molecular Basis of Bacterial Metabolism. Berlin, Heidelberg: Springer Verlag, 1990; 114-127.
4 Dimroth P. BioEssays 1991; 13: 1-6.
5 Dimroth P. FEBS Lett 1980; 122: 234-236.
6 Dimroth P, Thomer A. Biol Chem Hoppe-Seyler 1986; 367: 813-823.
7 Dimroth P, Thomer A. J Biol Chem 1990; 265: 7221-7224.
8 Dimroth P, Thomer A. Arch Microbiol 1989; 151: 439-444.
9 Hilpert W, Dimroth P. Nature (London) 1982; 296: 584-585.
10 Hilpert W, Dimroth P. Eur J Biochem 1983; 132: 579-583.
11 Buckel W, Semmler, R. Eur J Biochem 1983; 136: 427-434.
12 Hilpert W, Schink B, Dimroth P. EMBO J 1984; 3: 1665-1670.
13 Laubinger W, Dimroth P. Eur J Biochem 1987; 168: 475-480.
14 Laubinger W, Dimroth P. Biochemistry 1988; 27: 7531-7537.
15 Laubinger W, Dimroth P. Biochemistry 1989; 28: 7194-7198.
16 Tokuda H, Unemoto T. J Biol Chem 1982; 257: 10007-10014.
17 Tokuda H, Unemoto T. J Biol Chem 1984; 259: 7785-7790.
18 Unemoto T, Tokuda H, Hayashi M. The Bacteria 1990; 12: 33-54.
19 Dibrov PA, Kostyrko VA, Lazarova RL, Skulachev, VP, Smirnova IA. Biochim Biophys Acta 1986; 850: 449-457.
20 Krulwich TA, Ivey DM. The Bacteria 1990; 12: 417-447.
21 Hirota N, Imae Y. J Biol Chem 1983; 258: 10577-10581.
22 Maloy SR. The Bacteria 1990; 12: 203-224.
23 Blaut M, Müller V, Gottschalk G. The Bacteria 1990; 12: 505-537.
24 Efiok, BJS, Webster DA. Biochem Biophys Res Commun 1990; 173: 370-375
25 Dimroth P, Thomer A. Eur J Biochem 1983; 137: 107-112.
26 Schwarz E, Oesterhelt D. EMBO J 1985; 4: 1599-1603.
27 Schwarz E, Oesterhelt D, Reinke H, Beyreuther K, Dimroth P. J Biol Chem 1988; 263: 9640-9645.
28 Laussermair E, Schwarz E, Oesterhelt D, Reinke H, Beyreuther K, Dimroth P. J Biol Chem 1989; 264: 14710-14715.
29 Dimroth P, Hilpert W. Biochemistry 1984; 23: 5360-5366.
30 Hilpert W, Dimroth P. Eur J Biochem 1991; 195: 79-86.
31 Amann R, Ludwig W, Laubinger W, Dimroth P, Schleifer KH. FEMS Microbiol Lett 1988; 56: 253-260
32 Ludwig W, Kaim G, Laubinger W, Dimroth P, Hoppe J, Schleifer KH. Eur J Biochem 1990; 193: 395-399.
33 Kaim G, Ludwig W, Dimroth P, Schleifer KH. Nucl Acids Res 1990; 18: 6697.
34 Vik SB, Cain BD, Chun KT, Simoni RD. J Biol Chem 1988; 263: 6599-6605.
35 Cain BD, Simoni RD. J Biol Chem 1988; 263: 6606-6612.
36 Hoppe J, Sebald W. Biochim Biophys Acta 1984; 768: 1-27.
37 Fillingame RH. The Bacteria 1990; 12: 345-391.
38 Laubinger W, Deckers-Hebestreit G, Altendorf K, Dimroth P. Biochemistry 1990; 29: 5458-5463.
39 Boyer PD. FEBS Lett 1975; 58: 1-6.
40 Nagle JF, Morowitz HJ. Proc Natl Acad Sci USA 1978; 75: 298-302.
41 Boyer PD. Trends Biochem Sci 1988; 13: 5-7.

© 1992 Elsevier Science Publishers B.V. All rights reserved.
Molecular mechanisms of transport. E. Quagliariello, F. Palmieri, eds.

Probing the molecular mechanism of a P-type H^+-ATPase by means of fluorescent dyes.

G. Nagel, E. Bashi, F. Firouznia, and C.L. Slayman

Department of Cellular and Molecular Physiology, Yale School of Medicine,
333 Cedar Street, New Haven, CT 06510, U.S.A.

INTRODUCTION

Within the past 5 years the major emphasis in studies of membrane proteins, and especially transport proteins, has shifted from detailed biochemical and physiological description to elucidation of primary structures, and families of structures, via the powerful techniques of molecular biology. A burgeoning catalogue of primary structures, together with the growing X-ray crystallographic catalogue of soluble proteins (and the reaction center proteins of bacterial chromatophores; Ref. 1), is being used to predict secondary and tertiary structures, and it is apparent that rigorous crystallographic data will soon be coming available for membrane proteins generally. The central importance of this dazzling information explosion, of course, is the insight it should provide into how the molecular machines carrying out transport actually do work.

But a major obstacle in the way of detailed "structure-function" studies is a general lack of sensitive and easy functional assays. Measured increases, decreases, or even blocks of enzymatic activity or transport are crude assays; and the kinetic analysis required to define protein conformational intermediates--the classic tool of transport biophysicists--is far too laborious for use with more than a few single-site mutants of any one protein. A great practical need exists, therefore, for specific reporter molecules which can bind to transport proteins and signal conformation changes while themselves minimally biasing the transport reactions.

Several years ago Klodos and Forbush (2) observed that certain fluorescent styryl dyes, originally synthesized as membrane-voltage probes (3), seemed to act as specific reporters for the so-called $\mathbf{E}_2$ state of the sodium pump (Na^+,K^+-ATPase). Nagel and co-workers (4,5) followed with similar observations on the fungal plasma-membrane proton pump (H^+-ATPase). Later, Läuger (6) and Bühler et al. (7) argued that these dyes are electrochromic indicators of local field changes near the transport protein. The purposes of the present report, then, are to describe the basic behavior of the styryl dye RH-160 in reaction with fungal H^+-ATPase, including spectral observations which are *not consistent* with a pure electrochromic indicator mechanism; and to demonstrate use of the dye in deciphering the reaction pathway for the ATPase.

METHODS

Experiments were conducted with detergent-washed fragments of *Neurospora* plasma membrane, prepared as described by Bowman et al. (8). Such preparations consist of a mixture of flat sheets and broken vesicles, about five-fold enriched in H^+-ATPase, and containing a small admixture of protein and carbohydrate particles. The standard experimental buffer contained 100 mM MES-Tris at pH 6.7 and 1 μM RH-160. The excitation spectrum for the dye, added to asolectin liposomes or *Neurospora* membrane fragments, peaks at

510 nm; and the emission spectrum peaks near 700 nm. Total fluorescence emission was therefore measured through a 640-nm cut-off filter. Excitation spectra were determined by scanning (at 5 nm/sec) from 400 to 600 nm; and kinetic measurements were obtained with dual-beam excitation, using 550 nm for the test wavelength and 450 nm for the reference, on an SLM DW-2000 spectrophotometer with a differential fluorescence adaptor. Analysis was carried out via a graphics program written by Dr. Bliss Forbush III.

RESULTS

A generic structure for the common styryl dyes is shown in Figure 1, along with spectra obtained with RH-160 in *Neurospora* membranes under several conditions. [For RH-160, the chain-length subscripts **L**, **M**, and **N** are 4,2, and 3, respectively, and **R** = SO_3^-. Among the dozen dyes thus far tested with fungal ATPase, all showing effects equivalent to RH-160 have negative groups (SO_3^- or $PO_3^=$ at **R**); these include RH-295, RH-527, and RH-528.] Addition of Mg-ATP to the enzyme-dye mixture causes ca. a 3% decrease of apparent excitation efficiency centered at 500 nm, and subsequent addition of the $\mathbf{E_2}$-trapping agent orthovanadate (9) produces an apparent 6-9% enhancement of excitation efficiency centered at 540 nm, which can be seen most clearly in the rescaled difference spectrum

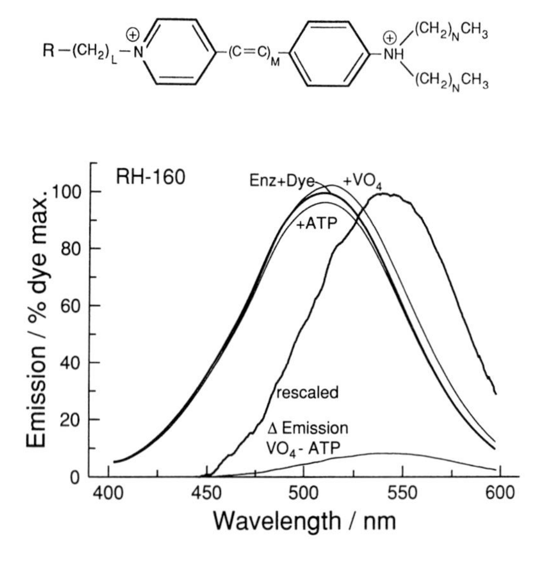

Figure 1. Generic structure and excitation spectra for styryl RH- dyes. See text for R, and values of L, M, N in RH-160. All spectra corrected for the dye-blank with *Neurospora* membrane fragments, and taken after serial addition of 1 μM RH-160, 1 mM Mg^{++} (not shown), 5 mM Mg-ATP, 100 μM orthovanadate. Each spectrum is the average for 3 or more trials. Difference spectrum rescaled to dye maximum, to show shape and noise.

(VO_4 - ATP) of Fig. 1. The fact that the vanadate-induced difference spectrum, following reaction of the enzyme-dye complex with magnesium and ATP, is almost entirely positive does *not* accord with an electrochromic indicator mechanism, which should produce an axial shift in the spectrum, with little alteration of excitation efficiency (10).

For the *Neurospora* plasma-membrane ATPase, the rate of rise (d**F**) of the vanadate-induced signal was found to be proportional to the steady-state level of $\mathbf{E_2}$ existing just prior to vanadate addition (4,5). That fact has made it possible to monitor $\mathbf{E_2}$ levels under various reaction conditions, and thereby to infer other steps in the reaction pathway. We shall concentrate here on measurements of changes in d**F** with concentration changes in three reagents: Mg-ATP, H^+, and salt (KCl).

ATP effects

The pH optimum for the *Neurospora* ATPase is near 6.7, and at that pH with low or moderate salt concentrations, the $K_{0.5}$ for ATP is 1.5-2.0 *m*M (8,11), compared with 1-100 μM for the animal P-type ATPases under dynamic conditions (12). For purified enzyme and purified ATP, the plot of ATPase velocity against increasing Mg-ATP concentration is very nearly hyperbolic, increasing monotonically and tending toward saturation at concentrations above 10 mM (13). The fluorescence change, d**F**, however does not rise monotonically, but peaks near 1.5 mM ATP--as reported by Nagel et al. (14) and illustrated in Figure 2--then declines at higher ATP concentrations. This surprising result suggests the occurrence of low-affinity ATP binding. But since the reduction of d**F** occurs *without reduction of ATPase activity*, the effective low-affinity binding must occur after the principal hydrolytic steps of the overall enzymatic reaction, withdrawing the enzyme to a vanadate-unreactive form.

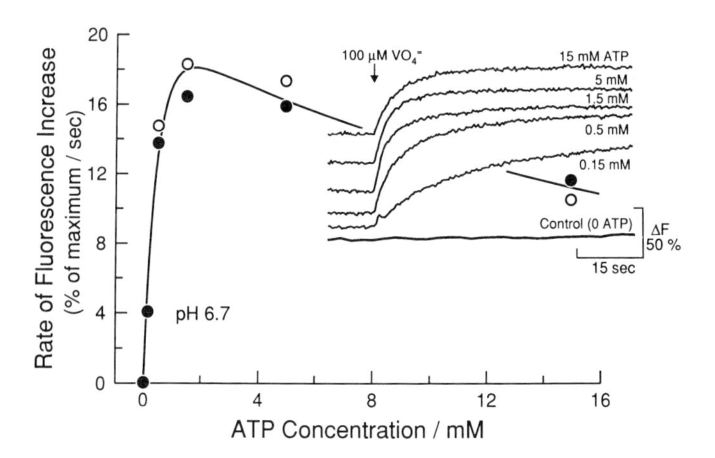

Figure 2. ATP-dependence of the vanadate-induced fluorescence increase (d**F**) by RH-160 plus *Neurospora* plasma-membrane H^+-ATPase. Note inhibition above 1.5 mM. Experiment run with duplicate samples. Filled circles represent initial slopes of traces shown by Inset. Smooth curve drawn by eye.

Proton effects

In the presence of millimolar ATP, the pH optimum for vanadate-induced fluorescence is near 6.7, and the overall pH-curve for **dF**, shown in Figure 3 (filled circles) is similar to that for ATP hydrolysis (8). But vanadate also induces a slow increase of RH-160 fluorescence in the absence of ATP, and that reaches a maximum of about 15% of the control value (for 5 mM ATP), but with a much higher pH optimum: near 8.4 instead of 6.7 (Fig. 3, open circles). At still higher pH's, **dF** with zero ATP is larger than with millimolar ATP, but the curves decline monotonically toward zero at pH values above 9.0. The increase of **dF** with falling H^+ concentration (rising pH, up to 8.4) cannot result from forward cycling of the enzyme, both because substrate ATP is absent and because substrate protons are declining. Instead, the effect must result from increasing enzyme accessibility to vanadate via the "back" reaction.

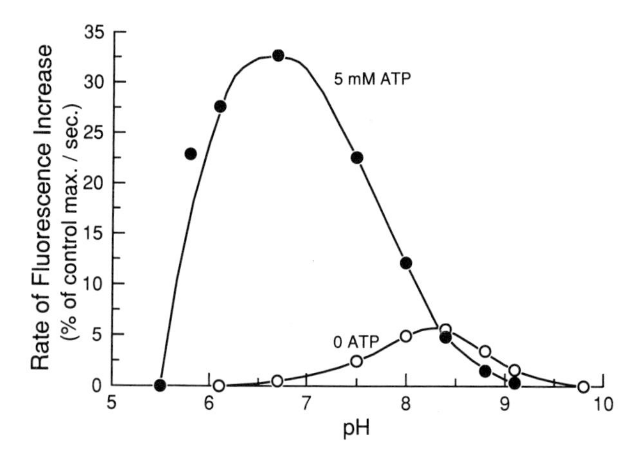

Figure 3. pH-dependence of vanadate-induced fluorescence changes, with and without ATP. Removal of ATP depresses **dF** 5-6 fold, but shifts pH optimum up nearly two units.

Salt effects

It has long been known that plant and fungal plasma-membrane ATPases are stimulated by elevated salt (15) . The H^+-ATPase of *Neurospora,* in particular, is maximally stimulated by 50-100 mM added salts: 25% for NaCl, 75% for KCl, and 100% for NH_4Cl (8). Although some authors have interpreted this kind of result from fungi (16) and especially from plants (17), to mean that alkali metal cations--usually K^+--are cosubstrates with H^+ (like K^+ with Na^+, for the animal sodium pump), no one has demonstrated stoichiometric co- or counter-transport of the two species. In fact, the extreme electrogenicity of proton pumping in *Neurospora* (18) effectively rules out coupling of any transported ions other than the single H^+, to ATP hydrolysis. That leaves open the question, however, of whether alkali metal cations might *substitute* for H^+ under certain conditions, just as H^+ can substitute for either K^+ or Na^+ in the animal sodium pump (19,20), or as Na^+ can substitute for H^+ in the gastric proton pump (21).

A direct approach to this question, via transport measurements in *Neurospora* membrane vesicles, had failed (C.P. Cartwright & C.L. Slayman, unpublished experiments), so we turned to the indirect method of searching for substrate-like enhancement of RH-160 fluorescence, with results as shown in Figure 4. Experiments were carried out at high pH (8.2) in order to reduce substrate competition from H^+ ions. Elevation of KCl in the reaction mixture, just prior to addition of 100 μM vanadate, only decreased d**F**. That decrease was modest, 20-40%, over the concentration range for stimulation of ATPase activity (50-100 mM), but reached 95% at molar salt, which also inhibits ATP hydrolysis by ca. 50% (E. Bashi & C.L. Slayman, unpublished results).

Thus, although elevating salt concentration in the buffer can stimulate ATPase activity and presumably speed cycling of the transport reaction, this maneuver--like inhibition of d**F** by high ATP or low pH--must also act by withdrawing enzyme from the vanadate-binding state, $\mathbf{E_2}$, not (substrate-like) by pushing enzyme into that state.

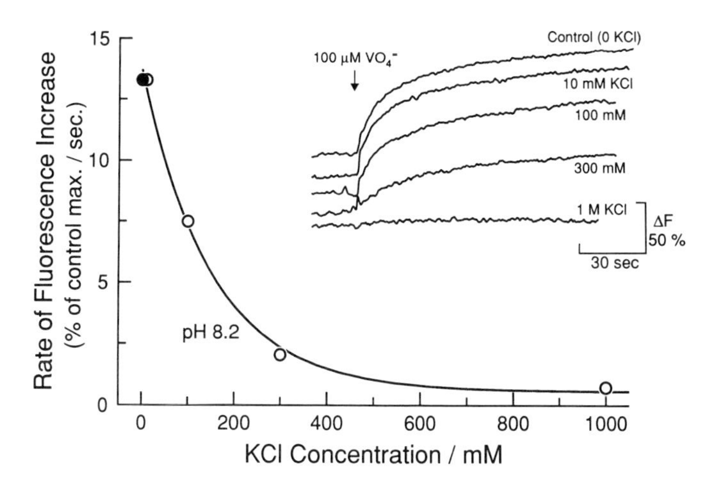

Figure 4. Inhibition of vanadate-induced fluorescence increase by potassium salt. Plotted points represent initial slopes of Inset curves (filled circle = control). Smooth curve fitted to a single exponential decay, with characteristic constant of 6.2 M^{-1}.

DISCUSSION

The three sets of results are most easily summarized and interpreted by reference to a specific reaction scheme for the H^+-ATPase, such as that drawn in Figure 5. It shows six principal states (conformations) of the enzyme, deduced largely by analogy with the better-known reaction scheme for the Na^+,K^+-ATPase (22).

Read clockwise in the top line:

$\mathbf{E_1}$, uncomplexed enzyme, having a high affinity for at least one of the substrates, H^+ or ATP

$H^+\mathbf{E_1}$ATP, fully complexed, but not yet phosphorylated, and

H^+E_1~P, the phosphorylated state, immediately "pretransport".

And read clockwise in the bottom line:

H^+E_2-P, the immediate "posttransport" state, still associated with H^+ and phosphate, but loosely

E_2-P, the deprotonated enzyme, and

E_2, the uncomplexed enzyme having low affinity for both substrates.

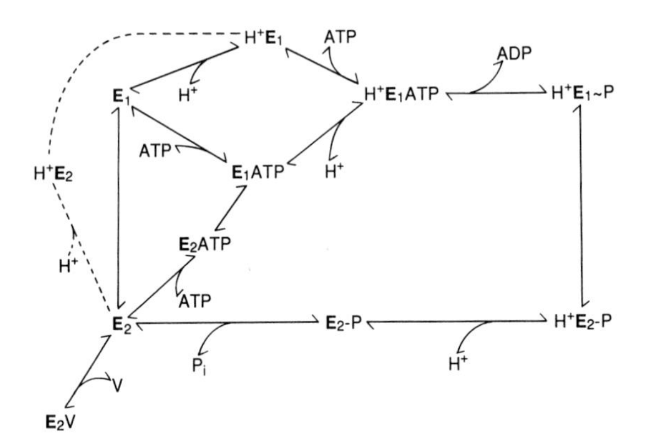

Figure 5. Reaction scheme for *Neurospora* plasma-membrane H^+-ATPase, to account for behavior of vanadate-induced d**F**, with changes of ATP, pH, and K^+. Basic scheme assumed from the animal Na^+,K^+-ATPase (22) with multiple-ion steps deleted, and from electrical-kinetic analysis of *Neurospora* H^+-pump (24,25). H^+E_2 and E_2ATP postulated specifically to account for results in Figs. 2-4 above.

In the presence of magnesium ions (23), orthovanadate can bind tightly with E_2, thus inactivating the enzyme. We do not know whether E_2 itself, or only E_2V is responsible for the characteristic fluorescence shift with RH-160, but the fact that the shift occurs for the Na^+,K^+-ATPase without vanadate suggests involvement of more than one E_2- form.

The enhancement of vanadate-induced d**F** by low concentrations of ATP is assumed to represent acceleration of the forward reaction (from E_1 clockwise through H^+E_1ATP to E_2) by a rate-limiting substrate. The other events described in Figs. 2-4 must involve the "return" segment of the cycle (clockwise from E_2 through E_1) since their direction of change is always inverse to effects on the overall rate of reaction cycling (i.e., the rate of ATP hydrolysis).

The easiest way to account for reduction of d**F** by high concentrations of H^+ at zero ATP is by reaction of H^+ with E_1 to form the "mainline" intermediate, H^+E_1, which is a deadend without ATP, and which would indirectly remove enzyme from the vanadate-reactive state E_2. Of course, direct reaction of H^+ with E_2 (dashed curve in Fig. 5), to produce a separate unreactive state, H^+E_2, is also possible; and a slow transition from H^+E_2 to H^+E_1 could also account for the decline of both ATP hydrolysis (8) and the (ATP-dependent) vanadate-induced **dF** below ~pH 6.7 (Fig. 3). Indeed, formation of a corresponding $\mathbf{E_2}$-intermediate with ATP seems the simplest way to account for the inhibitory effect of high ATP concentrations on **dF** (Fig. 2). The overall suggestion, therefore, is that $\mathbf{E_2}$ (or, more exactly, $\mathbf{E_2}$ complexed with RH-160) becomes vanadate insensitive after reaction with either ATP or H^+, but that the specific intermediate $H^+\mathbf{E_2}$, converts only slowly to a catalytically active form ($H^+\mathbf{E_1}$). Elevated salt (Fig. 4) would speed removal of $\mathbf{E_2}$ into either $\mathbf{E_1}$ or $\mathbf{E_2}$ATP.

Evidently, study of the interactions between varied pH, ATP concentration, and salt, in relation to the vanadate-induced **dF**, should further delimit the reaction steps between $\mathbf{E_2}$ and, say, $H^+\mathbf{E_1}$ATP, and such experiments are now underway.

Acknowledgements

Supported by Research Grant FG02-85ER13359 from the U.S. Department of Energy, and by a Brown-Coxe Fellowship from Yale University (to G.N.). Initial work supported by Research Grant GM-15858 from the U.S. National Institutes of Health.

REFERENCES

1 Deisenhofer J, Michel H. Ann Rev Biophys Biophys Chem 1991; 20:247-266.
2 Klodos I, Forbush B III. J Gen Physiol 1988; 92:46a, item 97.
3 Grinvald A, Hildesheim R, Farber IC, Anglister L. Biophys J 1982; 39:301-308.
4 Slayman CL, Klodos I, Nagel G. In: Dainty J et al., eds. Plant Membrane Transport: The Current Position. Amsterdam: Elsevier, 1989; 467-472.
5 Nagel G, Slayman CL, Klodos I. Biophys J 1989; 55:338a, item Tu-Pos224.
6 Läuger P. In: Kaplan JH, DeWeer P, eds., The Sodium Pump: Structure, Mechanism, and Regulation. New York: Rockefeller Univ. Pr. 1991; 303-315.
7 Bühler R, Stürmer W, Apell H-J, Läuger P. J Membr Biol 1991; 121:141-161.
8 Bowman BJ, Blasco F, Slayman CW. J Biol Chem 1981; 256:12343-12349.
9 Cantley LC, Cantley LG, Josephson L. J Biol Chem 1978; 253:7361-7368.
10 Loew LM. In: Loew LM, ed. Spectroscopic Membrane Probes. Boca Raton: CRC Press, 1988; Vol. 2: 139-152.
11 Slayman CL, Long WS, Lu CY-H. J Membr Biol 1973; 14:305-338.
12 Slayman CL, Zuckier GR. Ann N Y Acad Sci 1989; 574:233-245.
13 Bowman BJ, Slayman CW. J Biol Chem 1977; 252:3357-3363.
14 Nagel G, Klodos I, Xu J-C, Slayman CL. Biophys J 1990; 57:349a, item Tu-Pos473.
15 Hodges TK. In: Lüttge U, Pitman MG, eds., Transport in Plants II, Part A, Cells: 260-283.
16 Villalobo A. Can J Biochem Cell Biol 1984; 62:865-877.
17 Leonard RT. In: Martonosi A, ed., Membranes and Transport, Vol 2. New York: Plenum, 1982; 633-637.
18 Gradmann D, Hansen U-P, Long WS, Slayman CL, Warncke J. J Membr Biol 1978; 39:333-367.
19 Polvani C, Blostein R. J Biol Chem 1985; 260:16757-16763.
20 Blostein R. J Biol Chem 1985; 260:829-833.
21 Polvani C, Sachs G, Blostein R. J Biol Chem 1989; 264:17854-17859.

22 Skou JC, Norby JG, Maunsbach AB, Esmann M, eds. The Na,K-Pump. Part A: Molecular Aspects. New York: Alan R Liss, 1988; 655 pp.
23 Smith RL, Zinn K, Cantley LC. J Biol Chem 1980; 255:9852-9859.
24 Hansen U-P, Gradmann D, Sanders D, Slayman CL. J Membr Biol 1981; 63:165-190.
25 Slayman CL. J Bioenerg Biomembr 1987; 19:1-20.

TRANSPORT SYSTEMS IN BACTERIA

© 1992 Elsevier Science Publishers B.V. All rights reserved.
Molecular mechanisms of transport. E. Quagliariello, F. Palmieri, eds.

Sodium coupled transport in bacteria

B. Tolner, M.E. van der Rest, G. Speelmans and W.N. Konings.

Department of Microbiology, University of Groningen, Kerklaan 30, NL-9751 NN Haren, The Netherlands.

Introduction

In bacteria, energy consuming processes, such as solute transport, can utilize electrochemical ion gradients as a driving force. Although H^+ is often the central coupling ion in bacterial energy transduction, recently more and more evidence is accumulated that also Na^+ can perform this function. Na^+-ions are essential for growth of many marine, halophilic, alkaliphilic and rumen bacteria which live in Na^+ rich environments and are an important growth factor for methanogenic bacteria. Na^+ not only plays a role in Na^+-coupled solute-symport systems but also in pH homeostasis and in the activity of many enzymes (Table 1).

Table 1 Occurrence of sodium coupled transport

Environment	High salt	High pH	Extreme	Non extreme
Bacteria	Halophiles Marine bacteria Rumen bacteria	Alkaliphiles Alkalitolerants	Thermophiles Methanogenes	Enterobacteria
Mode of $\Delta\mu_{Na^+}$ generation	Na^+/H^+ antiport Respiratory chain Decarboxylase ATP-hydrolysis	Na^+/H^+ antiport	Na^+/H^+ antiport ATP-hydrolysis	Na^+/H^+ antiport Decarboxylase

The advantage of using Na^+-dependent transport systems is clear under specific culture conditions such as high environmental pH, high external sodium concentrations or elevated temperatures. The electrochemical gradient of sodium ions can be generated by primary Na^+-pumps, such as Na^+-pumping decarboxylases, ATP-ase and electron transport systems or with H^+/Na^+ antiport systems [1] (Fig. 1). Also in neutrophilic bacteria Na^+-dependent transport systems have been recognized. The advantage of having Na^+-dependent systems instead of H^+-dependent systems in these organisms is not always clear. Recently, several transport systems have been reported to transport a solute in symport with 2 different cations, H^+ and Na^+ [2,3,4,5,6].

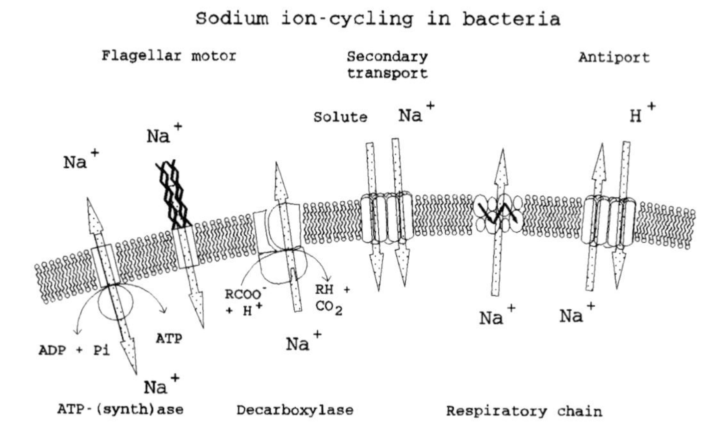

FIG. 1. Putative elements of a sodium cycle in a cell. Indicated are gradients generated by primary Na^+-pumps (Na^+-pumping decarboxylases, ATP-ases and redox-converting enzymes) and a H^+/Na^+ antiport system. Also the role of sodium in secondary transport processes, ATP-synthesis and the flagellar motor is shown.

This has been observed for the L-.glutamate transport systems from *Escherichia coli* ($GltS_{Ec}$) [2,3], *Bacillus stearothermophilus* ($GltP_{Bs}$) [4], *Bacillus caldotenax* ($GltP_{Bc}$) [5] and for a citrate transport system from *Klebsiella pneumoniae* [6]. Also transport systems have been reported which can use either H^+ or Na^+ as a coupling ion, such as the *E. coli* melibiose transport system [7].

Amino acid transport in thermophilic bacteria.

The transport of amino acids has been studied in three thermophilic bacteria; the anaerobic bacterium *Clostridium fervidus* and the aerobic bacteria *B. stearothermophilus* and *B. caldotenax*.
Clostridium fervidus is an anaerobic, thermophilic bacterium with an optimum temperature of growth of 68 °C, which can ferment peptides, amino acids and carbohydrates [8]. In membrane vesicles of *Cl. fervidus* uptake of neutral (serine, leucine, glycine, alanine), basic (lysine, arginine), acidic (glutamate, aspartate) and aromatic (phenylalanine) amino acids is driven by a Na^+ concentration gradient ($\Delta\mu_{Na^+}$) and an electrical potential ($\Delta\Psi$). A pH gradient never functions as a driving force for amino acid uptake [9]. As an example the uptake of serine driven by different artificial gradients at pH 7.0 is shown (Fig 2). The same results were found at pH 6.0 and 8.0. The transport systems for all amino acids have a low affinity for Na^+ (in the mM range).

Amino acid uptake driven by artificial gradients has a transient character which makes it impossible to determine the relation between the driving forces and the uptake of amino acids. Since in membranes of fermentative bacteria no suitable proton motive force generating systems are present, a foreign proton pump, such as cytochrome-*c* oxidase from the thermophilic bacterium *B. stearothermophilus*, has been introduced to generate a stable proton motive force [9]. At pH 6.0 the Na^+/H^+-exchanger monensin converts the pH gradient generated by cytochrome-*c* oxidase into a $\Delta\mu_{Na^+}$ and this enhances the rate and steady state level of uptake of L-serine (Fig 3).

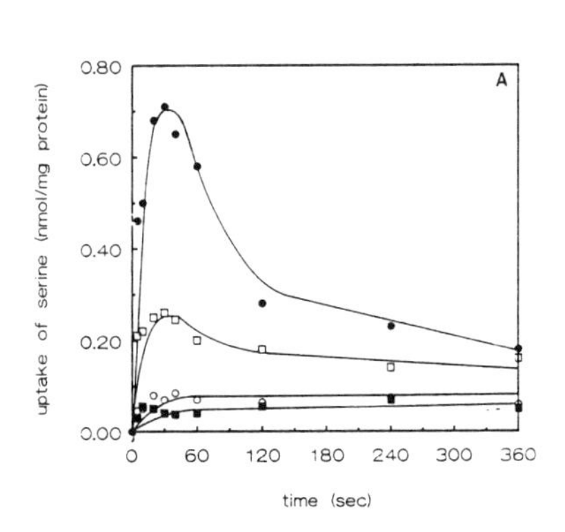

Figure 2 Uptake of L-serine driven by artificial gradients: a $\Delta\bar{\mu}_{Na^+}$ (□); $\Delta\bar{\mu}_{Na^+}$ and $\Delta\Psi$ (●); a ΔpH in the presence of sodium (>); no gradient (0). Uptake experiments were performed at pH 6.0 and 40°C.

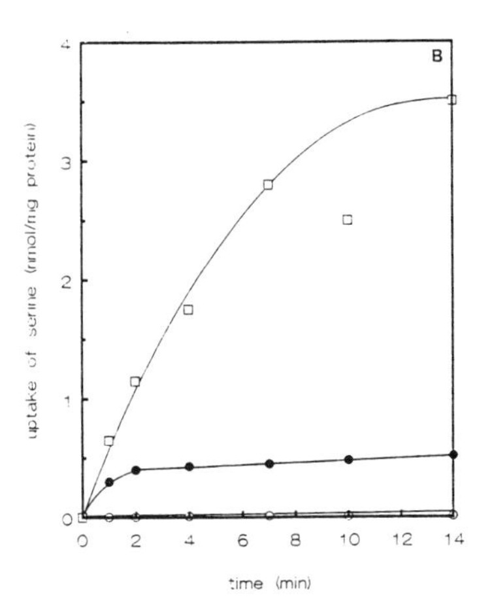

Figure 3. L-serine transport in membrane vesicles fused with cytochrome *c* oxidase containing liposomes in the absence (0) or presence (●) of reduced TMPD and Monensin (□).

The utilization of the *B. stearothermophilus* cytochrome *c* oxidase is restricted by the limitation of the acceptor concentration (O_2) at high temperatures. The incorporation of a light-driven Δp generator such as the reaction centre of the thermophilic phototrophic bacterium *Chloroflexus aurantiacus* can overcome this limitation (G. Speelmans, unpublished observations).

Amino acid transport was also studied in membrane vesicles of *B. stearothermophilus*. *B. stearothermophilus* is an aerobic thermophilic bacterium that grows optimally at 63°C. In this organism the uptake of different amino acids such as alanine or leucine also occurs in symport with sodium ions [4]. The basic amino acid l-lysine, however, appears to be transported by an electrogenic uniport mechanism [4].

Transport of the acidic amino acid L-glutamate was found to occur in symport with Na^+ and H^+ in a 1:1:1 stoichiometry [4, 10]. The glutamate transport protein is very thermostable. At 70°C a reduction of the activity by 50% occurs in 10 min [10]. The gene encoding this transport system has been cloned and sequenced [11]. Another thermophile *B. caldotenax* does have a higher optimum for growth (70° vs. 63 °C of *B. stearothermophilus*), and grows much faster on glutamate as sole source of energy, nitrogen and carbon than *B. stearothermophilus* (t_d: 30 min vs. 5 h for *B. stearothermophilus*). The L-glutamate transport gene of *B. caldotenax* was also isolated aand characterized (5). The cloning strategy of these genes was based on the complementation of an *E. coli* K12 strain, since K12 wild-type strains grow only poorly on glutamate as sole source of energy, nitrogen and carbon, due to a repressed uptake system for this solute [12]. Putative promoter, rho-independent terminator and ribosome binding site sequences could be assigned to both sequences.

The -35 and -10 promoter regions of the *glt*$_{Bs}$ and *glt*$_{Bc}$ genes (-35: TTGACA; -10: TAGAAT), and the spacing of 17 bp between the two regions, is similar to that of the consensus promoter sequences (-35: TTGACA; -10: TATAAT) that are recognized by the sigma^{43} and sigma^{70} factors of the holoenzym forms of RNA polymerase of *B. subtilis* [13,14] and *E. coli* [13,15], respectively. This suggests the existence of a sigma factor in *B. stearothermophilus* which is similar to its counterparts sigma^{43} in *B. subtilis* and sigma^{70} in *E. coli* that are involved in transcription of housekeeping functions.

The *gltP*$_{Bs}$ and *gltP*$_{Bc}$ genes encodes a protein of 421 residues, which corresponds with a molecular weight of 45 kDa. Alignment of these two amino acid sequences reveals extensive homology (96.7% identity). The amino acid sequences of the Na^+/H^+/glutamate transport proteins of *B. stearothermophilus* (GltP$_{Bs}$) and *B. caldotenax* (GltP$_{Bc}$) were also compared with the amino acid sequences of the H^+/glutamate transport protein of *E. coli* K12 (GltP$_{Ec}$) [16], and the Na^+/glutamate transport proteins (GltS$_{Ec}$) of *E. coli* B [2] and *E. coli* K12 [3]. Extensive similarity was found between the thermophilic Na^+/H^+/glutamate transport proteins and the H^+/glutamate transport system (GltP$_{Ec}$). The overall homology between these three sequences was 88.6%, of which 57.2% were identical and 31.4% were conserved changes.

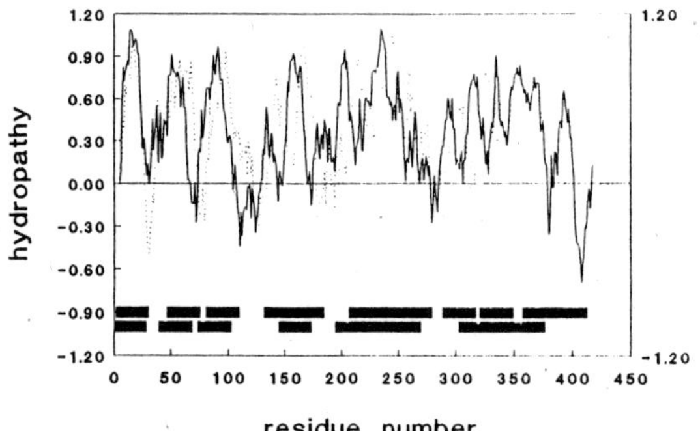

Figure 4. Hydropathy profiles of the $GltP_{Bs}$ and $GltP_{Ec}$ glutamate transport proteins. The hydrophobicity was calculated according to the method of Eisenberg *et al.* [19] with a window of 21 amino acids. Solid line: $GltP_{Bs}$; dashed line: $GltP_{Ec}$. The position of the putative membrane spanning segments in $GltP_{Bs}$ and $GltP_{Ec}$ are indicated by the solid bars.

These three proteins, furthermore, have the same substrate specificities (glutamate and aspartate) and are inhibited by cysteate (data not shown). However, the $GltP_{Ec}$ differs from $GltP_{Bs}$ and $GltP_{Bc}$ in ion-specificity (H^+ vs. Na^+ and H^+ for $GltP_{Bs}$ and $GltP_{Bc}$) and thermostability.
No similarity was found between the glutamate transport proteins of the thermophilic bacilli and any other protein, except for some local similarity with other Na^+ dependent transport proteins. The conserved amino acid sequence which has been proposed to be involved in Na^+ recognition or binding (__-G-__-A----L---GR-__) [2] and which was found in the Na^+/glutamate transport protein of *E. coli* [2], the Na^+/proline transport protein of *E. coli* [18] and the Na^+/glucose co-transporters of rabbit and human intestines [19, 20], can also be found in the $GltP_{Bs}$ and $GltP_{Bc}$ proteins (__-G_{38}-__-A_{62}----L_{67}---$G_{71}K_{72}$__) except for one mismatch (conserved change). Interestingly, apart of one mismatch the SOB-motif can also be found in the $GltP_{Ec}$. Since the $GltP_{Bs}$ and $GltP_{Ec}$ show extensive similarity and mainly differ in their mechanism of energy coupling the mismatch might explain the inability of $GltP_{Ec}$ to use Na^+ as coupling ion.
The method of Eisenberg *et al.* [17] predicts for both thermophilic proteins 10 membrane spanning regions. These regions are at similar locations when compared to the location of 10 of the 12 membrane spanning segments predicted for the *E. coli* H^+/glutamate transport protein (Fig. 4). These three proteins are very homologous, have a similar hydropathy prophiles, are all inhibited by cysteate (data not shown) and transport the same substrates (glutamate and aspartate). The number of membrane spanning segments is likely to be the same for all three proteins and most probably 12.

Citrate transport in the mesophilic bacterium Klebsiella pneumoniae.

K. pneumoniae is a mesophilic bacterium that can grow aerobically and anaerobically on citrate as a sole carbon and energy source. Anaerobic growth of *K. pneumoniae* on citrate is dependent on a citrate transport system which has been cloned by Schwarz and Oesterhelt [21]. Dimroth and Thomer [22,23] showed that the citrate transport system via this system is sodium dependent. The genes of the oxaloacetate decarboxylase, one of the enzymes involved in the fermentation of citrate during anaerobic conditions have been isolated together with the gene for the citrate transport system [6,21,24]. Oxaloacetate decarboxylase is a biotin-containing membrane protein that consists of three subunits. Dimroth discovered that this enzyme acts as a Na^+ pump [1]. The free energy of decarboxylation is used to translocate Na^+ ions out of the cell. Under purely fermentative conditions this decarboxylation thus leads to the generation of an electrochemical gradient of sodium ions which subsequently drives Na^+-citrate symport. Since anaerobic degradation of citrate by *K. pneumoniae* yields only one ATP by substrate level phosphorylation, the use of the electrochemical gradient of Na^+ for uptake of citrate is energetically very attractive. Under aerobic conditions citrate is taken up by *K. pneumoniae* in symport with H^+ by a completely different transport system [25]. In solution citrate is composed of different protonated species. The uptake of citrate via the proton dependent citrate carrier of *K. pneumoniae* was shown to be strongly pH dependent [26]. H-citrate^{2-} was found to be accumulated via this transport system in symport with 3 protons. Uptake of citrate by the anaerobic transport system can be driven by imposed ΔpH, $\Delta\Psi$ or $\Delta\tilde{\mu}_{Na^+}$ (Fig 5). In analogy with the aerobic transport system these observations suggest that also anaerobically H-citrate^{2-} is symported with 3 positive ions of which at least one is a Na^+.

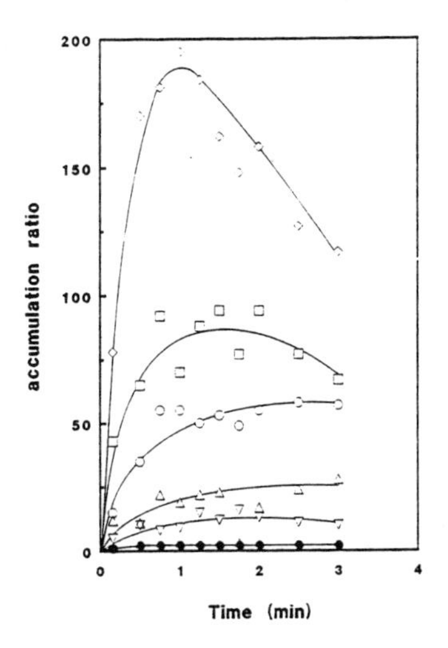

Figure 5. *Citrate uptake driven by artificially imposed ion gradients.* Driving forces: Δp and $\Delta\tilde{\mu}_{Na^+}$ (◇), ΔpH (○), ΔpH in the presence of Na^+ but no $\Delta\tilde{\mu}_{Na^+}$ (□), $\Delta\Psi$ and $\Delta\tilde{\mu}_{Na^+}$ (△), $\Delta\tilde{\mu}_{Na^+}$ (▽), no gradient (●).

The gene coding for the anaerobic Na^+-dependent citrate transport system has been cloned and sequenced [6]. The open reading frame of 1338 nucleotides codes for a protein of 446 amino acids. The molecular mass of the protein calculated from the amino acid sequence is 48 kDa. The amino acid composition is typical for membrane proteins with alternating hydrophobic and hydrophilic segments. The citrate carrier has an overall similarity of 76% with a citrate transport protein from *Lactococcus lactis* subsp *lactis* var *diacetylactis* [27]. The ion-dependency of this latter system has not yet been established. The hydropathy profiles of both carriers do not resemble the model proposed by Baldwin and Henderson [28] for a family of transport proteins, with 6 α-helices on either side of a central hydrophilic loop. Both carriers are asymmetric proteins with 7 α-helices in front and 5 after the hydrophilic loop. The Na-citrate transport protein of *K. pneumoniae* is not homologous with the glutamate transport proteins discussed above. However, the SOB-motif found in the glutamate carriers of *B.stearothermophilus* and *B.caldotenax* is also found in the citrate carrier of *K.pneumoniae*.

Concluding remarks

The thermophilic neutrophilic bacteria *B.stearothermophilus*, *B.caldotenax* and *Cl.fervidus* can use electrochemical gradients of sodium-ions for secondary transport processes. Amino acid transport in these bacteria is Na^+-dependent. It can be advantageous for a thermophilic bacterium to convert energy via sodium ion cycling. *B.stearothermophilus* adapts the phospholipid composition so that the membrane fluidity is optimal at growth temperature [29]. However, in doing so the membrane becomes very leaky for protons at high temperatures what leads to an inefficient transduction of redox (respiration) energy into electrochemical energy of protons at the growth temperature. Since the cytoplasmic membrane is less permeable for Na^+, a more efficient energy conversion with Na^+ would be possible. The aerobic respiration in *B.stearothermophilus* and *B.caldotenax* however, leads to the generation of electrochemical gradient of protons. The sodium ion gradient is subsequently generated with a Na^+/H^+ antiport system and not by a primary sodium pump. The advantage of transporting solutes in symport with sodium ions is therefore less obvious. One possible advantage could be that a sodium gradient at these elevated temperatures is less sensitive to fluctuations of the environment. *K.pneumoniae* uses a sodium gradient for different reasons. At 37°C there is no apparent use for secondary Na^+-transport systems. Fermentation of citrate yields only one ATP. The generation of electrochemical energy of protons by ATP-hydrolysis and transport of citrate via a H^+/citrate symport system would establish a futile cycle. However, oxaloacetate decarboxylase, a crucial enzyme in the fermentation pathway of citrate, converts the free energy of the decarboxylation reaction into an electrochemical gradient of Na^+. Thus, under purely fermentative conditions, a sodium gradient is established without ATP hydrolysis. This sodium gradient is subsequently used in citrate uptake. In this way the fermentation of citrate contributes to metabolic energy.

Literature cited

1 Dimroth P. Microbiological Reviews. 1987; 51: 320-340.
2 Deguchi Y, Yamato I, Anraku Y. J.Biol.Chem. 1990; 265: 21704-21708.
3 Kalman M, Gentry DR, Cashel M. Mol.Gen.Genet. 1991; 225: 379-386.
4 Heyne RIR, de Vrij W, Crielaard W, Konings WN. J.Bact. 1991; 173: 791-800.
5 Tolner B. (1991) Unpublished data.
6 v/d Rest ME, Siewe RM, Abee T, Schwarz E, Oesterhelt D, Konings WN. manuscipt in preparation.
7 Tsuchiya T, Lopilato J, Wilson TH. J.Membr.Biol. 1978; 42: 45-59.
8 Patel BKC, Monk C, Littleworth H,Morgan HW, Daniel RM. Int.J.Syst.Bacteriol. 1987; 37: 123-126.
9 Speelmans G, de Vrij W, Konings WN. J. Bact. 1989; 171: 3788-3795.
10 de Vrij W, Bulthuis RA, van Iwaarden PR, Konings WN. J.Bact. 1989; 171: 1118-1125.

11 Tolner B, Poolman B, Konings WN. Manuscript in preparation.
12 Halpern YS, Lupo M. J.Bact. 1965; 90: 1288-1295.
13 Helmann JD, Chamberlin MJ. Ann.Rev.Biochem. 1988; 57: 839-872.
14 Moran CP, Lang N, LeGrice SFJ, Lee G, Stephens M, Sonenshein AL, Pero J, Losick R. Mol.Gen.Genet. 1982; 186: 339-346.
15 Hawley DK, McClure WR. N.A.R. 1983; 1: 2237-55.
16 Tolner B, Poolman B, Konings WN. Manuscript in preparation.
17 Eisenberg D, Schwarz E, Komarony M, Wall R. J.Mol.Biol. 1984; 179: 125-142.
18 Nakao T, Yamato I, Anraku Y. Mol.Gen.Genet. 1987; 208: 70-75.
19 Hediger MA, Coady MJ, Ikeda TS, Wright EM. Nature 1987; 330: 379-381.
20 Hediger MA, Turk E, Wright EM. PNAS USA 1989; 86: 5748-5752.
21 Schwarz E, Oesterhelt D. EMBO J. 1985; 4: 1599-1603.
22 Dimroth P, Thomer A. Biol.Chem.Hoppe-Seyler 1986; 376: 813-823.
23 Dimroth P, Thomer A. J.Biol.Chem. 1990; 265: 7221-7224.
24 Laussermair E, Schwarz E, Oesterhelt D, Reinke H, Beyreuther K, Dimroth P. J.Biol.Chem. 1989; 264:14710-14715.
25 David S, v/d Rest ME, Driessen AJM, Simons G, de Vos WM. J.Bact. 1990; 172: 5789-5794.
26 v/d Rest ME, Schwarz E, Oesterhelt D, Konings WN. E.J.B. 1990; 189: 401-407.
27 v/d Rest ME, Abee T, Molenaar D, Konings WN. E.J.B. 1991; 195: 71-77.
28 Baldwin SA, Henderson PJF. Ann.Rev.Physiol. 1989; 51:459-471.
29 Reizer J, Grosswitz N, Barenholz Y. Biochim.Biophys.Acta 1985; 815: 268-280.

© 1992 Elsevier Science Publishers B.V. All rights reserved.
Molecular mechanisms of transport. E. Quagliariello, F. Palmieri, eds.

The K^+-uptake systems TrkG and TrkH From *Escherichia coli*: A pair of unusual transport systems involved in osmoregulation

Andreas Schlösser, Angela Hamann[a], Manfred Schleyer, and Evert P. Bakker

Abteilung Mikrobiologie, Universität Osnabrück, Postfach 4469, D-4500 Osnabrück, Germany

[a], Present address, Lehrstuhl für Technische Mikrobiologie, Universität Dortmund, Postfach 500500, D-4600 Dortmund 50, Germany

SUMMARY

Escherichia coli contains a pair of constitutive, high rate, low affinity K^+-uptake systems, TrkG and TrkH. These systems play a role in the adjustment of the turgor pressure after the cells have been subjected to a hyperosmotic or a hypoosmotic shock. TrkG and TrkH have a similar structure. TrkH is the natural *E. coli* system. The *trkG* gene appears to have entered the cell with the lambdoid prophage *rac* and may originate from an Enterobacterium with a genome low in G+C.

Cell K^+ and osmoadaptation

Cells of *Escherichia coli* that grow in a minimal salt medium of 200 mOsm contain a K^+ concentration in their cytoplasm of about 300 mM (Fig. 1). The most important function of this cell K^+ is its contribution to the turgor pressure of the cells [1-3]. Bacteria possess turgor pressures of the order of -0.1 to -0.4 MePa [4]. During osmoadapation the cells must vary the concentation of solutes in the cytoplasm in order

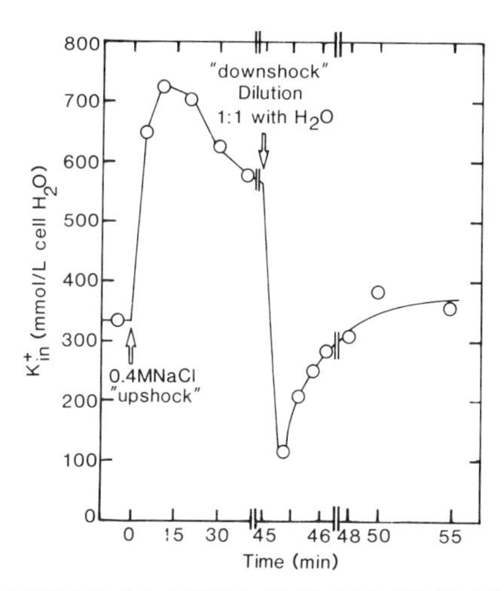

Fig. 1: Cells of *E. coli* strain LB322 (*ΔkdpABC5 kefB*::Tn*10 kefC*121-1 [7]) were grown at 37°C in a minimal mineral medium of 0.2 Osm with glucose as a carbon source. At zerotime the osmolarity of the medium was increased to 0.8 Osm by the addition of 0.4 M NaCl. At t = 45 min. the medium was diluted 1:1 with water of 37°C.

to maintain their turgor pressure within this range. Fig. 1 illustrates the role of K^+ in this process. When growing *E. coli* cells are put under accute osmotic stress by increasing the osmolarity of the medium from 200 to 800 mOsm (hyperosmotic shock or "upshock") the cells will respond by more than doubling their internal K^+ concentration. Concomitantly, the cells synthesize glutamate, which serves as the major counterion to the extra K^+ taken up by the cells [5]. Thus, by accumulating K^+ and glutamate the cells restore their turgor pressure.

At around twenty minutes after upshock *E. coli* starts to change the pattern of its internal solutes. Part of the K^+ and glutamate is replaced by trehalose (Fig. 1, and [5-7]). Moreover, in the presence of betaine or proline in the medium, the cells will replace their extra K^+, glutamate and trehalose by these zwitterions. High concentrations of the "compatible solutes" betaine, proline or trehalose have a much less severe influence on the structure of the cell water than does the small anorganic cation K^+ [3,4,8]. Thus, at high osmolarity compounds like betaine or trehalose allow the enzymes in the cytoplasm a relatively normal function [4,8].

The response of *E. coli* to a hypoosmotic shock ("downshock") is also biphasic. Accute stress was induced by decreasing the osmolarity of the medium from 0.8 to 0.4 Osm. Under these conditions the turgor pressure of the cells increased to -4 MePa and the cells have to release solutes in order to avoid bursting. Within seconds after osmotic downshock the cells lost 70-90% of their K^+ (Fig. 1), glutamate and trehalose [7], leading to a complete collapse of cell turgor. Remarkably, the release of these solutes is specific, since after downshock i) sucrose does not enter the cytoplasm, and ii) the cells loose neither alanine nor ATP [7]. It is not clear whether this extremely rapid solute release occurs via stretch-activated channels [9,10] or via other transport systems.

Within a minute after downshock the cells start to reaccumulate K^+ and glutamate (Fig. 1 and [7]). By this process *E. coli* restores its turgor pressure to a small negative value and is able to resume growth within 10 min after osmotic downshock [7].

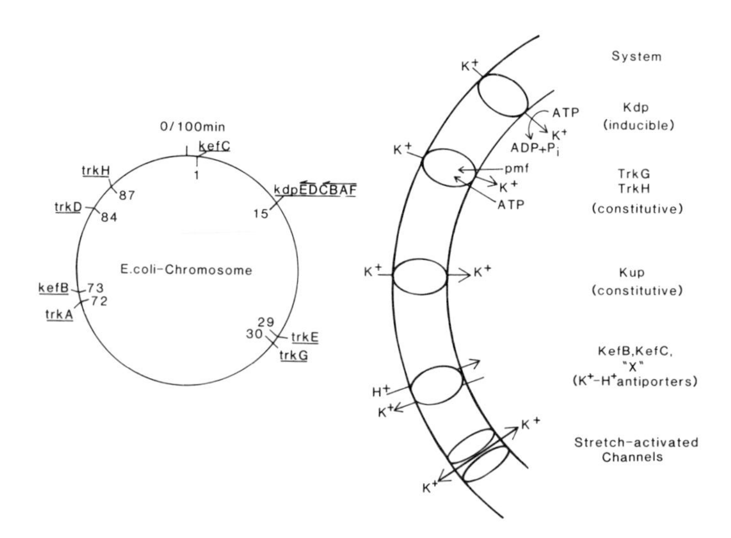

Fig. 2: K^+-transport genes and K^+-transport systems in *E. coli*. In the left-hand part the numbers (in min) indicate the positions of the genes on the *E. coli* chromosome according to the genetic map [13].

K^+-Transport genes and K^+-transport systems

The K^+ transport processes in Fig. 1 are catalyzed by separate K^+-uptake and K^+-efflux systems. *E. coli* possesses several of each [1,2,11]. The genes that encode K^+-transport proteins are shown in Fig. 2A. The *kdp* genes are organized in two operons. The other K^+-transport genes lay separate on the chromosome (Fig. 2A; [2,12,13]).

The genes *kefB* and *kefC* encode the K^+ efflux systems KefB and KefC, respectively. KefB and KefC have very similar properties and function as K^+-H^+ antiporters (Fig.2B, [11]). The experiment of Fig. 1 was carried out with a strain in which KefB and KefC were inactive. Hence, the K^+ efflux shown in it must have occurred via an additional system. It is not known whether this system is an antiporter ("X" in Fig 2B) or a strech-activated channel (see above).

K^+ uptake can be mediated by at least four systems. The Kdp system consists of the products of the *kdpABC* genes. They are only expressed by cells with a low turgor pressure. This condition is sensed by the *kdpD*-gene product that is located in the cell membrane. KdpD is thought to transmit a signal to the *kdpE*-gene product that may serve as a regulator of transcription of the *kdpFABC* operon ([2,14] and M. Walderhaug and W. Epstein, in preparation). The Kdp system is a K^+-translocating ATPase with a high affinity for K^+ (K_m about 2 μM [2,14,15]). At low turgor pressure Kdp is more active than at high turgor pressure [16]. Hence, the two phases of K^+ uptake shown in Fig. 1 could also have been mediated by Kdp, if a *kdp*$^+$ *trkA* strain had been used.

Besides Kdp *E. coli* possesses three constitutive K^+-uptake systems, TrkG, TrkH and Kup (or TrkD, i.e. the *trkD*-gene product [2,12,17]). These systems transport K^+ with low affinity (K_m between 0.1 and 2 mM). The V_{max} value of the TrkG and TrkH systems is at least 5 times higher than that of Kup. In addition Kup becomes inactive at medium osmolarities above 0.3 Osm. Thus, under most conditions of growth of wild-type cells the K^+-uptake shown in Fig. 1 occurs via the TrkG and TrkH systems.

The TrkG and TrkH Systems

Besides *trkD* Epstein and colleagues have identified four other *trk* genes, *trkA, trkE, trkG* and *trkH* [2,12,18]. Until very recently, the products of these four genes were thought to contribute to the activity of a single K^+-uptake system, Trk (e.g. [2]). However, Dosch et al. have proposed that Trk consists of two very similar systems, TrkG and TrkH [12]. Recent data support this notion (see below). The TrkG system is thought to consist of the integral membrane protein TrkG and the peripheral membrane protein TrkA. In addition, the *trkE*-gene product increases the affinity of the system [12]. The TrkH system has a similar composition, except that TrkH replaces TrkG, and that it shows a stronger requirement for TrkE [12].

The TrkG and TrkH systems have unusual properties

Before the distinction between the TrkG and TrkH systems was made Epstein and coworkers had established that the Trk system shows a number of remarkable features [2]. The same is likely to be true for both the TrkG and the TrkH systems.

First, the TrkG/TrkH systems require for activity both a high transmembrane proton motive force (pmf) and a high ATP concentration inside the cells [19]. Subsequent work has suggested that ATP plays a regulatory role rather than that it serves as an energy source for transport [20]. This leaves the pmf as the driving force for K^+ uptake. The membrane potential is often too low to energize K^+ uptake according to a uniport mechanism [21]. Several groups have therefore proposed that bacterial Trk-like K^+-uptake systems function as K^+-H^+ symporters [22-25]. For none of these systems it has, however, been shown that K^+ uptake is accompanied by alkalinization of the medium.

The second remarkable feature of the TrkG/TrkH systems is their strong regulation by the turgor pressure. This property is not unique, since Kdp [16] and the proline-uptake system ProP (also PPII [26]) show similar features. In swollen cells the TrkG/TrkH systems transport K^+ with a velocity that is only a few % of that of cells in which the turgor pressure has collapsed (e.g [17]). Physiologically, this down regulation of K^+-uptake activity makes sense, since in this manner the cells prevent rapid futile cycling of K^+ across the cell membrane, and thereby the waste of energy.

Third, the subunit composition of the TrkG/TrkH systems may be more complex than that of other secondary transport systems, which consist of one single type of subunit. The evidence that the TrkG/TrkH systems are secondary porters is circumstantial (see above). However, TrkG/TrkH translocate K^+ with a turn-over rate of at least 10^3 . complex^{-1}. s^{-1} [27]. Such a number is typical for a secondary porter and too high for a primary pump. If then the TrkG/TrkH systems indeed consist of more than one type of subunit, this must mean that the additional subunits confer regulation to the porters.

The TrkA protein

Some mutations in *trkA* inactivate the "Trk system" completely. Since mutations in other single *trk* genes caused smaller effects, it was originally thought that the TrkA protein forms the core of the "Trk complex". Studies with antibodies have shown, however, that TrkA is a peripheral membrane protein bound to the inner side of the cell membrane and that a large portion of the protein also occurs in the fraction of the soluble proteins [27]. Membrane binding of TrkA is necessary for K^+ transport, since in a strain that carries point mutations in both the *trkG* and *trkH* genes and does not transport K^+ via the Trk systems, TrkA occurred in soluble form only. Membrane binding of TrkA and K^+-uptake by the double mutant was restored by transformation of this strain with a multiple-copy plasmid that either carried the wild type *trkG* or *trkH* gene [27], suggesting that the TrkG or TrkH proteins are involved in the anchouring of TrkA to the membrane.

The amino acid sequence deduced from the nucleotide sequence of *trkA* supports the view that TrkA is not an integral membrane protein. Among its 458 residues it does not contain a stretch of hydrophobic amino acids long enough to traverse the membrane as an α-helix [28]. The Diagon plot of the TrkA sequence suggests that the protein forms an internal dimer [28]. At the N-terminus of each half TrkA contains a Rossmann fold:

GAGQVG (residues 7 to 12), and
GNIGAG (residues 241 to 246).

This sequence occurs in a variety of enzymes that interact with nucleotides, ranging from ATPases [29] to dehydrogenases that employ NADH as a coenzyme [30]. We have attempted to show that the membrane-bound form of TrkA binds [^{32}P]-8-azido-ATP or is phosphorylated by [γ-^{32}P]-ATP. However, these experiments gave negative results, which may be due to the low copy number of TrkA (20-50 copies/cell) and the fact that only a small portion of TrkA is bound to the membrane [27].

The *trkG* gene is not an intrinsic *E. coli* gene

The *trkG* gene is located within the region of the lambdoid prophage *rac* [12,13]. The transcript of *trkG* is much longer than one would have expected on the basis of the size of the gene, which may suggest that it is a prophage transcript. The nucleotide sequence of *trkG* indicated that it is not an intrinsic *E. coli* gene: the G+C content of *trkG* is 37%, which is much lower than that of other *E. coli* genes (average 51%]. In addition, the codon preference plot of *trkG* shows that the gene contains random rather than typical *E. coli* codons [31].

The TrkG protein

TrkG is an integral membrane protein and contains between 10 and 12 transmembrane helices [31]. Fig. 3A gives a scheme of the folding of the protein in the membrane, assuming that the protein crosses the membrane ten times. This folding pattern puts the positive charges (closed circles) at one side of the membrane, most likely the cytoplasm [32]. According to the 10 helix model TrkG contains large hydrophilic loops both at the cytoplasmic and at the periplasmic side of the membrane (Fig. 3A).

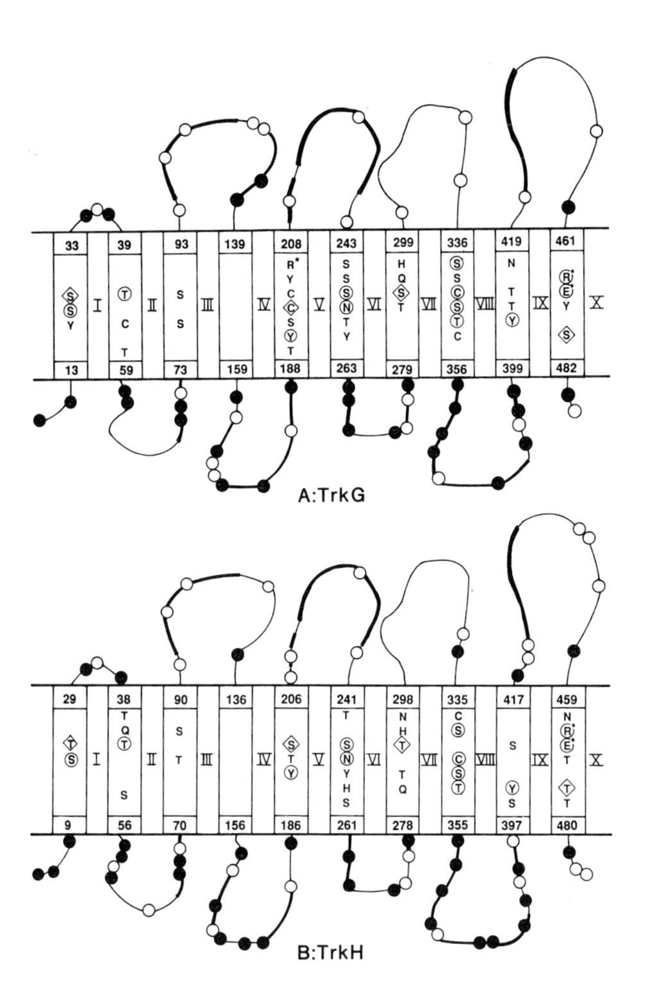

Fig. 3: Comparison of the potential membrane folding of TrkG (A) and TrkH (B) according to a 10-helix model. The closed and open circles indicate positively (K,R) and negatively charged residues (D,E), respectively. For further details see the text.

The peptide FKDDKLTPRLADY (residues 172 to 183) located within the loop between helices IV and V was synthesized by Dr. E. Meyer, Max Planck Institut für Biochemie, Munich, Germany and was used to generate antibodies. With these antibodies we could only detect TrkG in membranes derived from cells containing *trkG* in a multiple copy plasmid, but not in wild type cells. Comparison of the intensities of radioactively labeled TrkG protein in cells that overproduce it to different extents suggest that the wild type contains less than 200 copies of TrkG/cell.

The TrkH protein

Ken Rudd (NIH, Bethesda, USA) and Wolf Epstein (Chicago) have notified us that TrkG (485 residues [31]) is 40% identical to the product of an open reading frame of 421 residues (ORF421 [33]). However, ORF421 is likely to contain one nucleotide too many close to its 3`-end. Without it ORF421 becomes ORF483. The comparison between TrkG and the ORF483-gene product show that the they contain 40% identical and 18% conservatively exchanged amino acids throughout their complete sequence. Epstein (private communication) has established that ORF483 is indeed *trkH*. Its G+C content is 51% (14 % higher than *trkG*!) and its codon-preference plot suggests that *trkH* is a weakly expressed *E. coli* gene.

Comparison between TrkG and TrkH

Fig. 3 compares the folding of TrkG (A) and TrkH (B) through the membrane according to the 10-helix model. It shows that the distribution of the charged residues at the two sides of the membrane is very similar for the two proteins. The residues R and E inside helix X are also present in both proteins. However, the positioning of two potentially charged residues inside one trans-membrane will make this helix rather polar. Hence, it could also be that helix X and the C-terminus of the two proteins are located at the periplasmic side of the membrane. We will test the model of Fig. 3 with gene-fusion experiments [34].

In Fig. 3 the thick lines within the loops between the helices indicate regions in which the similarity of the two proteins is more than 70%. Such regions are found at both sides of the membrane. They may be involved in K^+ binding, and, at the inside, in the anchoring of the TrkA protein to the membrane (see above). In particular the loops between helices IV and V, and between helices VIII and IX may be important in the latter process.

Trans-membrane helices IV, X, I and XII show with 90, 71, 66, and 66%, respectively more than average similarity. Helix IV only contains apolar residues. In the other helices the approximate positions of the polar residues and tyrosine is indicated by the one letter symbols. Identical positions inside the helices at which the same or a similar polar residue is located in TrkG and TrkH are indicated by circles and diamonds, respectively. Remarkably, these positions are clustered around the middle of the membrane. With localized mutagenesis experiments we plan to investigate whether these residues play a role in the K^+-translocation process.

***trkE* and its gene product**

The nucleotide sequence of *trkE* has not yet been determined. Genetic data suggest that TrkE is a component of the TrkG/TrkH systems, rather than that it forms the fifth K^+-uptake system [12]. It remains, however, a puzzling observation that the TrkG system can do without an active TrkE protein [12,27]. Studies with a strain that contains a *trkE-lacZ* fusion suggest that TrkE is a membrane protein [27].

Model for the TrkG and TrkH systems

Fig. 4 gives a scheme that fit the present data about the structure of the TrkG and TrkH systems. The close similarity between the TrkG and TrkH proteins (Fig. 3) support the notion that these proteins form the core of two almost identical transport systems [12]. Apparently, *trkH* is the intrinsic *E. coli* gene and *trkG* has been brought into the cells via the prophage *rac*. The source of *trkG* is unclear, but its G+C content and the close similarity with *trkH* may suggest that *trkG* originates from a enterobacterium with a genome of low G+C content, possibly *Proteus*.

Although we have not proven this, we assume that TrkA confers regulation by ATP to the porter proteins TrkH and TrkG. This effect could occur via protein-protein interaction of TrkA with TrkG and TrkH and nucleotide binding to TrkA (Fig. 4). Alternatively, TrkA might be a protein kinase and phosphorylate the membrane

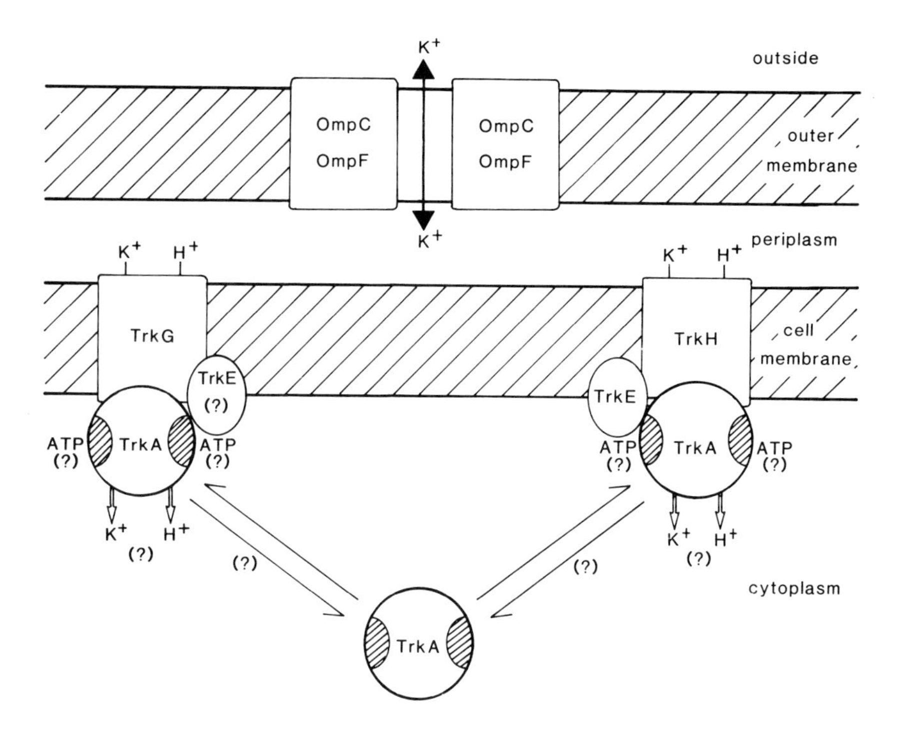

Fig. 4: Model for the structure of the TrkG and TrkH systems. See the text for details and for the meaning of the question marks. The dashed parts in TrkA indicate the two Rossmann folds at which a nucleotide may bind.

component(s) of the complex. The first model explains why in cells with a functional TrkG or TrkH system a portion of TrkA occurs in the membrane fraction. The latter model explains more easily why a large % of the TrkA protein is found in the fraction of soluble proteins [27].

Finally, we don't know how the TrkG and TrkH systems sense changes in the turgor pressure of the cells. We assume, however, that this occurs via helix-helix interactions inside the membrane. A comparison with a more simple turgor-sensing system like KdpD may give some clues.

Acknowledgments

We thank W. Epstein and K. Rudd for helpful discussion and for communicating unpublished results to us, and Eva Limpinsel for expert technical assistence. This work was supported by grants from the DFG (SFB171, Teilprojekt C1), the Fonds der Chemischen Industrie, and the European Community, contract SC-1-0334-C(A)

REFERENCES

1 Bakker EP, In: Bakker EP, ed. Alkali Cation Transport Systems in Prokaryotes. Boca Raton: CRC Press, 1992; in press
2 Walderhaug MO, Dosch DC, Epstein W, In Rosen BP, Silver S, eds. Ion Transport in Prokaryotes. New York: Academic Press, 1987; 84-130
3 Epstein W. FEMS Microbiol Rev 1986; 39: 73-78
4 Brown AD, Microbial Water Stress Physiology. Principles and Perspectives. Chicester: J Wiley and Sons, 1990
5 Dinnbier U, Limpinsel E, Schmid R, Bakker EP. Arch Microbiol 1988; 150: 348-357
6 Larsen PI, Sydness LK, Landfald B, Strom AR. Arch Microbiol 1987; 147: 1-7
7 Schleyer M. 1990: M. Sc. Thesis, University of Osnabrück
8 Wiggins PM. Microbiol Rev 1990; 54: 432-449
9 Martinac B, Buechner M, Delcour AH, Adler J, Kung C. Proc Natl Acad Sci USA 1987; 84: 2297-2301
10 Zoratti M, Ghazi A, In: Bakker EP, ed. Alkali Cation Transport Systems in Prokaryotes. Boca Raton: CRC Press, 1992; in press
11 Booth IR, In: Bakker EP, ed. Alkali Cation Transport Systems in Prokaryotes. Boca Raton: CRC Press, 1992; in press
12 Dosch DC, Helmer GL, Sutton SH, Salvacion FF, Epstein W. J Bacteriol 1991; 173: 687-696
13 Bachmann BJ. Microbiol Rev 1990; 54:130-197
14 Epstein W, In: The Bacteria, Volume XII. New York: Academic Press, 1991; in press
15 Rhoads DB, Waters FB, Epstein W. J Gen Physiol 1976; 67:325-341
16 Rhoads DB, Epstein W. J Gen Physiol 1978; 72:283-295
17 Bossemeyer D, Schlösser A, Bakker EP. J Bacteriol 1989; 173: 2219-2221
18 Epstein W, Kim BS. J Bacteriol 1971; 108: 639-644
19 Rhoads DB, Epstein W. J Biol Chem 1977; 252: 1394-1401
20 Stewart LMD, Bakker EP, Booth IR. J Gen Microbiol 1985; 131: 77-85
21 Bakker EP. FEMS Microbiol Lett 1983; 16: 229-233
22 Silver S, In: Rosen BP, ed. Bacterial Transport. New York: Marcel Dekker, 1978; 222-308
23 Bakker EP Harold FM. J Biol Chem 1980; 255:433-440
24 Wagner EF, Schweiger M. J. Biol. Chem 1980; 255: 534-539
25 Michels M, Bakker EP. J Bacteriol 1987; 169: 4335-4341
26 Milner JL, Grothe S, Wood JM. J Biol Chem 1988; 263: 14900-14905
27 Bossemeyer D, Borchard A, Dosch DC, Helmer GC, Epstein W, Booth IR, Bakker EP. J. Biol. Chem. 1989; 264: 16408-16410
28 Hamann A. 1991: Ph. D. thesis, University of Osnabrück
29 Walker JE, Saraste M, Runswick WJ, Gray NJ. EMBO J 1982: 1: 945-951
30 Rossmann MG, Liljas A, Brändén C-I, Banaszak LJ. The Enzymes 1975; Volume XII, Part A: 61-102
31 Schlösser A, Kluttig S, Hamann A, Bakker EP. J Bacteriol 1991; 173: 3170-3176
32 von Heijne G. EMBO J 1986; 5: 3021-3027
33 Nakahigashi K, Inokuchi H. Nucleic Acids Res 1990; 18: 6439
34 Manoil C, Beckwith J. Science 1986; 233: 1403-1408

© 1992 Elsevier Science Publishers B.V. All rights reserved.
Molecular mechanisms of transport. E. Quagliariello, F. Palmieri, eds.

Regulated transport systems in bacteria

I.R. Booth, R.M. Douglas, A.W. Munro, A.J. Lamb, G.Y. Ritchie, S-P. Koo and G.P. Ferguson

Department of Molecular and Cell Biology, University of Aberdeen, Marischal College, Aberdeen, AB9 1AS, SCotland, U.K.

INTRODUCTION

The transport of amino acids and sugars into bacterial cells is generally effected by transporters that are not subject to regulation of their activity at the level of the membrane. This situation reflects the role of the majority of transport systems in housekeeping functions within the cell (i.e. the capture and the re-capture of nutrients). In contrast, transport systems that are primarily involved in the cellular homeostases, such as regulation of cytoplasmic pH, turgor pressure and osmotic balance are subject to control over their activity. Such transport systems frequently have cations as their substrate but others are involved in the transfer of organic solutes across the membrane. There is good evidence that such transport systems are activated by perturbation of the cell as a consequence of changes in the environment. Such systems pose problems not only in their mechanisms of transport but also in the means by which the regulation of their activity is achieved. The objective of this presentation will be to review our understanding of gated transport systems and in particular to discuss our current views on the regulation of the KefC potassium efflux system.

TRANSPORT SYSTEMS AND HOMEOSTASIS

In microorganisms the response to an imposed stress can usually be divided into two distinct phases: the initial response involving existing enzymes and transport systems and a subsequent change in the pattern of gene expression to allow the synthesis of components that will allow the homeostasis to be maintained in the face of continued stress. This two tier system is well understood in the context of metabolism. The shortage of an amino acid or base elicits the release from feedback of the primary branch-point enzyme that exerts strong control over carbon flux into the appropriate biosynthetic pathway. Such allosteric modulation of cytoplasmic enzymes is well-characterised. If the supply of the nutrient is still insufficient to match the demands of the cell then new gene expression will be elicited to create more biosynthetic capacity. The key feature of this response to stress is that the initial response is achieved by allosteric modulation of protein components that change the balance of carbon flux within the cell.

In some cases the secondary response may involve the increased expression of transport components specific for the nutrient that is in short supply but usually these systems are not themselves subject to regulation of their activity. This is not to imply that they do not play a role in the regulation of the pool size of their transported solute. It is widely recognised that the major role of many transport proteins is in the maintenance of cell pools. Thus, they scavenge for their solute in the environment but they also re-capture material that has leaked from the cell or which is excreted during biosynthetic processes (e.g. peptides during cell wall synthesis; 1). This re-capture plays an important role in the energy economy of the cell (2) and since it is central to the stabilisation of the pool size it is a component of the regulation of metabolite synthesis.

In contrast to the essentially passive role of the above type of transporter in cellular homeostasis other transport systems are regulated in their activity such that they respond specifically to appropriate stresses (3,4). These transport systems are subjected to modulation of their activity although in many instances the identity of the effector is unclear. We have previously characterised such transport components as having the following properties: the transport system is relatively quiescent until cells are stressed (the system is rarely completely inactive - this reflects biological imperfections in the control over protein activity but also that cells are rarely free of stress hence the term "relatively quiescent"). The major change in transport activity is provoked by a stress (which may cause activation or inhibition) and the system remains in the new state until the stress is alleviated. One consequence is that the pool size of the metabolite transported is regulated by the applied stress. This is particularly evident during osmoregulation in the enteric bacteria when the pool of potassium and of compatible solutes is regulated in the main through their transport (4,5). However, in pH homeostatic systems the solute may be cycled through the cell to effect regulation with little or no change in the cytoplasmic pool.

In the enteric bacteria we know of at least nine transport systems that exhibit regulation of activity consistent with a role in homeostasis. These systems are the potassium uptake systems, TrkA, Kup and Kdp, three potassium efflux systems (KefB, KefC and KefA), the major sodium-proton antiport NHA, and the betaine uptake systems ProP and ProU (3-8). There is a further efflux system, which regulates the accumulation of betaine and proline in the enteric bacteria (9).

THE GLUTATHIONE-GATED POTASSIUM EFFLUX SYSTEMS

There are at least three potassium efflux systems in *E. coli*, KefB, KefC (3,10) and a third system (now called KefA; W. Epstein, personal communication) which is evident in strains that lack KefB and KefC activity (6). The KefB and KefC systems are responsible for potassium efflux elicited by the addition of NEM and similar sulphydryl reagents (see below). The *kefB* and *kefC* (formerly *trkB* and *trkC*; 11) loci were identified as mutations that cause a growth defect when the potassium content of the growth medium is low. The mutations were shown to affect the retention of potassium (12). Analysis of suppression of the growth defect by nonsense mutations and transposon insertions (10) established that the lesions identified by Epstein's group are missense mutations that activate the normally quiescent systems. However, the mutations affecting *kefC* and *kefB* have been shown not to prevent further activation of potassium efflux by NEM (see below). The *kefB* and *kefC* genes have been shown to encode two separate potassium efflux systems that are inhibited by glutathione and are activated by specific glutathione metabolites (6). The KefC system has been characterised in most detail at both the physiological and the molecular level (13-15).

The role of glutathione in potassium retention

A role for glutathione in the control of potassium retention was first proposed by Kepes and colleagues upon their discovery that sulphydryl reagents, such as N-ethylmaleimide (NEM), provoke the rapid loss of potassium from *E.coli* cells (16,17). The potassium loss could be reversed by the addition of a reducing agent leading to the renewed accumulation of potassium (16,18). It was suggested that NEM modified a small soluble thiol that was essential for potassium retention rather than affecting a membrane protein directly. Subsequently, it was observed that mutants that are deficient in the synthesis of glutathione, the major non-protein thiol of the the bacterial cell, leak potassium. Such mutants also lacked the NEM-stimulated potassium efflux observed in the parent(17). In this way a link was established between glutathione and the NEM-elicited potassium efflux. However, the mutants used in this study were isolated after chemical mutagenesis and subsequent study suggested that they also possessed defects in potassium transport *per se* (Epstein, personal communication). When well-defined transposon mutants affecting glutathione synthesis (*gshA::Tn10kan)* were investigated potassium leakage in the absence of glutathione was observed to be rapid but very limited (13) and the cells retained a substantial potassium pool in the absence of glutathione. The cloning of the structural gene for KefC allowed the role of glutathione to be tested more rigorously. A cell lacking glutathione but with multiple copies of the *kefC* gene should exhibit a major leak of potassium; however, this was found not to be the case. Extensive potassium loss was evident only when the glutathione deficiency was combined with mutations affecting the potassium uptake systems and the multicopy plasmid bearing the *kefC* gene. Once again the cells still retained a substantial potassium pool. This loss of potassium could be prevented by incubation with glutathione. Consequently, it is apparent that although glutathione does act negatively on the efflux system the activity of the system in the absence of glutathione is quite low.

Supplementation with glutathione restored both the potassium pool and NEM-elicited efflux in all glutathione-deficient mutants (13,17). The absence of rapid and extensive potassium loss (comparable with that which occurs in the presence of NEM) in the glutathione-deficient strain is not consistent with the model proposed by Kepes in which glutathione solely acts as a negative regulator (17). Further, we have shown that both iodoacetate and iodoacetamide completely convert the glutathione pool to metabolites but do not elicit significant potassium efflux (13). These reagents also block the action of NEM (13). Iodoacetate-treated cells exhibit normal NEM-elicited potassium efflux if they are re-incubated with glutathione prior to NEM treatment. These data emphasise the importance of the synthesis in the cytoplasm of specific glutathione metabolites in order that potassium efflux be effected (13). These studies can only be explained if a specific glutathione metabolite, derived from the reaction of NEM with glutathione, can act as an activator of the efflux system.

In cells extracts of NEM-treated cells the glutathione pool was found to be converted to N-ethylsuccinimido-S-glutathione (ESG) and this could be prevented by prior incubation with iodoacetate which converts the glutathione to S-carboxymethylglutathione (CMG). ESG and CMG could be broken down by cells during incubation with reducing agents re-generating free glutathione (13,15). It is the breakdown of ESG and the release of glutathione that leads to the prevention of further potassium loss. The re-capture of potassium is assumed to take place through the uptake systems (18) although TrkA is inhibited by NEM in a non-reversible manner. Presumably it is the residual activity of these systems that functions to recover the potassium lost through the efflux systems.

Potassium efflux can be activated by other sulphydryl reagents such as 1-chloro-2,4-dinitrobenzene (CDNB), PCMB, PCMBS, phenylmaleimide and diamide (13). Potassium efflux elicited with CDNB cannot be reversed with reducing agents; presumably, the conjugate of glutathione and CDNB (2,4-dinitrophenyl-S-glutathione; DNG) is very stable and cannot be broken down by the cell even in the presence of reducing agents (although see below and ref. 25). Despite the dinitrophenyl group DNG is not permeant through the membrane to any significant extent and consequently DNG cannot be used as an externally-added activator of the efflux process.

Diamide oxidises glutathione to the disulphide form and efflux provoked by this compound has been suggested to reflect redox control over KefC activity (19). However, HPLC analysis of the products formed between diamide and glutathione has revealed some glutathione adducts, presumably linked through the sulphydryl group (unpublished data). Thus the activation may reflect the adduct rather than the oxidation of glutathione. Other oxidising agents that provoke potassium efflux (eg. hydrogen peroxide) do not utilise KefC activity. Thus it is unlikely that redox events are essential to the regulation of the KefC potassium efflux system.

Structure of the KefC system

The structural gene for the KefC system has recently been cloned and the DNA sequence determined (14). The cloned fragment that restores NEM-elicited potassium efflux to a mutant strain carrying null lesions in the chromosomal *kefB* and *kefC* genes encodes a putative polypeptide of 620 amino acids. On SDS-polyacrylamide gels a membrane protein of approximately 60kDa has been detected which is somewhat smaller than the predicted molecular mass of 79kDa. The membrane location of the putative gene product has been confirmed through the formation of KefC'-'LacZ protein hybrids which attach the β-galactosidase to the cytoplasmic membrane. Even quite short fragments of the amino-terminus are sufficient to attach β-galactosidase to the membrane.

Analysis of the amino acid sequence of the putative KefC protein has revealed that only the amino-terminal 2/3rds is sufficiently hydrophobic for the protein to enter the membrane. The carboxy-terminal region of the protein is essential for KefC activity but is predominantly hydrophilic in nature and is unlikely to integrate into the membrane domain. There is no strong amino acid homology to other proteins but there are sequence similarities with a number of glutathione-utilising enzymes such as dehalogenase (20) and glyoxalase (21). These sequence similarities start within the hydrophobic domain and extend into the "hydrophilic domain" close to the carboxy-terminus. Both dehalogenase and glyoxalase are soluble proteins and are known to possess hydrophobic pockets (22). Given these properties we are attracted by the model of the KefC system in which only the region close to the amino acid terminus (first 200 amino acids) are involved in the transport of potassium and the rest of the protein forms a regulating domain. We believe that glutathione and its adducts bind to the regulatory domain to effect the opening and closing, respectively, of KefC.

The mechanism of KefC

Analysis of the potassium efflux in the *kefC* mutants and of the events occurring after NEM addition has led to the proposal that KefC is a regulated potassium-proton antiport. NEM addition was observed to result in acidification of the cytoplasm while the membrane potential was observed to increase (18). In the case of the mutants it was observed that the effect of the mutations was to provoke a lower potassium gradient than could be sustained by the membrane potential. These are both circumstantial evidence for a potassium-proton antiport as the mechanism for KefC. However, it has become evident that the system is almost certainly multimeric with co-operativity between the subunits (unpublished data) a property often associated with channels. Thus, the cloned *kefC* gene can suppress the leakage of potassium associated with the missense mutations affecting the chromosomal *kefC* locus. When NEM is added to the mutant carrying the cloned *kefC* gene potassium is lost from the cell almost instantaneously. The degree of suppression of the mutant phenotype is determined by the level of expression of the cloned gene (R.M. Douglas, unpublished data). These data can be explained if the mutated and the wild-type KefC protein interact in the membrane to form a complex that has intermediate properties. Preliminary data with cross-linking reagents has shown that KefC can be cross-linked to give a high molecular mass protein. Oligomeric structure would be consistent with channel formation although it would not exclude antiport function.

The role of glutathione-gating in cell physiology

KefC activity has been detected in all the Gram negative bacterial species examined but was absent from Gram positive species with the exception of *Staphylococcus aureus* (15). KefC activity was defined as NEM-elicited potassium efflux that could be prevented by pre-treatment of cells with iodoacetate. In all cases the cells possessed glutathione and could synthesise ESG and CMG. Similarly, the NEM-elicited potassium efflux could be reversed by incubation with reducing agents and inhibition of NEM-elicited efflux by iodoacetate could be prevented by incubation of the treated cells with glutathione. Considerable variation in KefC activity was observed in the different organisms although the significance of this observation is unclear. A notable factor is that there was no correlation between the magnitude of the glutathione pool and KefC activity revealed by NEM-activation. Indeed some

Gram positive bacteria are able to accumulate significant glutathione pools but have no detectable KefC activity.

By Southern hybridisation, using the *E.coli kefC* gene as probe, homology at the DNA level was very poor and indicated that even within the enteric bacteria there was little sequence conservation between the genes for KefC (15). It was also clear that even closely related enteric bacteria exhibited only limited sequence homology and that the conserved regions lay at the 5' end of the gene. Interestingly, the *kefB* gene does not appear to be homologous to *kefC* even in *E.coli.*

The widespread distribution of KefC activity among the Gram negative bacterial kingdom argues for a function for the transport system that has led to its retention by a range of divergent species. However, mutants of *E.coli* lacking both KefC and KefB activity were isolated with ease (10) and under the growth regimes investigated no advantage or disadvantage has been observed for mutants lacking KefB and KefC. This may simply indicate that other compensating changes can occur and it may be that the construction of strains lacking other potassium efflux system may shed some light on the value of KefC to *E.coli*.

Glutathione is utilised by cells as part of the mechanism for de-toxifying methylglyoxal a natural metabolite produced as a side-product of glycolysis (23,24). Added methylglyoxal activates KefC indicating that there could be a link between the production of this toxic metabolite and the role of glutathione-regulated potassium efflux. In addition to its role in the de-toxification of methylglyoxal glutathione is also involved in the metabolism of a number of toxic compounds such as CDNB and reduces the toxicity of this compound to the cell (25). The primary route for CDNB metabolism is via glutathione-S-transferase but there may be a second route for de-toxification which operates in the absence of glutathione (unpublished data). Thus glutathione may occupy a significant niche in the resistance of the cell to certain toxic metabolites. Conceivably, the ability of such metabolites to activate KefC and lead to spontaneous potassium loss may be an important function for the transport system. Transient potassium loss during CDNB-induced stress may serve as a general inhibitor of cell growth to allow for the repair of DNA damage caused by toxic compounds.

In conclusion, we are unclear what the function of the KefC system is in cell physiology and it is clear that a greater understanding may evolve from analysis of mutants lacking other potassium efflux systems and/or the metabolic systems for glutathione metabolism.

CONCLUSION

The KefC system shows properties typical of a regulated transport system. Its activity is very low until the metabolism of glutathione is perturbed leading to the formation of activators. For many of the activators there are mechanisms for their destruction thus removing the signal activating KefC. The system displays many features in common with channels of higher organisms and thus poses an interesting problem in terms of the actual mechanism of potassium efflux.

Fortunately the cloning of the structural gene opens the way for experiments that will directly test this property of the system. It is conceivable that KefC performs an important function in cell physiology and that the regulated flux of potassium is central to this function. Despite significant progress in understanding this system there is still much to be discovered both in terms of the function of the system and in the mechanism and regulation of its activity.

ACKNOWLEDGEMENT

The authors are indebted to the Science and Engineering Research Council, the Agricultural and Food Research Council and Unilever plc for support for their research. The authors gratefully acknowledge the support of their colleagues in Aberdeen, both past and present, and also Ken Douglas, Wolf Epstein, John Findlay, Chris Higgins and Olaf Pongs.

REFERENCES

1 Goodell EW, Higgins CF. J Bacteriol 1987; 169:3861-3865.
2 Ames GF-L. Ann Rev Biochem 1986; 55:397-425.
3 Booth IR. Microbiol Revs 1985; 49:359-378.
4 Booth IR, Cairney J, Sutherland L, Higgins CF. J Appl Bacteriol Symp Suppl 1988; 35S-49S.
5 Epstein W. FEMS Microbiol Rev 1986; 39:73-78.
6 Bakker EP, Booth IR, Dinnbier U, Epstein W, Gajewska A. J Bacteriol 1987; 169:3743-3749.
7 Padan E, Maisler N, Taglicht D, Karpel R, Schuldiner S. J Biol Chem 1989; 264:20297-20302.
8 Bossemeyer D, Sclosser A, Bakker EP. J Bacteriol 1989; 171:2219-2221.
9 Koo S-P, Higgins CF, Booth IR. J Gen Microbiol 1991; 137: in press
10 Booth IR, Epstein W, Giffard PM, Rowland GC. Biochimie 1985; 67;83-90.
11 Epstein W, Kim BS. J Bacteriol 1971; 108:639-644.
12 Rhoads DB, Waters FB, Epstein W. J Gen Physiol 1976; 67:325-341.
13 Elmore MJ, Lamb AJ, Ritchie GY, Douglas RM, Munro A, Gajewska A, Booth IR. Mol Microbiol 1990; 4:405-412.
14 Munro AW, Ritchie GY, Lamb AJ, Douglas RM, Booth IR. Molec. Microbiol 1991; 5:607-616.
15 Douglas RM, Roberts JA, Munro AW, Ritchie GY, Lamb AJ, Booth IR. J. Gen. Microbiol 1991; 137:1999-2005.
16 Meury J, Lebail S, Kepes A. Europ J Biochem 1980; 113:33-38.
17 Meury J, Kepes A. EMBO J 1982; 1:339-343.
18 Bakker EP, Mangerich WE. FEBS Lett 1982; 140:177-180.
19 Meury J, Robin A. Eur J Biochem 1985; 148:113-118.
20 La Roche SD, Leisinger T. J Bacteriol 1990; 172:164-171

21 Murata K, Inoue Y, Rhee H-I, Kimura A. Can J Biochem 1989; 35: 423-431.
22 Douglas KT, Al-Timara A. in *Thioredoxin and Glutaredoxin Systems: Structure and Function* (edited by A. Holmgren) Raven Press, New York. 1986; 155-164.
23 Ackerman RS, Cozzarelli NR, Epstein W. J Bacteriol 1974; 119:357-362.
24 Murata K, Tani K, Kato J, Chibata I J. Gen Microbiol 1980; 120: 545-547.
25 Summer K-H, Goggelmann W. Mutation Res 1980; 70:173-178.

© 1992 Elsevier Science Publishers B.V. All rights reserved.
Molecular mechanisms of transport. E. Quagliariello, F. Palmieri, eds.

Mechanism of lysine uptake and secretion by *Corynebacterium glutamicum*

Stefan Bröer and Reinhard Krämer

Institut für Biotechnologie I, Forschungszentrum Jülich, 5170 Jülich, F.R.G.

SUMMARY

Secretion of lysine in *Corynebacterium glutamicum* strains, deregulated in feed-back inhibition of aspartate kinase by lysine, is catalyzed by a specific export carrier system. Efflux due to unspecific leakage or permeation through pores could be ruled out. Also secretion by functional inversion of lysine uptake could be excluded. The lysine uptake system of *C.glutamicum* was characterized as an exchange system with very low activity, using alanine, valine or isoleucine as countersubstrates. Lysine export is an electrogenic process, modulated by the membrane potential, the pH-gradient and the lysine gradient. The kinetic mechanism can be described as cotransport of lysine together with two OH^--ions. This carrier system seems to be well adapted to export and shows several particular features different from uptake systems, with respect to affinity, symport-ion and translocated charge.

INTRODUCTION AND OVERVIEW

In contrast to solute uptake in general, solute secretion, although beeing a wide-spread phenomenon, is poorly understood. There is increasing evidence that bacteria possess a variety of systems excreting e.g. organic acids, carbohydrates, or toxic compounds [1]. Coryneform bacteria have been used for amino acid production since more than 30 years. Expecially *Corynebacterium sp.* and *Brevibacterium sp.* are used for producing mainly glutamic acid and lysine. By the use of these processes, more than 600 000 t of amino acids per year are produced.

Various mechanisms have been postulated for the excretion of amino acids in *C.glutamicum.* (i) It is still widely accepted that excretion of these solutes may occur as a result of a physically changed (leaky) membrane (leak model) [2]. (ii) Lysine efflux has been suggested to be caused by osmotically regulated pores in the plasma membrane (pore model) [3]. (iii) A general explanation for solute efflux, also applied for glutamate secretion in *C.glutamicum*, postulates the functional inversion of the corresponding uptake systems, due to a change in energetic conditions and possibly triggered by membrane alteration (inversion model) [4]. (iv) We provided evidence for the presence of specific secretion carriers for isoleucine, glutamate and lysine in *C.glutamicum* (secretion carrier model) [5-7].

When studying amino acid secretion, it is important to investigate also the corresponding uptake systems for at least two reasons. Besides possibly interfering with the observed activity of secretion, at least in kinetic terms, the detailed knowledge of the uptake mechanism is indispensable for evaluating the above mentioned inversion model. So far we have studied both uptake and efflux of three amino acids, namely glutamate (anionic substrate), lysine (cationic) and isoleucine (zwitterionic). In all three cases we found, besides the expected specific uptake systems, also specific secretion carriers for these solutes (Tab. 1).

Table 1
Carrier systems for uptake and secretion of glutamate, lysine and isoleucine in *C.glutamicum*

amino acid	transport direction	transport mechanism
glutamate	uptake	primary active [8,9] presumably binding protein dependent
	efflux	secretion by specific carrier [10]
lysine	uptake	antiport carrier (see below)
	efflux	specific secretion carrier (see below)
isoleucine	uptake	Na^+-coupled, secondary uptake [11]
	efflux	secretion carrier [7] presumably secondary mechanism

Secretion of lysine is difficult to investigate in the wild-type strain *C.glutamicum* ATCC 13032, which, under normal conditions, does not export lysine in significant amounts. In general, in lysine producing strains, aspartate kinase, the key enzyme in regulation of lysine anabolism, is insensitive to feedback regulation. An high internal lysine pool and effective lysine secretion is the result of this mutation.

MATERIALS AND METHODS

Growth of organisms. The strains used were *C.glutamicum* ATCC 13032 (wild type) and the mutant DG 52-5 derived from it which is insensitive to feedback inhibition by lysine and threonine. Cells were aerobically grown in mineral medium with glucose as carbon and energy source and harvested in the late exponential phase [7].

Kinetic measurements of lysine uptake and secretion were performed using silicone oil centrifugation for separation of external and internal space. Transport of lysine was in general measured by monitoring internal and external amino acid concentrations using HPLC after precolumn derivatization. In the case of uptake, also radioactively labeled lysine was used [12]. For detailed calculation of lysine efflux kinetics see ref. 7 and 13.

Components of the protonmotive force. The cytoplasmic volume was determined using labeled taurine and water. The membrane potential was measured either by distribution of ^{14}C-tetraphenylphosphonium bromide or by ^{86}Rb in the presence of valinomycin. The pH-gradient was determined using permeant weak acids (benzoic acid) or bases (methylamine). Different diffusion potentials were generated in intact cells using permeation of acetate (creating a pH-gradient) and K^+ in the presence of valinomycin (membrane potential) [7].

ATP determination was carried out by the luciferase test using both external and internal control values.

RESULTS AND DISCUSSION

Lysine uptake

Uptake of labeled lysine into *C.glutamicum* cells follows Michaelis-Menten kinetics with a K_m of 10 μM and a V_{max} of only 0.2 nmol$\cdot$min$^{-1}\cdot$mg dw^{-1}. Determination of energetic parameters revealed that neither membrane potential, nor pH-gradient nor ATP had any influence on uptake of labeled lysine [13]. When monitoring internal and external lysine concentrations, actually no change at both sides was observed. This is in agreement with a homologous lysine/lysine exchange observed under these conditions. Heterologous exchange could not be detected with any tested amino acid or related substance, only lysine itself and the lysine-analogue S-aminoethylcysteine (AEC) provoked efflux of labeled lysine which was loaded into the cells previously.

On the other hand, lysine-auxotrophic strains of *C.glutamicum* must take up lysine per definition. This could in fact be observed during growth, as measured by HPLC. Under particular experimental conditions [12] net uptake of lysine could also be measured in washed *C.glutamicum* cells. In contrast to the above mentioned transport acitvity (homologous antiport), this transport now was electrogenic and absolutely dependent on the presence of a membrane potential. Detailed analysis revealed that the mechanism of lysine uptake under these conditions can be described as heterologous

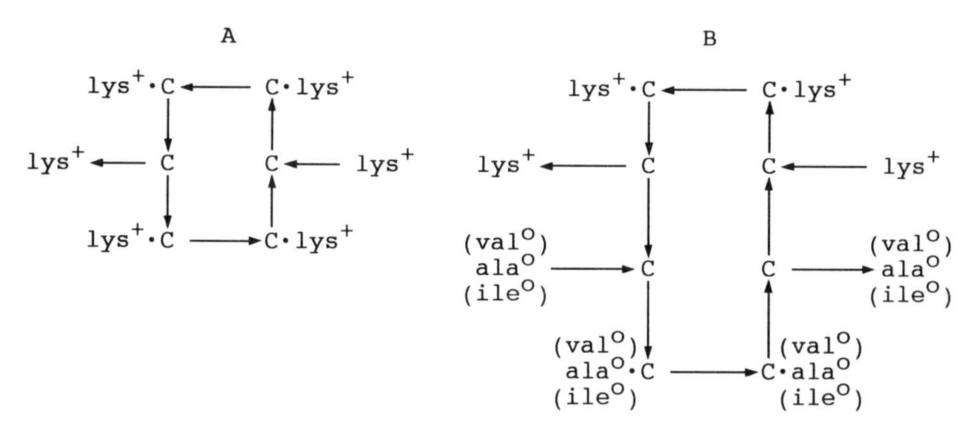

Figure 1. Transport cycle of the lysine uptake system. A: In general, only homologous lysine exhange is observed. B: Under particular conditions, e.g. in lysine auxotrophic mutants, the system switches to heterologous, electrogenic antiport of lysine against alanine, valine or isoleucine.

antiport against an uncharged counterion showing surprisingly low specificity for the antiport substrate. Since we used intact cells for these measurements, the counterions could only be tested from the outside (reverse function of the carrier) under completely deenergized conditions, otherwise the membrane potential would inhibit outside movement of lysine. These experiments revealed that antiport of lysine takes place against alanine, valine or isoleucine with rates comparable to those of the homologous lysine/lysine exchange reaction. Due to the charge difference of the antiport substrates this mode of the lysine uptake carrier is now electrogenic.

The two modes of exchange reaction of the lysine carrier, homologous and heterologous antiport, are shown in Fig. 1. As mentioned above, heterologous antiport against uncharged amino acids was only observed under particular conditions. The reason for switching from one mode to the other is not simply due to different internal concentrations of possible counterions, but can be regulated by the cell [12]. The nature of this regulation is unknown so far.

Lysine secretion: Definition of a specific carrier system

For investigating lysine secretion, the strain DG 52-5 with feedback insensitive aspartate kinase was used. The experiments provided several convincing arguments ruling out the possibility of lysine efflux mediated by unspecific diffusion or by osmotically regulated pores, as suggested earlier [3].

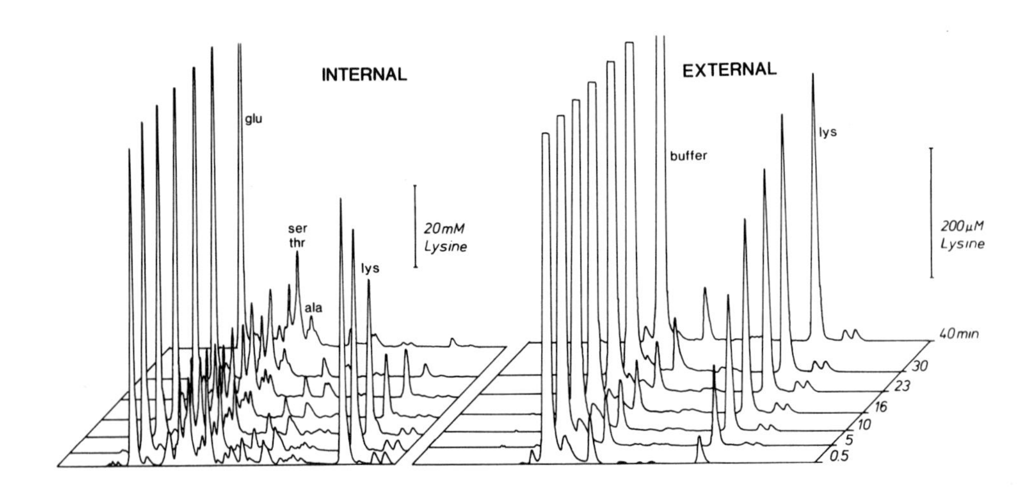

Figure 2. Specificity of lysine secretion in strain DG 52-5. HPLC-analysis of internal (left panel) and external (right panel) amino acid content. Besides glutamate and lysine, a variety of other amino acids is observed internally. At the external side, only lysine and the buffer used (two peaks) are observed in the HPLC-analysis in significant amounts.

(i) Lysine secretion is specific. Fig. 2 shows diagrams of HPLC-measurements of the internal and external amino acid composition during lysine secretion. Although a complex pattern is seen inside the cells including very high glutamate concentrations, only lysine appears at the exterior in significant amounts.
(ii) Lysine secretion is highly temperature dependent. Between 20°C and 35°C an activation energy of 54 kJ/mol was observed, below 18°C the activation energy was 260 kJ/mol. Above 50°C lysine secretion stopped completely due to inactivation of the carrier protein [7]. Especially the latter observation is an excellent argument against any possibility of unspecific efflux.
(iii) During secretion, all gradients including membrane potential (up to 190 mV at pH 9), K^+-gradient, and substrate gradients were fully maintained.

Also the pore model could be ruled out by the fact that the export system can accumulate the transported cation lysine at the external (positively charged) side of the membrane. An at least 30fold accumulation was observed under particular conditions. Higher ratios could not be measured due to technical reasons, i.e. problems in determination of low internal at high external concentrations.

The mechanism of the uptake system (homologous or heterologous antiport) excludes the use of this system for lysine export under our experimental conditions where no appropriate antiport substrate was present. Also a hypothetical functional conversion from an antiport into a uniport (efflux) system was ruled out since we observed constant V_{max} values of uptake during the export reaction. These values were identical to the activity measured in the absece of secretion. It has to be mentioned, however, that the uptake activity did not significantly interfere with secretion because the V_{max} of the uptake reaction is at least 50 times lower than that of export.

The above mentioned data could also be summarized in argueing for the presence of a specific secretion carrier system. We additionally measured the further basic kinetic constants of the efflux system. The K_m at the cis (internal!) side was determined to 20 mM, the maximum export rate was 12.5 nmol$\cdot$min$^{-1}\cdot$mg dw^{-1}, the pH-optimum was found around pH 8 [7].

Lysine secretion: Mechanism

The dependence of lysine export on the various energetic parameters was studied in intact cells by varying more or less independently the electrical potential, the pH-gradient, the lysine gradient, and the internal ATP-concentration. It turned out that (i) internal ATP was not correlated at all to the acitivty of lysine efflux, (ii) the strongest influence was exerted by the membrane potential, (iii) also the pH-gradient and the lysine gradient modulated the secretion activity, however, only under particular conditions of weakened driving forces for export [13]. Since we could not measure true *accumulation ratios* which would be the most elegant method to determine the quantitative relation of export and its driving forces, the correlation of *export rates* to the different energetic parameters was studied. In Fig. 3, this relation is shown for the membrane potential and the pH-gradient. Similar results can be obtained when varying the lysine gradient [13].

It has to be pointed out, that strong dependence of the transport rate on external pH (pH-gradient), as shown in Fig. 3B, could only be seen, if the external lysine concentration wa high. Under the conditions of Fig. 3A (external lysine 0.1 mM), no significant dependence on external pH would be observed (see also explanation in legend to Fig. 4).

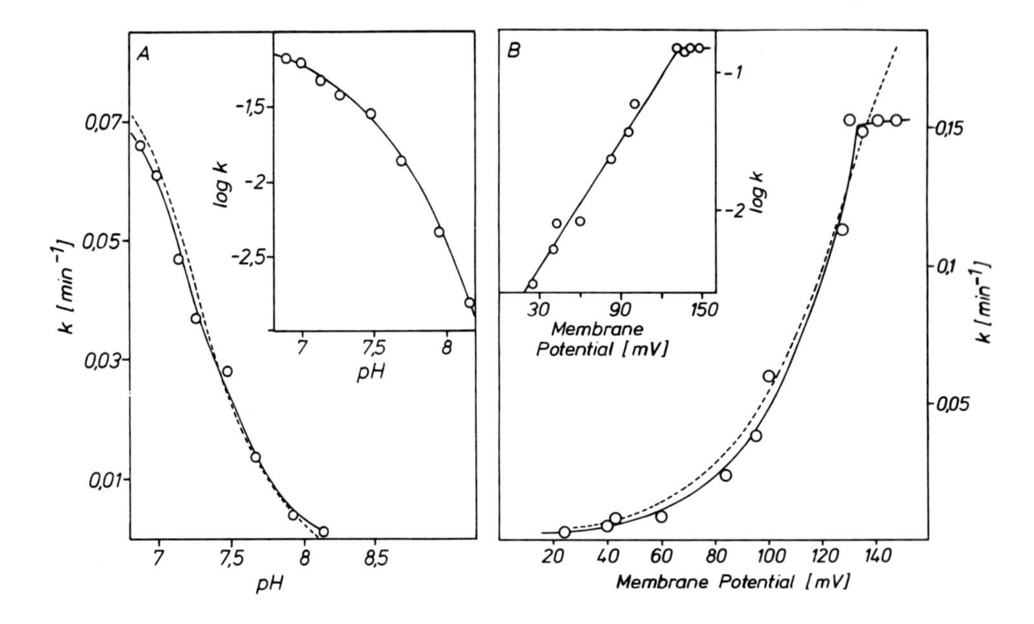

Figure 3. Correlation of export rate (first order rate constants of the export reaction, linearily correlated to the export velocity) to membrane potential (A) and pH-gradient (B), respectively. Membrane potential was varied by different K^+-concentrations in the presence of valinomycin. The internal pH is more or less constant around 7.5-7.8 in these experiments. External lysine concentration was 0.1 mM (A) and 92 mM (B). Inserts: logarithmic plots of the same data.

Taken together, these and further findings [13] could be explained in a mechanistic sheme as shown in Fig. 4. Lysine is exported to the cytosol in symport with two OH^--ions. The substrate loaded transport complex is uncharged (charge-compensated), whereas the unloaded reorienting carrier is positively charged.

The major reason why H^+ cannot replace OH^- in this transport mechanism is due to the fact that protons, if used as coupling ions, had to be transported within the partial reaction in the lower part of the transport cycle (Fig. 4). However, the influence of membrane potential and pH-gradient was never correlated in export measurements, whereas the pH-influence was completely interchangeable with the influence of the lysine gradient. Thus, the ion coupled to the lysine movement must be transported within the partial reaction in the upper part of the scheme, i.e. it must, at least formally, be OH^-.

The kinetic equations describing the scheme shown in Fig. 4 can be derived using simple kinetic equations [13]. The resulting equation for the export velocity comprises several affinity constants for the internal and external substrates (OH^- and lys^+), as well as rate constants for the two transport steps. The rate constant for the reorientation step (unloaded carrier) is modulated by the membrane potential. The equation furthermore contains the actual internal and external substrate concentrations (lys^+ and pH). All essential values could be determined experimentally. When inserted into the derived equation, the dependence of the secretion rate on the modulating parameters, i.e. membrane potential and pH-gradient, can be caluclated, The result of this calculation is shown in Fig. 3 by dotted lines. Obviously, model and experimental data coincide reasonably well.

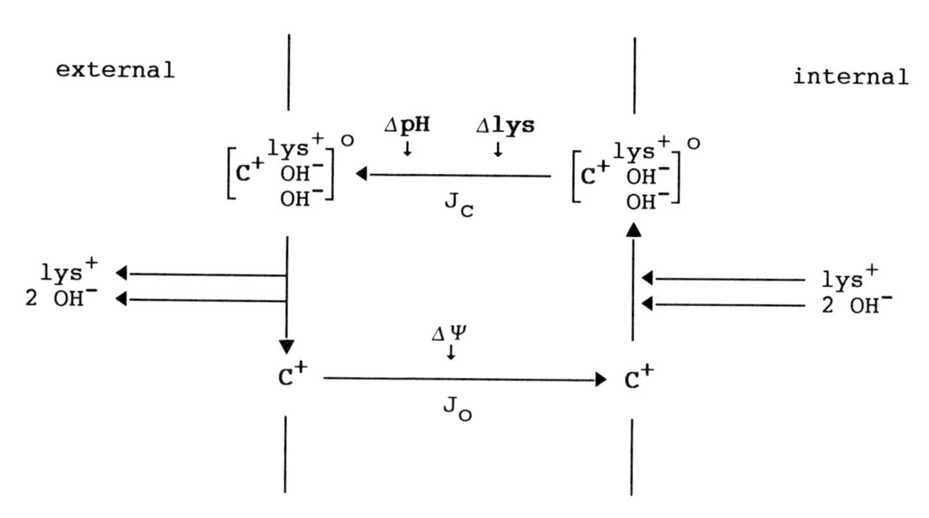

Figure 4. Kinetic model of lysine export by the secretion carrier of *C.glutamicum*. The two substrates lys^+ and OH^- (two molecules) bind to the positively charged carrier at the inside, leading to an uncharged translocation complex. This particular step of the overall transport cycle is influenced by the pH-gradient and the lysine gradient. After dissociation of the substrates at the outside, the positively charged, unloaded reorientation complex orientes back to the inside. Due to the net postive charge, this particular step is driven by the membrane potential (inside negative). The two basic fluxes (permeabilities) J_C, for the translocation step, and J_O for the reorientation step were found to be very assymmetric, i.e., J_C is at least 50 times larger than J_O. That results in an predominant control of the whole cycle by the influence of the membrane potential on this limiting step. The binding and the dissociation steps are assumed to be not rate limiting.

In principle, the lysine secretion system in *C.glutamicum* is a classical secondary system, modulated by the electrical potential and the gradients of the transport ions, lysine and OH^-. Besides showing features typical for a secondary system, e.g. limitation by the basic rate constant of the reorientation step (permeability $P_C \gg P_O$), it is characterized by several properties which make it ideally suited for export purposes.

(i) The affinity of the system at the cis side which in this case is the internal side, is very low (20 mM), thus preventing loss of lysine under normal conditions.

(ii) The coupling ion is OH^- and not H^+, which makes possible a secondary system

essentially driven by membrane potential (together with argument iii!).
(iii) The unloaded carrier is charged. In import systems in general, the substrate-loaded complex is assumed to be charged.

REFERENCES

1 Driessen AJM, Konings WN. In: Krulwich TA, ed. The Bacteria Vol XII. New York: Academic Press, 1990; 449-478
2 Demain AL, Birnbaum J. Current Topics Microbiol Immunol 1968; 46: 1-25
3 Luntz MG, Zhdanova NI, Bourd GI. J Gen Microbiol 1986; 132: 2137-2146
4 Clement Y, Escoffier B, Trombe MC, Laneelle G. J Gen Microbiol 1984; 130: 2589-2594
5 Ebbighausen H, Weil B, Krämer R. Appl Microbiol Biotechnol 1989; 31: 184-190
6 Hoischen C, Krämer R. Arch Microbiol 1989; 151: 342-347
7 Bröer S, Krämer R. Eur J Bioch 1991; 202 (in press, first paper)
8 Krämer R, Lambert C, Hoischen C, Ebbighausen H. Eur J Bioch 1990; 194: 929-936
9 Krämer R, Lambert C. Eur J Bioch 1990; 194: 937-944
10 Hoischen C, Krämer R. J Bacteriol 1990; 172: 3409-3416
11 Ebbighausen H, Weil B, Krämer R. Arch Microbiol 1989; 151: 238-244
12 Bröer S, Krämer R. J Bacteriol 1990; 172: 7241-7248
13 Bröer S, Krämer R. Eur J Bioch 1991; 202 (in press, second paper)

PHOSPHOTRANSFERASE SYSTEMS

© 1992 Elsevier Science Publishers B.V. All rights reserved.
Molecular mechanisms of transport. E. Quagliariello, F. Palmieri, eds.

The enzymesII of the PTS as carbohydrate transport systems: what the evolutionary studies tell us on their structure and function.

J.W. Lengeler, J. Bockmann, H. Heuel, F. Titgemeyer.

Universität Osnabrück, Fachbereich Biologie/Chemie, Postfach 4469, W-4500 Osnabrück (FRG).

INTRODUCTION

Bacteria have evolved as an adaptation to the frequent and abrupt changes in their environment different types of carbohydrate transport systems, each system optimized for its substrate and for specific growth conditions. These are complex, membrane bound, and substrate specific protein channels containing between one and six subunits. More than fifty different systems have been analysed thus far, but none of these is understood at the molecular level because mostly classical physico-chemical methods like NMR spectroscopy and X-ray cristallographic studies using purified proteins have failed thus far. This leaves basically a set of biological methods the use of which begins to show molecular details of these complex systems and processes: 1) Isolation of mutants with specific transport defects and mutations of defined residues by localised mutagenesis to any wanted amino acid. 2) Construction of intra- and intergenic protein hybrids in fused and free form, as well as of deletions of specific parts of the proteins to define independent and large protein domains for e.g. substrate binding, ion channeling, or phosphotransferase activity. 3) Fusions of parts of complex membrane bound proteins to enzymes active only in the periplasm or in the cytoplasm to differentiate between transmembrane helices and extramembraneous loops. Such fusions complement data obtained through conventional biochemical tests able to establish the topology of membrane bound enzymes.

Recently, a large series of DNA sequences for genes encoding carbohydrate transport systems, e.g. 22 for enzymeII-complexes of the bacterial phosphotransferase system have been obtained. From the analysis of these sequences, a set of rules emerges which helps further to understand such complex membrane bound systems as will be shown below.

BACTERIAL PHOSPHOENOLPYRUVATE (PEP)-DEPENDENT CARBOHYDRATE: PHOSPHOTRANSFERASE SYSTEMS (PTS).

Many carbohydrates (mono- and disaccharides, polyhydric alcohols) are taken up and accumulated in eubacteria as the phosphate ester by a series of phosphotransferase systems which, as far as is known, all function according to the general scheme outlined in Fig. 1 [1,2]: Two general and soluble proteins, called traditionally EnzymeI (EI) and Histidine protein (HPr), catalyse the first step in which EI autophosphorylates at the expense of PEP and transfers the phosphoryl group from a histidine residue to a histidine of HPr. Phospho-HPr in turn phosphorylates one of a series of substrate specific and membrane bound protein complexes called EnzymesII (EII). Each contains three structurally distinct domains (EIIA, B, and C) which constitute a functional unity. They may be fused to a single polypeptide of about 675 residues length, split into two, or even three distinct proteins, and arranged in almost any possible sequence at the genetic and primary amino acid level, further corroborating the functional autonomy of each domain.

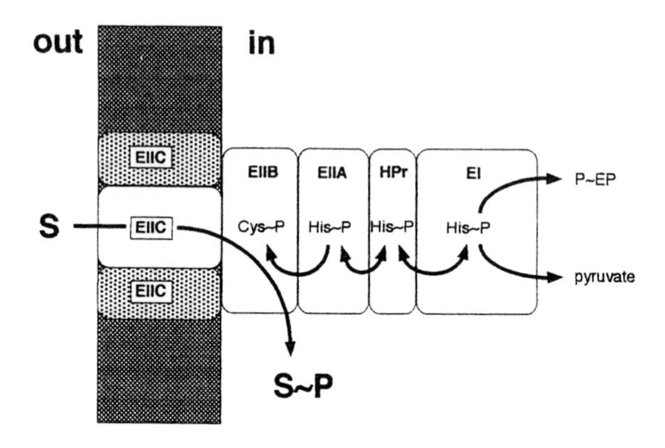

Figure 1. General scheme of a phosphotransferase system (for abbreviations and further explanations see text).

Domain EIIA, named before EnzymeIII, is phosphorylated by phospho-HPr again at a histidine residue such that the phosphorylation cascade up to this step remains fully reversible. Through phospho EIIA is phosphorylated domain EIIB, interestingly at a cysteine residue and a slightly lower energy level than phospho-histidines. EIIA and EIIB are hydrophilic proteins or domains located at the cytoplasmic site of the inner membrane, by contrast to the third and integral membrane bound domain EIIC. This hydrophobic domain of about 350 amino acid length forms the transmembrane channel proper with a specific substrate binding site. EIIC alone, as shown by a series of deletion experiments, still inserts correctly in the membrane and forms a substrate specific channel, while both EIIA and B are required for the concomitant phosphorylation of the substrate during transport [for reviews and references see 3 to 7].

PROTEIN STRUCTURES AS DEDUCED FROM THE COMPARISON OF DNA AND PROTEIN SEQUENCES.

Most obvious in comparing the many PTS analysed thus far is the remarkable similarity of the biochemical steps involved. It is reflected in the obligatory presence of domains EI, HPr and EIIA to EIIC, which together are responsible for this vectorial phosphorylation. According to present models on the evolution of genes and polypeptides, systems as closely related as the PTS should have evolved by gene and gene fragment duplications, followed by mutational diversification of the copies. During evolution, these genes are split into the equivalent of functional enzyme domains and the fragments shuffled and reshuffled while passing through new organisms. A search for such conserved amino acid motifs and consensus sequences should thus be very helpful to define functional domains. For obvious reasons, the least conserved sequences and structures can be found as areas with amino acid exchanges in otherwise closely related molecules, and the strongest conserved parts as unchanged motifs in distantly related molecules; intragenic rearrangements of large amino acid groups, however, help to define conserved functional domains especially if supported by intergenic complementation tests [3].

In recent years, a series of rules have been found which help to improve strongly the accuracy of predictions for protein structures as deduced from primary amino acid sequences. The lactose permease LacY of enteric bacteria may be considered rightly as the paradigm of a transport system in which the existence of such predicted secondary structures has been tested by means of all biological and biochemical methods mentioned [references in 8 - 10]. The model assumes first that, as predicted from a hydropathy analysis according to the method of Kyte and Doolittle [11], the polypeptide contains 12 transmembrane domains connected by more hydrophilic loops. The hydrophobic domains are considered to be α-helices of about 20 amino acids length (3.6 residues per turn), bridging the phospholipid bilayer in a nearly perpendicular way, and not to be composed of ß-sheets or ß-barrels. Many transport systems seem to form a substrate-specific channel through the membrane. The side of an α-helix which faces the lipid must be hydrophobic, while the interior face might be hydrophilic and contribute to a hydrophilic core of such a channel. Such amphiphilic helices have a tendency to form oligomers or to be arranged, in the monomeric form, in parallel to the phospholipid surface, hydrophobic residues in, hydrophilic residues out of the membrane. They may be recognised by a characteristic asymmetric location of the polar and non-polar residues. If a ring of helices is large enough, its interior may be filled with other (hydrophilic) parts of the polypeptide. Protease treatments, monoclonal antibody binding tests, and fusion studies with such hydrophilic "intramembrane" parts might easily give ambiguous and even false results. The presence of hydrophobic helices alone is not sufficient to predict their integration in the membrane accurately. Rather, the charged amino acids that lie on either end of these helices are crucial in determining their orientation. Thus membrane spanning hydrophobic stretches generally have a net positive charge on the cytoplasmic site, their orientation depending on the balance between the number of positively charged amino acids at the amino terminus and those in the hydrophilic loop following its carboxy terminus. Furthermore, periplasmic hydrophilic loops have a clear tendency to be shorter than cytoplasmic loops [12].

AN INTEGRATED MODEL FOR THREE PROTON SYMPORT DRIVEN TRANSPORT SYSTEMS.

Predictions on protein structures can be improved furthermore if several related sequences for proteins with similar functions are available. When aligned, functionally important structures show up as highly conserved motifs or even stretches of amino acids. Except for the LacY permease from *E. coli* [8,9] and from *Klebsiella pneumoniae* [13], a plasmid encoded raffinose permease RafB [14], and a newly discovered sucrose permease CscB encoded in the chromosome of wild type strains of *E. coli* [our unpublished results] have been sequenced. All are of the proton symport type and show partially overlapping substrate specificity, while the protein sequences contain between

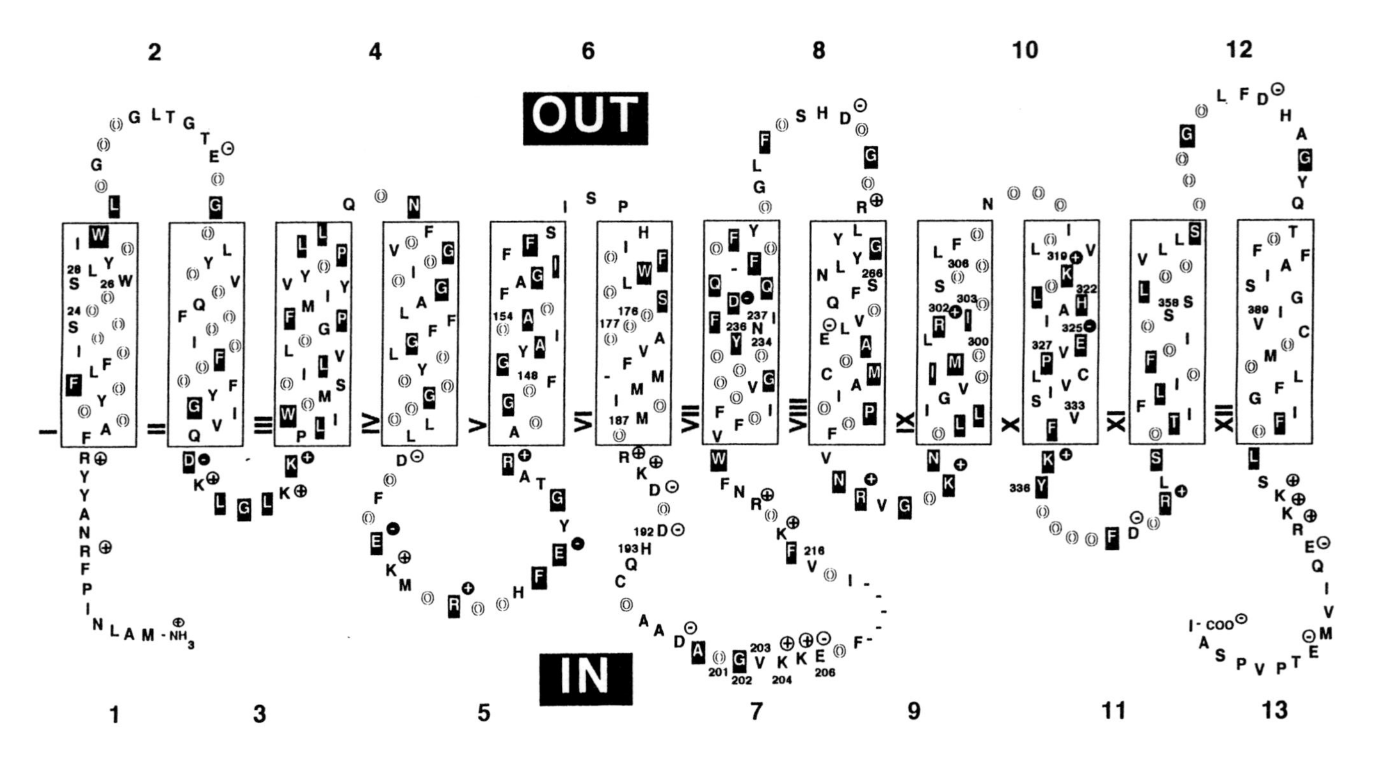

Figure 2. Secondary structure model for ion symport permeases. The model is according to that of King et al. [9], but based on the integrated protein sequence of the LacY lactose permeases from *E. coli* and *K. pneumoniae*, the plasmid encoded RafB raffinose permease, and a CscB sucrose permease from *E. coli*. Putative transmembrane α helices are shown as boxes connected by hydrophilic loops. Identical amino acids are boxed, conserved (as in CscB) indicated by capitals, non-conserved by O, and gaps by a -, while residues identified as important in LacY of *E. coli* are numbered (according to the LacY numbers). Loops 1 and 13, finally, are taken unchanged from the CscB sequence.

31 and 56% identical amino acids for about 405 overlapping residues. These were aligned and a consensus sequence deduced in which identical, conserved (classified as hydrophobic, neutral or hydrophilic according to Kyte and Doolittle [11], and non-conserved residues are indicated.

As shown in Fig. 2, the first conserved structures are putative transmembrane helices. Their length is in general strictly conserved as is the hydrophobicity of the amino acids involved. Amino acid insertions and deletions, by contrast, are characteristic for loops connecting secondary structures, e.g. the hydrophilic loops between transmembrane helices (TM). A difficulty in aligning less related sequences is that the nature, the number and the general location of essential amino acids may be conserved, but not necessarily their exact position. This is clearly visible for the hydrophobic and neutral residues within transmembrane structures, for the location of positively charged residues in cytoplasmic loops of membrane bound proteins, and for stretches of about 20 characteristic amino acids (often Pro, Ala, Gln, Ser and charged residues) with linker or hinge function [15]. These often connect structurally distinct, but autonomous domains, which interact functionally. Based on these criteria, a common model for the three permeases emerges according to which 12 putative transmembrane helices exist, 6 of which (# III,V,IX,X,XI,XIII) are highly conserved and hydrophobic, 4 (# IV,VI,VII, VIII) at an intermediate level, and 2 (# I, II) at a low level. The number of identical residues is highest for helix III, VII, V, IX, and X, perhaps indicating central roles as has been shown for helix X containing the "proton translocating pathway" [8].

The second conserved motifs are located on hydrophilic loops, mostly on these directed towards the inside. They correspond largely to the small loops containing the intracellular positively charged amino acids needed to hold the TM correctly in the membrane [12]. Interestingly the larger loops (# 1,7 and 13) vary in length and contain among the least conserved structures of the whole molecules. This includes loop 7 considered as critical for inducer exclusion and binding of $EIII^{Glc}$.

HYPOTHETICAL MODEL AS DEDUCED FROM AN EII^{Scr} CONSENSUS SEQUENCE ANALYSIS.

More than 20 DNA and protein sequences for various EIIs from different bacteria have become available recently. The similarities range from 95% identical residues to values at the limit of statistical significance. In view of the predictions from the common evolution hypothesis, it came as a surprise that at the primary amino acid level, the different EnzymesII seem organised not in one large, but in four distinct families [3]. These are: i) The D-glucopyranoside family with PTS specific for sucrose, ß-glucopyranosides, D-glucopyranoside, maltose, N-acetyl-glucosamine, and probably trehalose, all generating D-glucopyranoside 6-phosphate during transport. Of these, the former two form a subfamily more closely related to each other than to the other members mentioned which form a second subfamily. ii) The D-mannitol/D-fructose PTS family generating the corresponding 1-phosphates during transport. iii) The D-mannose/ketose family with PTS specific for D-mannose, D-fructose and L-sorbose, each accepting D-fructofuranoside and generating the corresponding 6-phosphate during transport. iv) The lactose/cellobiose family with PTS specific for ß1,4-disaccharides. While the percentage of identical amino acids ranges from 20 to 40% for members of a family, similarity is reduced for members from different families to few and locally restricted conserved amino acid motifs or lacking at all.

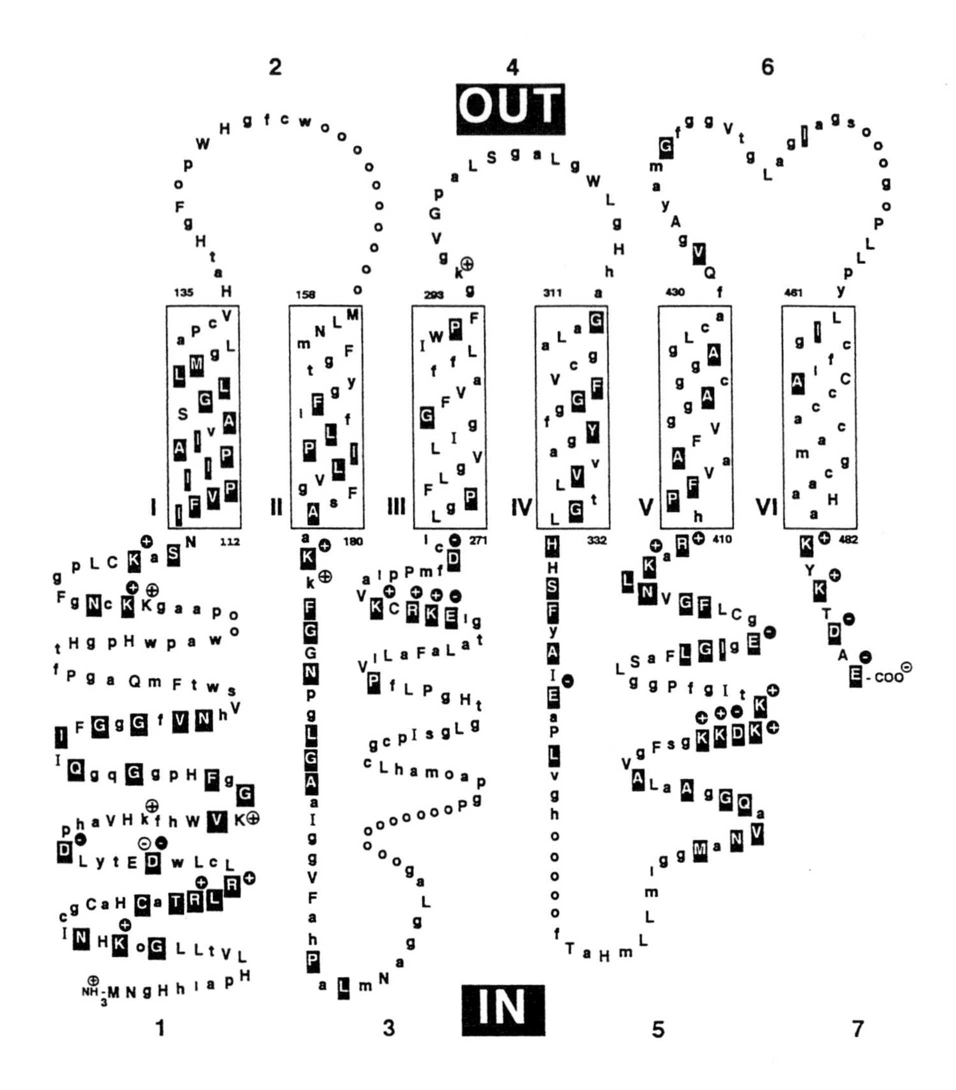

Figure 3. Secondary structure model of EII^{Scr} transport systems. The model is based on an integrated sequence (see text) derived from the 5 published sequences for EII^{Scr}. Six putative transmembrane helices are boxed and numbered. Residues identical in all sequences are boxed, conserved (capital letters) and non-conserved (small letters) residues averaged according to Kyte and Doolittle [11] and the amino acid closest to this value given, gaps labeled with a O. The C-terminal loop is taken unchanged from the pUR400 and *K. pneumoniae* sequence, and intracellular positively charged residues lost by the averaging procedure are retained.

Six EII^{Scr} sequences have become available for sucrose pathways from an enterobacterial plasmid [16] and the chromosome of *K. pneumoniae* (95% identical residues) [our unpublished results], two systems from *Bacillus subtilis* [17,18], and one system each from *Vibrio alginolyticus* [19] and *Streptococcus mutans* [20], the latter four systems with from 33 to 55% identical residues among each other and the first two systems. Using the rules as described for the symport systems, a consensus sequence and the corresponding model was deduced as shown in Fig. 3. Identical amino acids (24%) are listed, while for conserved and non-conserved residues the average hydrophobicity value according to Kyte andDoolittle [11] was calculated, and the amino acid closest to this average on the scale listed in the consensus sequence. The Kyte-Doolittle hydropathy plot of this integrated sequence was finally used to identify putative hydrophobic structures.

According to this analysis, there are six putative transmembrane helices of which the first contains a remarkable 63%, the others between 10 and 30% identical residues. None contains a conserved charged amino acid. The prospected periplasmic loops are from about 20 to 30 amino acids long, two containing extended insertions/deletions. Their hydrophilic character is more visible through the absence of conserved and hydrophobic residues than by the presence of charged residues as the latter, from 3 to 6 per loop in each sequence, are lost by the averaging procedure. Most striking are three domains of about 80 to 115 residues length located according to the model and the "positive-net-charge" rule on the cytoplasmic side of the membrane. The first and amino terminal (residues # 1-115) of these corresponds to the cystein-domain EIIB which in the closely related EII^{Glc}/EII^{Nag} subfamily is located at the carboxy terminal end (fused to the equivalent of residue 490 in Fig. 3). As expected, Cys26 supposed to be a member of the phosphorylation cascade, is present in each sequence and surrounded by an extended consensus sequence, identified before as typical for the D-glucopyranoside PTS family [3]. This sequence contains a series of charged amino acids possibly also involved in the phosphotransferase reaction, and of structural amino acids. These might participate in the interaction of EIIB with HPr on the one side, and with EIIC on the other side. The second domain (residues # 180-272) begins with a strongly conserved stretch (residues # 186-205) of weak but sufficient hydrophobicity to be integrated in the membrane, followed by a non conserved area with large insertions/deletions (residues # 206-236). This arrangement would require that residues # 225-247 form again a transmembrane helix to be followed by a positively charged intracellular loop (residues # 248-272). It cannot be decided thus far which of the two possibilities is correct, although the hydropathy plots of the individual sequences seem to favour the first solution.

The third domain (# 333-413) contains, in contrast to domain 2 but similar to the EIIB, domain throughout, a series of identical residues, among them many charged amino acids. These include a histidine (# 333) and a GI.E motif (# 298-301), the latter flanked by a series of charged amino acids, conserved in most or all EIIs at the equivalent place. This domain thus seems critical in important catalytic steps such as substrate binding and substrate phosphorylation [3,4].

CONCLUSIONS

Biochemical and genetical topology methods allowed the proposition of a well founded model for the lactose permease, by far the best analyzed of all transport systems. Combination of this model with the sequence of the raffinose permease and a newly discovered sucrose permease, both also of the ion symport type, gave results, the equivalent of an extended mutagenesis study, which corroborate the model admirably in a series of molecular details as demonstrated in Fig. 2. When the rules used to construct this model were applied to an integrated protein sequence of six EII^{Scr} permeases, a putative model emerges as shown in Fig. 3. A major difference seems the presence of six instead of twelve transmembrane helices. There are, however, two weakly membrane seeking stretches in the middle hydrophilic domain, while helices # I and II (this study) and the more hydrophilic central structures of the ion symport model can still not be considered as unambiguous transmembrane α-helices. Both models have in common the high degree of conservation in true transmembrane helices and in catalytically important structures, often localised at the cytoplasmic face of the membrane.

Based on the presence of locally restricted, but conserved motifs for all EII sequences available, we have postulated before, that most, if not all, Enzymes II share considerable structure similarity [3,4]. As expected, the model for EII^{Bgl} based on the same rules as used for the EII^{Scr} model, is basically identical to the one shown in Fig. 3, with the modification that now the fused EIIA is found as a fourth intracellular domain. The same holds for an integrated model obtained for all EII^{Glc}, EII^{Nag}, and EII^{Mal} [references in 6,21,22] sequences available (data not shown). In the latter, as shown before, the EIIB domain is fused to the C-terminal end of EIIC and thus intermediate between EIIA, the last phospho-histidine domain, and the intracellular domain considered as essential in substrate binding and phosphorylation. A similar configuration can easily be visualised for EII^{Scr} and EII^{Bgl} through closing a transmembrane ring between helix I and VI, especially if it is remembered that the various domains are fused to this ring and among each other through highly mobile linker and hinge structures. A six transmembrane helix model, also predicting four (including EIIA) extended intracellular domains, has been found for EII^{Mtl} from *E. coli* and validified by over 40 unique *mtlA-phoA* fusions [G. Jacobson, pers. commun.]. This model, and even better the model for two EII^{Fru} [6], the other members of the family, fits exactly the model proposed in Fig. 3. 90% of the identical and 76% of all conserved residues between the EII^{Fru} and the EII^{Scr} sequences (15.8% overall similarity) are located in the predicted six transmembrane helices, the positively charged intracellular loops and catalytically important motifs such as the G..E motif. Finally, integrated sequences for the lactose-cellobiose, and even for the mannose-ketose family predict again the existence of six conserved transmembrane helices, and of four large intracellular loops, which alternate with short non-conserved periplasmic loops, while carrying themselves conserved and catalytically important structures [references in 3,6]. Consequently, all enzymesII, regardless of which family they belong to, seem to form one type of transmembrane and substrate specific channel also responsible for the vectorial phosphorylation of all PTS substrates. Such a result is clearly in favour of the hypothesis that all PTS originated from a common predecessor.

ACKNOWLEDGEMENTS

We would like to thank Eileen Placke for help in preparing the manuscript and the Deutsche Forschungsgemeinschaft for financial support through SFB171, TPC4.

REFERENCES

1 Postma PW, Lengeler JW. Microbiol Rev 1985; 49: 232-269.
2 Meadow ND, Fox DK, Roseman S. Annu Rev Biochem 1990; 59: 497-542.

3 Lengeler JW, Titgemeyer F, Vogler AP, Wöhrl BM. Phil Trans R Soc Lond B 1990; 326: 489-504.
4 Lengeler JW. Biochim et Biophys Acta 1990; 1018: 155-159.
5 Saier MH Jr. Res Microbiol 1990; 141: 281-286.
6 Reizer A, Pao GM, Saier MH Jr. J Mol Evol 1991; 33: 179-193.
7 Grisafi PL, Scholle A, Sugiyama J, Briggs C, Jacobson GR, Lengeler JW. J Bacteriol 1989; 171: 2719-2727.
8 Kaback HR. In: Krulwich TA ed. The Bacteria XII 1990; 151-202.
9 King CS, Hansen CL, Hastings Wilson T. Biochim et Biophys Acta 1991; 1062: 177-186.
10 Calamia J, Manoil C. Proc Natl Acad Sci USA 1990; 87: 4937-4941.
11 Kyte J, Doolittle RF. J Mol Biol 1982; 157: 105-132.
12 von Heijne G, Nilsson I. Cell 1990; 62: 1135-1141.
13 McMorrow I, Chin DT, Fiebig K, Pierce JL, Wilson DM, Reeve ECR, Hastings Wilson T. Biochim et Biophys Acta 1988; 945: 315-323.
14 Aslanidis C, Schmid K, Schmitt R. J Bacteriol 1989; 171: 6753-6763.
15 Wooton JC, Drummond MH. Prot Eng 1989; 2: 535-543.
16 Ebner R, Lengeler JW. Molec Microbiol 1988; 2: 9-17.
17 Fouet A, Arnaud M, Klier A, Rapoport G. Proc Natl Acad Sci USA 1987; 84: 8773-8777.
18 Zukowski MM, Miller L, Cosgwell P, chen K, Aymerich S, Steinmetz M. Gene 1990; 90: 153-155.
19 Blatch GL, Scholle RR, Woods DR. Gene 1990; 95: 17.23.
20 Sato Y, Poy F, Jacobson GR, Kuramitsu HK. J Bacteriol 1989; 171: 263-271.
21 Vogler AP, Lengeler JW. Mol Gen Genet 1991; in press.
22 Reidl J, Boos W. J Bacteriol 1991; 173: 4862-4876.

© 1992 Elsevier Science Publishers B.V. All rights reserved.
Molecular mechanisms of transport. E. Quagliariello, F. Palmieri, eds.

Modular design and multiple functions: The hexose transporters of the bacterial phosphotransferase system

T. Schunk[a], E. Rhiel[a], R. de Meyer[ab], A. Buhr[ac], U. Hummel[b], C. Wehrli[c], K. Flükiger[c] and B. Erni[abc]

[a]Department of Biology, University of Marburg, D-3550 Marburg, Germany

[b]Department of Microbiology, Biocenter, University of Basel, CH-4056 Basel, Switzerland

[c]Institute for Biochemistry, University of Bern, Freiestr. 3, CH-3012 Bern, Switzerland

INTRODUCTION

Domains are independently folding units of structure within a polypeptide chain. Often they are also distinct units of function. The rate and specificity of a sequence of chemical reactions can be increased if the catalytic units are spatially close and properly oriented with respect to the reaction intermediates. This can be accomplished either by organizing the functional units as domains in a single polypeptide chain or as polypeptide subunits in a multisubunit complex. It is not clear why and when the two apparently equivalent forms are chosen. Multisubunit arrangements may be prevalent in prokaryotes which have a small and non compartimentalized cell space. The coordinate expression and assembly of subunits is ensured by polycistronic transcripts. Eukaryotes in contrast have an at least 1000 fold larger, strongly compartimentalized cell space, and do not produce polycistronic messages. In this case the assembly of functional units in the correct stoichiometry might be ensured by translation of a multidomain polypeptide from a monocistronic message.

The bacterial phosphotransferase system comprises a group of proteins which function in carbohydrate transport, regulation of metabolism and chemotaxis [1]. The proteins transfer in a divergent cascade phosphorylgroups from two phosphoryldonors to a large number of phosphorylacceptors. A schematic view of the cascade is given in Fig. 1. The donors are the high energy metabolic intermediates phosphoenolpyruvate and acetylphosphate. The acceptors are different hexoses, disaccharides and hexitols. The phosphotransferase proteins can be grouped into five functional units according to their position and function in the phosphotransfer cascade. The first four units are hydrophilic proteins each containing one phosphorylation site (sites 1-4). The fifth unit is a membrane protein which contains the carbohydrate binding site. The fourth and fifth unit together transport carbohydrates by a mechanism that couples translocation to the phosphorylation of the substrates. The first three units have regulatory functions. They coordinate carbohydrate transport and metabolic control [1, 2]. Their regulatory activity appears to be determined by their phosphorylation state. They can act upon their targets either allosterically or by phosphorylation.

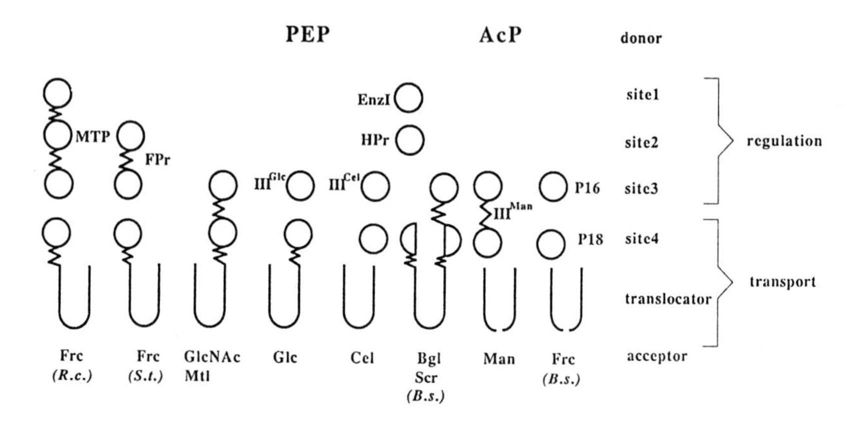

Figure 1. Domain structure of the phosphotransferase proteins. *Open circles* represent the domains which become transiently phosphorylated. Phosphoryltransfer occurs vertically and can occur diagonally between different sites of different systems. The *wiggles* indicate domain linkers. The *U* represent the transmembrane translocator units. The sugar specificities of the transporters belonging to a given type are indicated below.

Enzyme I is the first unit. It transfers phosphoryl groups from phosphoenolpyruvate or acetyl phosphate to the second unit, a 9 kD phosphorylcarrier domain termed HPr. Enzyme I and HPr are found in practically all bacteria. In *S. typhimurium* a unit homologous to HPr is found as a domain of a larger protein termed FPr [3]. The third functional unit transfers the phosphorylgroup to phosphorylation site 4 of the transporter complex. The units 3 and 4 have molecular weights between 14 and 22 kD. They exist as single polypeptdide chains or as units in a multidomain protein. The site 3 unit is found linked to the site 2 unit in the FPr protein of *E. coli* [3], and to the site 4 unit in III^{Man} of *E. coli* [4]. Mostly, site 4 unit are, however, linked with the transmembrane domain. An additional and so far unique combination of the three domains containing sites 1-3 in a single polypeptide has been found in *R. capsulatus* [5]. The phosphorylated residues in units 1-3 are histidines. The phosphorylated residue of site 4 is a cysteine in the mannitol transporter and most of the other transporters [6] but a histidine in the mannose transporter [4]. The transmembrane unit is composed of either one or two polypeptide chains.

When linked, the units are arranged in the following order starting from the N-terminus: (transmembrane site 4)-site 3-site 2-site 1. The sequential order of the site 4 unit and the transmembrane domain can be either way. In the fructose transporter of *E. coli* the site 4 unit could be a composite of two hydrophilic segments each attached to one end of the transmembrane domain. We would like to speculate that the apparently different arrangements of hydrophilic and hydrophobic segments in the different transporters are circularly permuted forms of a common tertiary structure [7].

Table 1
Domain linker sequences

Ala Pro linker		
III^{Man}	AAPAPAAAAPKAAPTPAK	site 3-site 4
II^{GlcNAc}	EATPATAAPVAKPQ	site 3-site 4
II^{Bgl}	KRQPAQGAPQEKTP	site 3-site 4
MTP(*R.c.*)	GAAAPEAATPGLTGAGA	site 1-site 2
MTP(*R.c.*)	GAAAPVAAPAETPADFA	site 2-site 3
LKTPGRED linker		
II^{Glc}	LKTPGRED	site 4-transmembrane
II^{GlcNAc}	LKTPGRED	site 4-transmembrane
II^{Glc} (*B.s.*)	LKTPGRED	site 4-transmembrane
MalX [15]	LKTPGRD	site 4-transmembrane
II^{Mtl}	EEDDIEAATRRMQDMK	site 4-transmembrane
Q linker		
II^{Glc} (*B.s.*)	RKPRPEPKTSAQEEVGQQVEEVIAEPLQNEIGEE	site 3-site 4
II^{Bgl}	RQPAQGAPQEKTPEVITPPEQGG	site 3-site 4
II^{GlcNAc}	ARGPVAAASAEATPATAAPVAKPQAVPNAVSIAE	site 3-site 4

Characteristic linker sequences can be identified in the interdomain regions of the multidomain proteins. Three types of liker have been found so far: Ala-Pro linkers, LKTPGRED linkers, and Q linkers. The Ala-Pro linker is characterized by the cumulation of Ala Pro repeats with interdispersed charged residues. This linker connects the site 3 and 4 domains of III^{Man} [4], the site 1, site 2 and site 3 domains of the MTP protein of *R. capsulatus* [5] and the site 3 and site 4 domains of II^{GlcNAc} and II^{Bgl} [8]. All the domains linked by the Ala Pro rich linker are phosphoryltransfer units. Similar Ala-Pro linkers are abundant in decarboxylases and dehydrogenases where they link the biotinylated and lipoylated domains respectively [9]. Occasionally they have integrated sites of posttranslational proteolytic processing [8]. The LKTPGRED linker connects site 4 with the transmembrane domain. It is found in II^{Glc} of *E. coli* and *B. subtilis*, and in II^{GlcNAc} of

E. coli [Hummel, manuscript submitted]. A sequence of similar amino acid composition is found in II^{Mtl} of *E. coli* [10] and domain linkers of similar amino acid composition were found by inspection of protein crystal structures [11]. The Q linkers which are rich in glutamine, glutamate, arginine and proline were originally discovered in proteins of prokaryotic two component regulatory systems [12]. In the phosphotransferase system a Q linker appears to connect site 3 and site 4 of the glucose transporter from *B. subtilis* [13]. The linkers in II^{GlcNAc} and II^{Bgl} connecting sites 3 and 4 can be classified either as Ala-Pro linkers or as Q linkers.

We wonderd why homologous functional units sometimes are expressed as single polypeptides and sometimes as domains of a larger polypeptide. Are these arrangements equivalent with respect to both transport and regulation? Or is their regulatory function affected if free diffusion is restricted by tethering of subunits? Are the linker sequences of general use for the construction of multidomain proteins from individual subunits? In order to investigate such questions multidomain proteins were constructed by fusion of PTS proteins shown in Fig. 2, and multidomain proteins were split by gene interruptions.

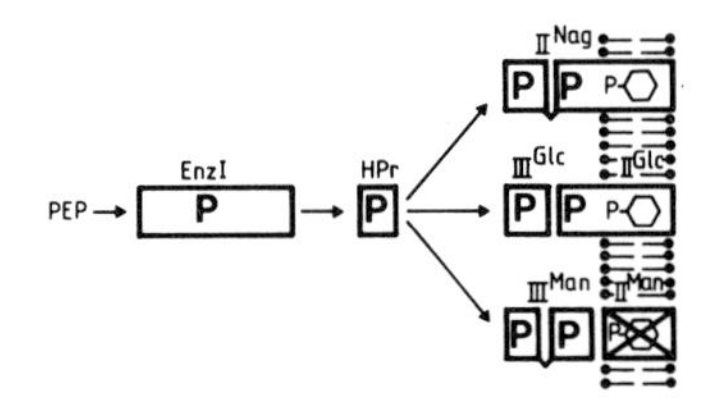

Figure 2. Components of the phosphotransferase system of *E. coli* used for the construction of fusion and hybrid proteins.

RESULTS AND DISCUSSION

Enzyme I-HPr fusions

The 9 kD phosphorylcarrier protein HPr is a substrate of enzyme I. HPr also stimulates the autophosphorylation of enzyme I and/or stabilizes its phosphorylated form, indicating that the two subunits form a functional complex [14]. HPr and enzyme I are encoded by the genes *ptsH* and *ptsI* which form an operon. Three fusion proteins have been constructed, an inframe fusion between HPr and enzyme I, and two fusions containing the Ala-Pro rich sequence taken from III^{Man} (Table 1) as a linker either once, or twice in tandem. All three fusion proteins are active *in vivo* as judged from the fermentation postive phenotype of transformed cells. The *in vitro* activities of cytoplasmic extracts and the purified proteins are given in Tables 2 and 3.

Table 2
Phosphotransferase activities of cytoplasmic cell fractions (A) containing free HPr and enzyme I (H,I), and HPr-enzyme I fusions with a Ala-Pro rich tandem hinge (HdhI) a single hinge (HhI) and no hinge (H·I) [nmol dGlc-6P/30 min/µg protein, 1 µg is approx. 7 pmol enzyme I], and of membranes (B) containing II^{Glc} and II^{Glc}-III^{Glc} fusions with and without hinges [nmol αMeGlc-6P/30 min/µg protein]. The PEP:hexose phosphotransferase assay is described in [4, 14].

A

source	activity
H,I	32.0
HdhI	2.8
HhI	2.0
H·I	1.8

B

source	activity
II^{Glc} (plusIII^{Glc})	8.3
IIdhIII^{Glc}	30.5
IIhIII^{Glc}	30.7
II·IIIGlc	5.1

Table 3
Phosphotransferase activitiy of purified HPr, enzyme I and HPr-enzyme I fusion protein HdhI (100% activity is 3.34 nmol dGlc-6P/30 min/pmol enzyme I).

proteins (molar ratio)	% activity
HPr plus enzyme I (2.3:1)	100
HPr plus enzyme I (1:1)	42
HdhI	5
HdhI plus enzyme I (1:1)	5
HdhI plus HPr (1:1)	42
HdhI after trypsin digestion	39

The phosphotransferase activity of the fusion proteins is 5% - 10% of the control and increases slightly with increasing length of the linker. The reduced activity of the fusion could be caused: (i) by the 1:1 stoichiometry of enzyme I and HPr in the fusion which limits the availability of HPr for phosphoryltransfer and reduces the over all reaction rate; (ii) by a conformational strain due to the linking. It should be more pronounced the tighter the linkeage is; (iii) by non-optimal interaction between the fused enzyme I and HPr domains.

The phosphotransferase activity of the fusion can be complemented by addition of extra HPr (Table 3). Extra enzyme I, in contrast, has no effect and HPr must therefore be the rate limiting component. Proteolytic cleavage of the fusion with a limited amount of trypsin afforded two fragments with the electrophoretic mobilities of enzyme I and HPr [Schunk, unpublished results]. The fragments have significantly higher phosphotransferase activity than the fusion proteins. It therefore appears, that tethering enzyme I and HPr does not affect enzyme I activity nor prevent it from interacting with free HPr, but that the bound HPr is no longer competent to optimally interact with the enzyme I moiety and/or with phosphorylation sites 3. When enzyme I and HPr protein concentrations were diluted below 3.5 nM, the other PTS components being kept constant, the HPr-enzyme I fusion was more active than a 1:1 mixture of free enzyme I and HPr. This indicates intramolecular phosphoryltransfer and it suggests that linking of domains can compensate for the negative effect of non-optimal domain alignment at low protein concentration. The rate constants are 7.25 $pmol^{-1}min^{-1}$ for intermolecular phosphoryltransfer between free enzyme I and HPr and 3.0 min^{-1} for intramolecular transfer between the linked domains. This difference is smaller than the observed 10 fold reduction of phosphotransferase activity of the fusion (Table 2). It suggests that tethering of HPr to enzyme I interferes with both, the HPr enzyme I and the HPr-site 3 interactions.

II^{Glc}-III^{Glc} fusions

The glucose transporter consists of two subunits, III^{Glc} (phosphorylation site 3) and II^{Glc} (phosphorylation site 4 and transmembrane unit). The association between III^{Glc} and II^{Glc} is so weak that a stable complex could not be isolated. Nonetheless, III^{Glc} not only serves as the phosphoryldonor to II^{Glc} but it also affects the catalytic performance of II^{Glc} [14]. III^{Glc} plays a pivotal role in catabolite repression. In this function it has to interact with other proteins (adenylcyclase, carbohydrate transporters [2]). It is not known whether these "target proteins" form a supramolecular complex with the II^{Glc}-III^{Glc} complex *in vivo* or whether III^{Glc} diffuses between the complex and its targets. To approach this question, III^{Glc} and II^{Glc} were linked by two Ala-Pro rich linkers of different length, and a third fusion without a linker was also constructed. All three fusions were active *in vivo* and *in vitro* (Table 2). The fusions containing the Ala-Pro rich hinges could be stimulated several fold by addition of purified III^{Glc} subunit. In contrast, the fusion without a linker could not be stimulated, as if tight linking of II^{Glc} and III^{Glc} prevented the access of soluble III^{Glc} to II^{Glc} [Schunk, unpublished results]. The II^{Glc}-III^{Glc} fusion containing the Ala-Pro rich linker could not be purified in an intact form because it was cleaved by an endogenous protease after solubilization and/or during isoelectric focusing. The resulting II^{Glc} fragment, however, retained full activity [Schunk, unpublished results]. It remains to be seen, how tethering of III^{Glc} to II^{Glc} and the fixed 1:1 stoichiometry influence catabolite repression.

II-P^{Man}-IIM^{Man} fusion

The mannose transporter consists of three subunits, III^{Man}, II-P^{Man} and II-M^{Man}. III^{Man} consists of two hydrophilic domains, P13 and P20, which contain phosphorylation sites 3 and 4 respectively and which are linked by an Ala-Pro rich peptide. II-P^{Man} and II-M^{Man} form the transmembrane unit, a tight complex which cannot be dissociated without loss of activity [16]. No active complex is formed when independently expressed, detergent solubilized subunits are mixed. The hydropathy profile of II-P^{Man} and II-M^{Man} suggests that II-P^{Man} might contain four transmembrane helices and II-M^{Man} a large cytoplasmic loop followed by two hydrophobic helices. A similar topology (4 helices, a large loop and two helices) can also be predicted for the single subunit PTS transporters. If we assume that all PTS transporters have a similar topology, the C-terminal end of II-P^{Man} and the N-terminal end of II-M^{Man} should be next to each other on the cytoplasmic surface of the membrane. To test this proposition, II-M^{Man} and II-P^{Man} were linked once without intervening sequence and once through an Ala-Pro linker peptide. Both fusion proteins were active

Table 4
Phosphotransferase activities of the II-P^{Man} II-M^{Man} complex (P, M) and of II-P^{Man}/II-M^{Man} fusions with (PhM) and without (P·M) the Ala-Pro rich hinge before (-) and after (+) digestion with the optimal amount of trypsin. 100 % activity corresponds to 380 nmol dGlc-6P/30 min/µg II-M^{Man} protein. The protein concentration was judged from a Coomassie stained polyacrylamide gel.

	membranes		partially purified	
	-	+	-	+
P, M	100%	100%	100%	100%
PhM	4%	30%	11%	35%
P·M	4%	5%	6%	5%

in vivo and could be purified by isolelectric focusing. The purified fusion without an intervening linker had approximately 5% of wild-type activity, the fusion with the linker 11% (Table 4). The latter could be selectively cleaved with trypsin in the hinge peptide. Cleavage was accompanied by an increase of the specific activity to 30% of the wild-type control. These results suggest, that the C-terminal end of II-P^{Man} and the N-terminus of II-M^{Man} are close in space and can be joined through a flexible Ala-Pro rich linker. From these results we also conlude that II-M^{Man} is present in a 1:1 molar ratio with II-P^{Man} in the complex and not in excess as suggested by autoradiography, Coomassie blue staining and UV scanning of SDS polyacrylamide gels [Rhiel, unpublished results].

Heterologous fusions with III^{Man}

Three domain fusions of the form P13(Ala-Pro hinge)X(Ala-Pro hinge)P20, X being a heterologous protein and P13 and P20 domains of III^{Man} were engineered in an attempt to produce sandwich fusion proteins from which the heterologous part could be recovered by proteolytic cleavage at the Ala-Pro rich hinges. Mouse dihydrofolate reductase, EcoRI restriction nuclease and a homeodomain of *Drosophila melanogaster* were chosen as guest domains. All fusions could be expressed and had phosphotransferase activity *in vivo* as judged from the mannose fermentation positive phenotype of transformed cells. However, only the fusion with DHFR remained soluble and could be purified. The other two fusion proteins formed inclusion bodies and so far could not be refolded in spite of heroic efforts to do so. Digestion of the purified P13-DHFR-P20 fusion with limiting amounts of trypsin afforded the three domains as major productss but in addition a great number of other proteolytic fragments [de Meyer, unpublished results]. From these results we tend to conclude, that the Ala-Pro rich linker might not be the universal "protein polylinker".

Hybrid proteins between the glucose and the N-acetylglucosamine transporter

The glucose and the N-acetylglucosmine transporter have approximately 40% amino acid sequence similarity. Chimaeric proteins between the two transporters were generated *in vitro* by introduction of restriction sites at homologous positions and exchange of restriction fragments, and *in vivo* by forced homologous recombination and screening for fermentation positive recombinants. Only one out of the eight *in vitro* recombinants was active. It contained the N-terminal transmembrane domain of II^{Glc} and the site 4 and site 3 units of II^{GlcNAc}. The fusion was at the LKTGRED linker (Table 1). Exactly the same fusion was also obtained by forced homologous recombination *in vivo*. The fusion is glucose specific and 100% active [Hummel, manuscript submitted]. The site 3 domain of II^{GlcNAc} complements the III^{Glc} subunit of the glucose transporter. The reciprocal fusion between the membrane domain of II^{GlcNAc} and the site 4 domain of II^{Glc} did not complement fermentation of GlcNAc but unexpectedly, supported fermentation of Glc. However fermentation of Glc occurs only in strains which express all the PTS phosphorylcarrier proteins enzyme I, HPr, III^{Glc} and in addition glucokinase but not in strains which lack either PTS proteins or glucokinase. It appears as if the chimaeric II^{GlcNAc}-II^{Glc} protein transports the substrates by a phosphorylation dependent mechanism but can no longer phosphorylate them. If this were so, both, Glc and GlcNAc would be transported by the II^{GlcNAc} hybrid. However Glc only could be phosphorylated by glucokinase and then be metabolized. Another fusion comprising the first 150 residues of II^{Glc} and the complementary part of II^{GlcNAc} was found to be active and glucose specific. This indicates that the sugar specificity determinant must be localized in the N-terminal half of the transmembrane unit, possibly within the first three transmembrane helices. Experiments are now in progress to further zoom in on the sugar specificity determinant of the glucose transporter.

REFERENCES

1 Meadow ND, Fox DK, Roseman S. Ann. Rev. Biochem. 1990; 59: 497-542.
2 Saier MH. Microbiol. Rev. 1989; 53: 109-120.
3 Geerse RH, Izzo F, Postma PW. Mol. Gen. Genet. 1989; 216: 517-525.
4 Erni B, Zanolari B, Graff P, Kocher HP. J. Biol. Chem. 1989; 264: 18733-18741.
5 Wu LF, Tomich JM, Saier MH. J. Mol. Biol. 1990; 213: 687-703.
6 Pas HH, Robillard GT. Biochemistry 1988; 27: 5835-5839.
7 Erni B. Int. Rev. Cytol. 1992; in press
8 Erni B. FEMS Microbiol. Rev. 1989; 63: 13-24.
9 Perham RN. Biochemistry 1991; 30: 8501-8512.
10 van Weeghel RP, Meyer G, Pas HH, Keck W, Robillard GT. Biochemistry 1991; in press.
11 Argos P. J. Mol. Biol. 1990; 211: 943-958.
12 Wootton JC, Drummond MH. Protein Engineering 1989; 2: 535-543.
13 Sutrina SL, Reddy P, Saier MH, Reizer J. J. Biol. Chem. 1990; 265: 18581-18589.
14 Erni B. Biochemistry 1986; 25: 305-312.
15 Reidl J, Boos W. J. Bacteriol. 1991; 173: 4862-4876.
16 Erni B, Zanolari B, Kocher HP. J. Biol. Chem. 1987; 262: 5238-5247.

ACKNOWLEDGMENT This work was supported by grants Er 147/1-1 and Er 147/2-1 from the Deutsche Forschungsgemeinschaft, 31-909287 and 31-29795.90 from the Swiss National Science Foundation and by a grant from the Sandoz Research Foundation.

© 1992 Elsevier Science Publishers B.V. All rights reserved.
Molecular mechanisms of transport. E. Quagliariello, F. Palmieri, eds.

CONTROL OF CARBOHYDRATE METABOLISM IN ENTERIC BACTERIA: QUALITATIVE AND QUANTITATIVE ASPECTS

P.W. Postma, G.J.G. Ruijter, J. van der Vlag and K. van Dam

E.C. Slater Institute for Biochemical Research, University of Amsterdam, Plantage Muidergracht 12, 1018 TV Amsterdam, The Netherlands.

INTRODUCTION

Uptake of carbohydrates by micro-organisms can be catalyzed in various ways. In most cases the carbohydrate is accumulated by the cell against its electrochemical potential in an unchanged form at the expenditure of energy. For instance, lactose is accumulated in *Escherichia coli* via the lactose permease at the expense of an H^+ moving down its electrochemical gradient, a process called lactose-H^+ symport. Accumulation of other carbohydrates requires the investment of ATP or movement of a Na^+ ion down its electrochemical gradient. A different class of carbohydrates is accumulated via the PEP: carbohydrate phosphotransferase system (PTS). The PTS catalyzes the transport and concomitant phosphorylation of numerous carbohydrates in both Gram-negative and Gram-positive bacteria (for a review, see ref. 1). The phosphoryl group which is transferred to the carbohydrates, is derived from phosphoenolpyruvate (PEP) and is sequentially moved via the various PTS proteins, each of which is phosphorylated on a histidine residue (Figure 1). All PTS carbohydrates utilize the general PTS proteins, Enzyme I and HPr. Carbohydrate specificity is provided by the various Enzymes II (e.g. II^{Mtl}, specific for mannitol) and Enzyme II/III couples (e.g. II^{Glc}/III^{Glc}, specific for glucose). At least 15 different Enzymes II/III are known in enteric bacteria (3).

Apart from its function in the uptake and phosphorylation of carbohydrates, the PTS plays a central role in the transport and metabolism of many non-PTS carbohydrates. This phenomenon is most prominent in enteric bacteria like *Escherichia coli* and *Salmonella typhimurium*. A classical example is the phenomenon of diauxic growth, studied first by Monod (2). When *E. coli* is offered a mixture of glucose and lactose, the cells will utilize first all glucose before metabolism of lactose starts after a short lag. The enzymes encoded by the *lac* operon are synthesized only after all glucose has been used up. In this article we will describe some of the factors that are important in this type of regulation.

REGULATION BY THE PTS - AN OVERVIEW

The PTS regulates the uptake and subsequent metabolism of many compounds in *E. coli* and *S. typhimurium* at two possible levels (for a review, see ref. 3): (i) the expression of genes or operons, encoding the enzymes required for the uptake and metabolism of these compounds; and (ii) the activity of some of these enzymes. A central role is played by III^{Glc} which catalyzes together with II^{Glc} the transport and phosphorylation of glucose (Figure 1). Like the other PTS proteins, III^{Glc} can exist in two forms, a phosphorylated form ($P \sim III^{Glc}$) and a non-phosphorylated form. Both forms can interact with non-PTS enzymes and alter

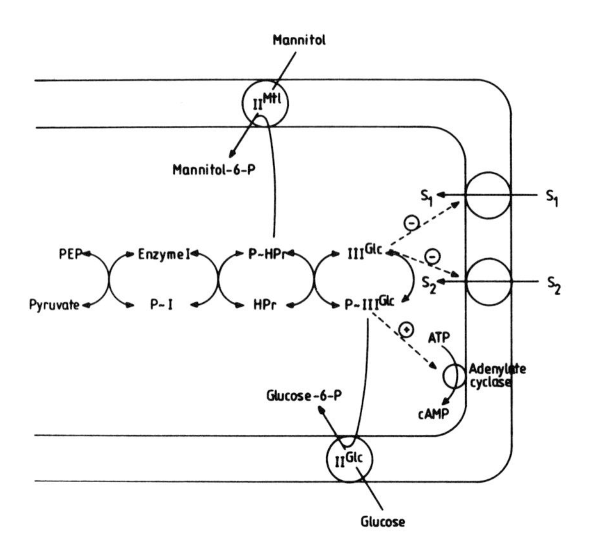

Figure 1. The PTS. In addition to the general proteins, Enzyme I and HPr, two permeases are shown, specific for glucose (II^{Glc}/III^{Glc}) and mannose (II/III^{Man}). PTS-mediated regulation is indicated by the inhibition (-) of certain non-PTS uptake systems (S_1 and S_2, representing lactose, melibiose, maltose or glycerol) and the activation (+) of adenylate cyclase by phospho-III^{Glc}.

their activity. Figure 1 shows this in a schematic way. In its non-phosphorylated form, III^{Glc} can bind to several non-PTS enzymes and inhibit their activity. These target proteins include the lactose and melibiose permease, a component of the maltose transport system (MalK) and glycerol kinase. In all cases, the direct effect is the inhibition of solute uptake and/or metabolism. This phenomenon is called inducer exclusion. Compounds like lactose, melibiose, maltose or glycerol cannot be taken up any more and thus cannot induce their respective operons. We will discuss below the evidence that in an intact cell III^{Glc} is indeed in a dephosphorylated state when a PTS carbohydrate like glucose is metabolized. Phospho-III^{Glc} (the form present in the absence of a PTS carbohydrate) is involved in the activation of adenylate cyclase, i.e. in regulating the level of intracellular 3',5'-cyclic AMP (cAMP) which is required for the expression of many metabolic operons.

ASPECTS OF PTS-MEDIATED REGULATION

Inducer exclusion. Evidence has been presented that non-phosphorylated III^{Glc} interacts with its various target proteins. III^{Glc} binds to the *E. coli* lactose permease with a K_D of approximately 10 μM. Phospho-III^{Glc} does not bind. Another interesting observation is the requirement of a substrate of the target protein for III^{Glc} binding. Thus, III^{Glc} binds

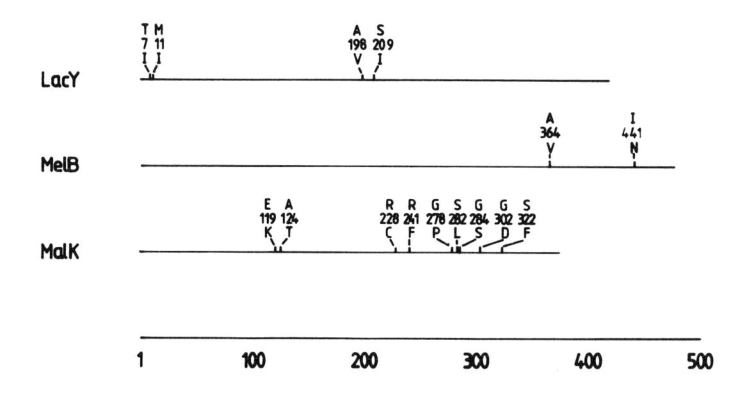

Figure 2. Mutations that result in resistance to inducer exclusion. LacY, MelB and MalK are the lactose permease, the melibiose permease and the MalK component of the maltose transport system, respectively. The one-letter symbols for amino acids are used. Data were taken from refs. 4 and 6 (LacY), 7 and 8 (MalK) and unpublished results of S.J. de Waard, M. Kuroda, T. Tsuchiya and P.W. Postma (MelB).

only to the lactose carrier in the presence of a β-galactoside (4). Similarly, III^{Glc} binds to and inhibits purified glycerol kinase only in the presence of glycerol (5). From these experiments it can be concluded that a conformational change is required in the target protein before III^{Glc} is able to bind to that protein. After reconstitution of the lactose carrier in liposomes, purified III^{Glc} is able to inhibit β-galactoside transport (4). The two proteins are sufficient to explain inducer exclusion in the case of the lactose permease.

These results suggest a specific interaction of III^{Glc} with a number of quite different target proteins, including integral membrane proteins (the lactose and melibiose permease), a peripheral membrane protein (the MalK component of the maltose transport system) and a soluble, cytoplasmic protein, glycerol kinase. Can we tell what kind of structure is recognized in these enzymes by III^{Glc}? A simple comparison of the linear amino acid sequences does not give any clue since the four proteins show little or no identity. Since III^{Glc} binds to and inhibits these proteins, it was realized earlier that mutated proteins could exist which are resistant to inhibition by III^{Glc}. Mutations have been isolated that render *E. coli* and *S. typhimurium* resistant to inducer exclusion, i.e. these mutants can metabolize lactose or melibiose in the presence of a PTS carbohydrate. In practice, these mutants are isolated by selecting for growth of *E. coli* or *S. typhimurium* on for instance melibiose in the presence of a non-metabolizable glucose analog such as 2-deoxyglucose or methyl α-glucoside. A number of such mutations has been recently isolated and identified by sequencing in the *lacY*, *melB* and *malK* genes, encoding the lactose, melibiose and maltose (MalK component) transport systems, respectively. It is assumed that in all cases III^{Glc} binding to the target proteins is defective but this has been verified only in one instance (4). A schematic representation of the isolated mutations is given in Fig. 2. At present, these mutations do not give any clear indication of which features might be conserved in those three proteins that constitute a III^{Glc} binding site. Although it has been pointed out (7,8) that some identity exists between residues 198-217 of LacY and 275-291 of MalK, regions which contain some of the mutations, other mutations clearly fall far outside this segment. In addition, the suggested identity is very weak. Clearly, many more mutations will be required to identify possible stretches of each polypeptide, involved in III^{Glc} binding. Most likely, only X-ray crystallography will be able to solve this specific interaction. The most promising complex in this respect is probably that of III^{Glc} and glycerol kinase.

Adenylate cyclase. The hypothesis that phospho-III^{Glc} is an activator of adenylate cyclase is mainly based on the observation that adenylate cyclase activity is inhibited in toluenized cells by PTS carbohydrates and that *crr* mutants, lacking III^{Glc}, have a low adenylate cyclase activity. Since mutants that contain only non-phosphorylated III^{Glc} (for instance *ptsHI* mutants that lack Enzyme I and HPr) have an equally low adenylate cyclase activity, it has been proposed that phospho-III^{Glc} is an activator of adenylate cyclase (Fig. 1), rather than III^{Glc} being an inhibitor, as is the case in inducer exclusion (1,3). Attempts to show that phospho-III^{Glc} can activate adenylate cyclase in an *E. coli* extract have been only partially successful. At most, a two-fold stimulation was observed (9). It is known, however, that the rate of cAMP synthesis can vary 100- or even 1000-fold, depending on the conditions. Even inhibition of cAMP synthesis by glucose in toluenized cells is five- to ten-fold. Possibly, factors other than phospho-III^{Glc} are required for full activation of adenylate cyclase.

The phosphorylation state of III^{Glc}. In the previous sections it has been assumed that the phosphorylation state of III^{Glc} in an intact cell is indeed changing in response to the presence of PTS carbohydrates in the growth medium. The available evidence (10) suggests that in cells growing on a non-PTS carbohydrate like galactose, III^{Glc} is mostly in the phosphorylated form; III^{Glc} is phosphorylated via phospho-HPr while no dephosphorylation via Enzyme II^{Glc} occurs. Addition of the non-metabolizable analog methyl α-glucoside (αMG) results in a dephosphorylation of phospho-III^{Glc}. This effect is even more pronounced in a mutant that lacks most of its Enzyme I, i.e. a mutant in which re-phosphorylation of III^{Glc} is slow. Thus, in a semi-quantitative way, the model shown in Figure 1 is supported. It is important to stress that the addition of any PTS carbohydrate can lead to dephosphorylation of phospho-III^{Glc}, even by those PTS carbohydrates that are not recognized directly by II^{Glc}. It has been shown that most of the reactions of which the PTS is made up, are reversible, i.e. if phospho-HPr is dephosphorylated completely due to donation of its phosphoryl group to another Enzyme II, e.g. II^{Mtl}, HPr can in its turn catalyze the dephosphorylation of phospho-III^{Glc}.

Levels of target proteins and III^{Glc}. The stoichiometric interaction between III^{Glc} and its target proteins, as discussed above, predicts that in (mutant) cells in which the amount of III^{Glc} is lowered or, alternatively, the amount of target proteins is increased, not all target proteins can be complexed by III^{Glc}. As a consequence, cells might escape from PTS-mediated regulation. It is important to know that in *E. coli* and *S. typhimurium* the level of III^{Glc} is relatively constant and independent of growth conditions. This will be of direct physiological importance because cells which have not been in contact with lactose before (and thus are not induced for the enzymes required for lactose metabolism) and which face the choice between glucose and lactose, will be unable to take up and metabolize lactose since the few lactose transport systems available will be inhibited by III^{Glc}. However, cells that have been grown previously in the presence of lactose, will be much more difficult to inhibit by glucose since the level of lactose permease is quite high in those cells.

An interesting consequence of the stoichiometric interaction is that synthesis of a second PTS-sensitive system can relieve inhibition of the first system. This has been observed experimentally (11). Inhibition of the maltose transport system by the glucose analog 2-deoxyglucose was lessened or abolished by induction of the glycerol uptake system, i.e. glycerol kinase. As expected, the presence of glycerol was also required for the relieve of inhibition of maltose transport by 2-deoxyglucose.

These examples show that several factors are important in PTS-mediated regulation. First, the target proteins have to be recognized specifically by III^{Glc}. Secondly, the amount of both III^{Glc} and the target proteins as well as the phosphorylation state of III^{Glc} are critical factors in regulation. In the next section we will describe some results that address the more quantitative aspects of PTS-mediated regulation, in particular the flux of phosphoryl groups through the various PTS enzymes.

CONTROL ANALYSIS OF PTS-MEDIATED TRANSPORT AND REGULATION

Introduction. The previous sections have stressed the importance of the phosphorylation state of III^{Glc} in PTS-mediated regulation. It has been shown in a semi-quantitative way that certain conditions in the growth medium result in dephosphorylation of phospho-III^{Glc}. However, for a better understanding of PTS-mediated regulation, a more quantitative description of the pathways that are responsible for phosphorylation and dephosphorylation is required. A useful model to analyse this in more detail has been proposed by Kacser and Burns (12) and Heinrich and Rapoport (13). It allows one to evaluate the fluxes through a pathway and the degree to which individual steps determine the overall rate through the pathway. In our case it can be used to determine which steps are controlling the fluxes of phosphoryl groups through the PTS pathways and eventually the phosphorylation state of III^{Glc}. To quantify to which extent one step of a pathway controls the overall flux through that pathway, the concept of control coefficient has been introduced.

$$S \xrightarrow{E_1} X_1 \xrightarrow{E_2} X_2 \; \ldots\ldots\ldots \; X_{N-1} \xrightarrow{E_N} P \qquad (1)$$

$$c_E^J = \frac{dJ/J}{dE/E} \qquad 0 \leq c_E^J \leq 1 \qquad (2)$$

$$\Sigma c_E^J = 1 \qquad (3)$$

(c_E^J = *control coefficient,* J = *flux*)

Consider the pathway of equation 1 in which a number of enzymes catalyze the conversion of S into P. The control coefficient, c_E^J, is the relative increase in pathway flux, divided by the relative increase in the amount of enzyme (equation 2), and can vary between 0 and 1. In the first case, an increase in the amount of enzyme E_i has no effect at all on the overall flux through the pathway, i.e. the enzyme E_i exerts no control. In the second case, a two-fold increase in the amount of enzyme is causing a two-fold increase in flux, i.e. E_i is completely controlling the flux. The summation theorem states that the sum of all control coefficients in a pathway equals 1 (equation 3).

Control of glucose uptake and metabolism. To establish which enzyme(s) controls the flux of phosphoryl groups through the PTS, one must be able to alter individually the amount of each of the components of the PTS and to determine the change in flux, keeping all others constant. For this purpose, we have constructed plasmids that contain the genes, encoding the enzymes of the glucose PTS, under control of the *tac* promoter. The plasmids contain in addition the *$lacI^Q$* gene. In the absence of an inducer like isopropylthiogalactoside (IPTG), expression of a gene under control of the *tac* promoter is almost completely repressed.

Figure 3 shows that increasing amounts of IPTG in the growth medium result in an increasing activity of II^{Glc} if the *E. coli ptsG* gene, encoding II^{Glc}, is inserted behind the *tac*

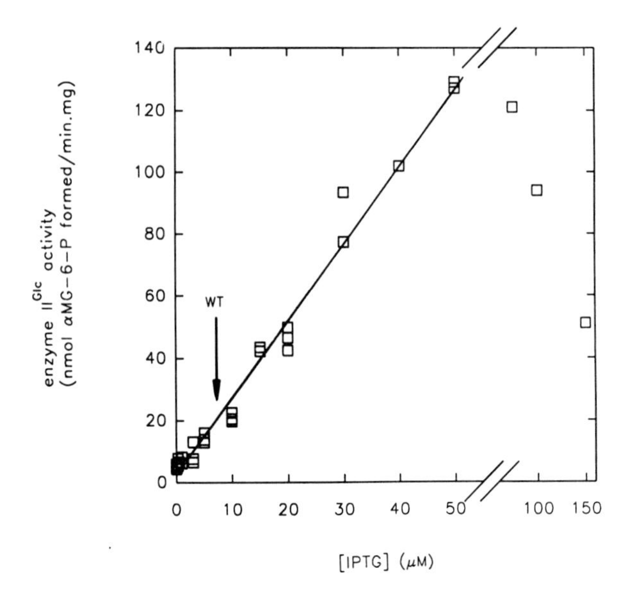

Figure 3. Activity of II^{Glc} in PPA211 (*ptsG ptsM*)/pTSG11. Cells were grown to exponential phase in minimal salts medium containing 0.2% glucose and varying amounts of IPTG. The II^{Glc} activity was determined enzymatically by in vitro phosphorylation of 0.5 mM [U-^{14}C]αMG as described in ref. 19. The arrow indicates the wild type level. Plasmid pTSG11 was gift of B. Erni, Universität Bern.

promoter (plasmid pTSG11). It can be calculated that the II^{Glc} activity can be varied between 20% and 600% of that encoded by the induced chromosomal gene. The low level of II^{Glc} in the absence of IPTG is due to a low basal expression level and the high copy number of the plasmid. This variation in II^{Glc} activity allows us to determine the control of II^{Glc} on the flux through pathways that include the II^{Glc}/III^{Glc} system. Figure 4 shows that glucose oxidation,

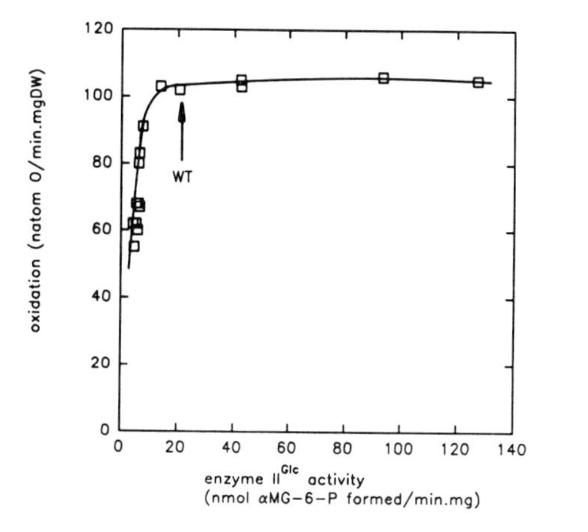

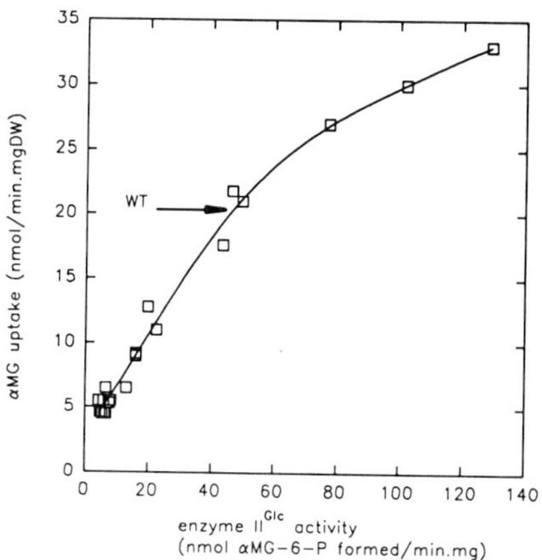

Figure 4. Dependence of the rate of glucose oxidation on the level of II^{Glc} in strain PPA211/pTSG11. Oxidation of 10 mM glucose was measured with a Clark-type electrode at 25 °C. The arrow indicates wild type level of II^{Glc}.
Figure 5. Dependence of the rate of αMG uptake on the level of II^{Glc} in strain PPA211/pTSG11. Transport of 0.5 mM [U-^{14}C]αMG was measured as described in ref. 18. The arrow indicates wild type level of II^{Glc}.

a rather complex pathway that consists of many steps, is not controlled by II^{Glc} in a wild type cell since an increase of the II^{Glc} activity above that resulting from the chromosomal gene (indicated by the arrow) does not lead to an increased rate of glucose oxidation. Figure 5 shows that different results are obtained when a shorter pathway is studied, i.e. the uptake and phosphorylation of the glucose analog αMG. An increase in the II^{Glc} activity of 2-fold results in an 1.3-fold increase of methyl α-glucoside uptake. In terms of control theory, the control coefficient $c^J_E = 0.65$. Separate measurements have shown that the activities of the other PTS enzymes involved in the αMG transport and phosphorylation steps, Enzyme I, HPr and III^{Glc}, did not change. Thus, the increase in the rate of αMG uptake is due to the increase in the II^{Glc} activity. Similar experiments have been done with the *crr* gene, encoding III^{Glc}, under control of the *tac* promoter. Increasing the III^{Glc} level above that present in the wild type strain containing only the chromosomal *crr* gene, did not result in an increase of the rate of αMG uptake nor of glucose oxidation. Thus, glucose metabolism is not controlled by III^{Glc} at wild type level.

Control of PTS-mediated regulation. PTS-mediated inducer exclusion can be measured by determining the inhibition of the uptake of a compound like maltose or glycerol by a PTS carbohydrate like αMG. Table 1 shows that in the wild type *E. coli* PPA234 glycerol uptake is inhibited 95% by αMG. Inhibition is not observed in a *ptsG* mutant, PPA211, which lacks II^{Glc}. Using strain PPA211 and plasmid pTSG11 that expresses II^{Glc} at defined levels, Table 1 shows that at II^{Glc} levels which are only 25% of that of the wild

Table 1
Correlation between II^{Glc} activity and inducer exclusion

Strain	IPTG (μM)	II^{Glc} activity (nmoles/min/mg)	Rate of glycerol transport (nmol/min/mg dry weight)		Inhibition (%)
			-αMG	+αMG	
PPA234/pJF118EH	0	22.5	15.5	0.7	95
PPA211/pJF118EH	0	0.5	15.7	15.7	1
PPA211/pTSG11	0	5.3	11.5	1.9	83
PPA211/pTSG11	10	20.9	12.0	1.0	92
PPA211/pTSG11	50	128	11.8	1.0	92

PPA234 (*ptsM glk*) and PPA211 (*ptsG ptsM glk*) were grown in glycerol minimal medium and subsequently for 1 h in glucose minimal medium. Transport of 0.5 mM [U-^{14}C]glycerol (specific activity 165 cpm/nmol) was measured as described elsewhere (18). αMG was added at a final concentration of 10 mM. pJF118EH is the vector. pTSG11 contains the *E. coli ptsG* gene under control of the *tac* promoter.

type level, inducer exclusion is still almost complete. At first sight this is a somewhat surprising result because the relatively high control coefficient (c=0.65) of II^{Glc} on αMG uptake suggests that at wild type levels of II^{Glc} and below, the phosphorylation of III^{Glc} via phospho-HPr is faster than the dephosphorylation via II^{Glc}. Consequently, in a wild type cell one expects III^{Glc} mostly in the phosphorylated form, even in the presence of glucose or its analogs like αMG. This does not agree with the inducer exclusion studies reported in Table 1 which show that III^{Glc} should be dephosphorylated. To resolve this apparent contradiction, one has to take the kinetics of the system (i.e. elasticities of the enzymes) into consideration. The K_m of II^{Glc} for phospho-III^{Glc} is between 3.4 μM (*S. typhimurium*, ref. 14) and 5 μM (*E. coli*, ref. 15). For II^{Glc} to operate at or close to V_{max} values, 20 μM phosphorylated III^{Glc} is sufficient. Since the intracellular III^{Glc} concentration is estimated to be 30-50 μM, an appreciable amount of III^{Glc} can be in the non-phosphorylated state. The concentration of glycerol kinase is approximately 15 μM (16) and the K_i of glycerol kinase for III^{Glc} is estimated to be 10 μM at pH 7 in *E. coli* (17). Assuming stoichiometric interaction between III^{Glc} and glycerol kinase, sufficient non-phosphorylated III^{Glc} remains to inhibit most glycerol kinase.

It can be concluded that despite the high control of II^{Glc} on glucose uptake, inducer exclusion of non-PTS solutes via non-phosphorylated III^{Glc} is still possible since sufficient non-phosphorylated III^{Glc} is available to complex the target proteins.

ACKNOWLEDGEMENTS

We thank B. Erni for the generous gift of plasmid pTSG11 and strain PPA211.

This work was supported by grants from the Netherlands Organization for Advancement of Research (NWO) under the auspices of the Netherlands Foundation for Chemical Research (SON).

REFERENCES

1 Meadow ND, Fox DK, Roseman S (1990) Annu. Rev. Biochem. 59:497-542
2 Monod J (1942) Recherches sur la Croissance des Cultures Bacteriennes. Ph.D. Thesis.
3 Postma PW, Lengeler JW (1985) Microbiol. Rev. 49:232-269
4 Nelson SO, Wright JK, Postma PW (1983) EMBO J. 2:715-720
5 de Boer M, Broekhuizen CP, Postma PW (1986) J. Bacteriol. 167:393-395
6 Wilson TH, Yunker PL, Hansen CL (1990) Biochim. Biophys. Acta 1029:113-116
7 Kühnau S, Reyes M, Sievertsen A, Shumann HA, Boos W (1991) J. Bacteriol. 173:2180-2186
8 Dean DA, Reizer J, Nikaido H, Saier MH (1990) J. Biol. Chem. 265:21005-21010
9 Reddy P,Meadow N, Roseman S, Peterkofsky A (1985) Proc. Natl. Acad. Sci. USA 82:8300-8304
10 Nelson SO, Schuitema ARJ, Postma PW (1986) Eur. J. Biochem. 154:337-341
11 Nelson SO, Postma PW (1984) Eur. J. Biochem. 139:29-34
12 Kacser H, Burns JA (1973) In: Davis DD, ed Rate control of biological processes. Cambridge: Cambridge University Press
13 Heinrich R, Rapoport TA (1974) Eur. J. Biochem. 42:89-95
14 Meadow ND, Roseman S (1982) J. Biol. Chem. 257:14526-14537
15 Grenier FC, Waygood EB, Saier MH (1986) J. Cell. Biochem. 31:97-105
16 Heller KB, Lin ECC, Wilson TH (1980) J. Bacteriol. 144:274-287
17 Novotny MJ, Frederickson WL, Waygood EB, Saier MH (1985) J. Bacteriol. 162:810-816
18 Postma PW (1977) J. Bacteriol. 129:630-639
19 Kundig W, Roseman S (1972) J. Biol. Chem. 246:1407-1418

© 1992 Elsevier Science Publishers B.V. All rights reserved.
Molecular mechanisms of transport. E. Quagliariello, F. Palmieri, eds.

MECHANISTIC DETAILS OF EII CATALYZED MANNITOL TRANSPORT

G. T. Robillard, J. S. Lolkema, R. P. van Weeghel, D. Swaving-Dijkstra and R. ten Hoeve-Duurkens

The BIOSON Research Institute, University of Groningen, Nyenborgh 16, 9747 AG Groningen, The Netherlands

INTRODUCTION

The enzymes II (EII) of the P-enolpyruvate-dependent phosphotransferase system (PTS) are carbohydrate transporters found only in prokaryotes. They transport hexoses and hexitols, but also pentitols and disaccharides at the expense of P-enolpyruvate, concomitant with phosphorylation. The entire process is characterized by a number of phospho-enzyme intermediates. These reactions and the associated phospho-enzyme intermediates are usually outlined as shown in Scheme I.

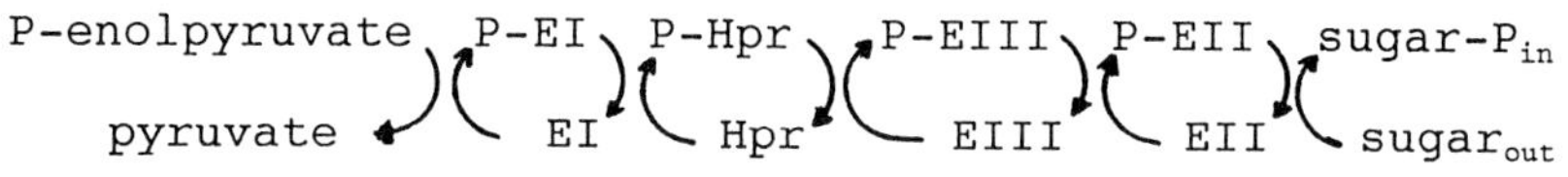

P-enolpyruvate P-EI P-Hpr P-EIII P-EII sugar-P_{in}

pyruvate EI Hpr EIII EII $sugar_{out}$

Scheme I

A flurry of nucleotide sequence activity in the past several years has shown that shuffling, splicing, fusion, duplication and deletion events have occurred in the various PTS operons during evolution resulting in virtually all combinations of the four proteins in Scheme I, EII/EIII fusions, EIII/HPr fusions and EIII/HPr/EI fusions. Only a complete fusion, EII/EIII/HPr/EI, is still missing. The purpose of this presentation is to provide an up-to-date picture of the structure and mechanism of the mannitol-specific EII.

EII ARCHITECTURE

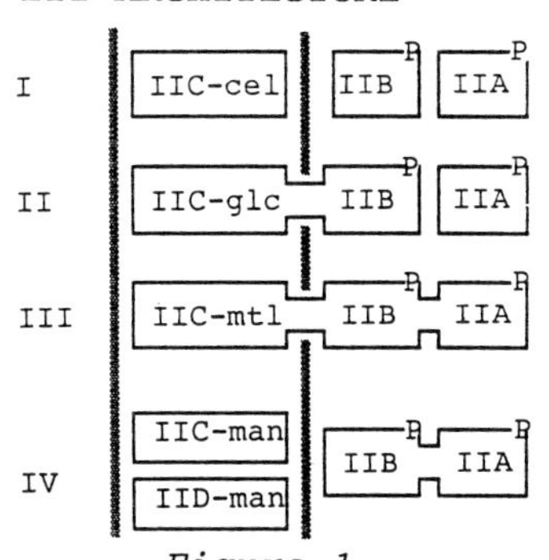

Figure 1

Sequence data in combination with functional studies reveal several classes of Enzymes II as shown in figure 1. These classes may very well change and expand as future studies define additional domains with new functions. The first class is represented by *E. coli* II^{Cel}. It consists only of a single hydrophobic peptide with no hydrophilic domain attached at either end [1,2]. The second class, represented by *E. coli* II^{Glc}, consists of a hydrophobic domain of approximately 360 residues followed by hydrophilic domain of approximately 100 residues [3]. The third class, represented by *E. coil* II^{Mtl}, consists again of a single hydrophobic domain of approximately

360 residues but with two covalently attached hydrophilic domains, equal, together, in size to the hydrophobic domain [4]. The A domain is proposed to function as a covalently attached EIII. The last class, represented by *E. coli* II^{Man}, consists of a membrane domain involving two distinct peptides, one very hydrophobic, and one somewhat less hydrophobic. The domains A and B are not covalently attached to the membrane domains but are separate cytoplasmic proteins [5,6].

The A Domain

As shown in Scheme I, the function of EIII is to transfer the phosphoryl group from P-HPr to EII. ^{32}P-phosphorylated peptides have been isolated from a number of EIII species and A domains of *S. carnosus* and *S. aureus* III^{Mtl} [7], *E. coli* II^{Mtl} [8], III^{Glc} [9], and III^{Man} [10]. In each case the phosphoryl group is carried on a histidine residue. Enough sequence similarity have been reported between these phosphorylated peptides and other Enzymes III or A domains to confirm that the A domains function as Enzymes III.

We have subcloned the A domain of *E. coli* II^{Mtl} after inserting a restriction site into a region of the structural gene corresponding to a flexible peptide which could function as a linker between domains [11, 12]. The purified domain restored 25% of wild-type mannitol phosphorylation activity when used in a mutant complementation assay together with a site-specific mutant, EII H554A, lacking the A domain H554 phosphorylation site. It's EIII-like function was also confirmed by substituting the purified domain for *S. carnosus* III^{Mtl} in an *in vitro* mannitol phosphorylation assay with *S. carnosus* II^{Mtl}. Twenty percent of the original activity was measured when an equal amount of the A domain was substituted for *S. carnosus* III^{Mtl}. Conversely, purified *S. carnosus* III^{Mtl} was able to replace the purified A domain in the *in vitro* mannitol phosphorylation assay with *E. coli* H554A II^{Mtl}.

The B Domain

The existence of a second hydrophilic domain was inferred from the difference in length of the cytoplasmic portions of class II and class III enzymes, from sequence similarities in II^{Glc}, II^{Bgl} and II^{Nag} and from the fact that these homologous regions are found on the C-terminal end of the hydrophobic domain in some cases and on the N-terminal end in others. Experimental evidence for a second hydrophilic domain came again from *E. coli* II^{Mtl} studies. Two phosphoryl groups are transiently incorporated per II^{Mtl} [13]. ^{32}P labelling located both sites in the hydrophilic portion of the enzyme, the H554 site on the A domain mentioned above and C384 [8] which is now considered to be part of the B domain.

The function of the B domain has been confirmed by subcloning and preliminary kinetic measurements. We subcloned the AB domain of *E. coli* II^{Mtl}, residues 348-637, after inserting a restriction site at a position corresponding to residue 348. The purified protein restored mannitol phosphorylation activity when measured with the A domain complementation assay mentioned above and the B domain complementation assay using a C384S inactive EII^{mtl} mutant [14]. The B domain alone, residues 348-488, was subcloned, over-produced and purified and was also active in the B domain complementation assay. These data confirm that the B domain exists as a separate structural and functional entity.

The C Domain

The C domain is proposed to be the carbohydrate binding domain. Experimental proof of this function has been provided only in the case of *E coli* II^{Mtl}. Mannitol binds with very high affinity; K_D 's of 35 to 150 nM have been reported [10, 13]. Deletion analysis revealed that removal of the C-terminal 40% of the protein, leaving only the N-terminal hydrophobic portion intact, did not significantly effect mannitol binding [15]. Lolkema et. al. cleaved off domains A and B by trypsin treatment of inside-out vesicles; the high affinity binding remained unaltered [10]. We have also subcloned the C domain, residues 1-347, and expressed it stably in membranes. These membranes show qualitatively the same mannitol binding properties as wild type enzyme [14].

Domain Interactions

The complementation experiments raise some intriguing issues about the association state of these proteins and the kinetics of their interactions. Do Enzymes II form stable homologous complexes in the membranes? If so, is it necessary to postulate the formation of stable heterologous complexes to explain, for example, the phosphorylation of the B domain of H554A II^{Mtl} by the A domain of II^{Mtl} or can the data be explained by assuming a transient complex formation between the domains? If stable homologous dimers are formed, do the subcloned domains such as the A, AB or B domain of *E. coli* II^{Mtl} displace inactivated domains to restore activity in the complementation assays?

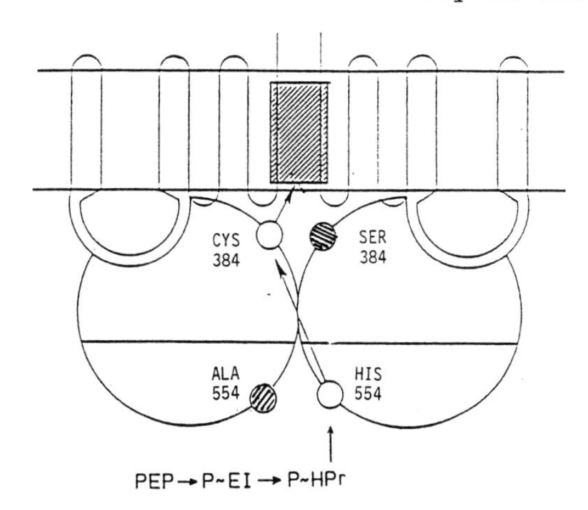

Figure 2

The A domain complementation assays have been carried out as a function of the A domain concentration in the presence of enough HPr, EI and P-enolpyruvate to keep the A domain fully phosphorylated in the steady-state. The kinetics exhibited saturation behavior with respect to the concentration of phospho-A domain. This result suggests that the transfer of the phosphoryl group to the

B domain of the H554A-II^{Mtl} is most likely preceeded by a formation of a complex. For this to occur, the inactive A domain carrying the H554A mutation might swing out of the way to allow for binding of the active A domain to the active B domain. On the other hand, there could be a complex between the active and mutant A domains and a phosphoryl group transfer across the subunit interface as shown in figure 2 [16].

The A domain-dependent saturation kinetics could, in principle, be explained in a less traditional way, namely that phosphoryl group transfer occurs upon collision of the two domains, rather than stable complex formation, and that some other kinetic step after the formation of the phospho-B domain is rate limiting.

Heterologous dimers have been demonstrated kinetically in the case of Mtl/Mtl-P exchange [12]. Detergent solubilized H554A-II^{Mtl} and C384S-II^{Mtl} were examined for exchange activity. The H554A enzyme catalyzed exchange at wild-type rates whereas the C384S enzyme was inactive. This is in keeping with the expectation that C384 is the phosphoryl group donor/acceptor site to Mtl/Mtl-P. When exchange kinetics were done at a fixed concentration of the H554A enzyme and increasing concentrations of the C384S enzyme, a C384S enzyme concentration dependent stimulation of up to 1.8 times the H554A rate was observed. This supports the proposal that association at least to a dimer is necessary for exchange activity in the II^{Mtl} system. In this view, the dimer is maximally active if there is at least one Cys 384 residue per dimer. Such a state is achieved by complementing each H554A subunit with a C348S subunit. A maximum two-fold stimulation would be expected, all other things being equal.

Stable heterologous complexes are not necessary to explain the limited P-enolpyruvate-dependent mannitol phosphorylation kinetic data now available from domain complementation assays; transient complexes between domains are sufficient. The challenge remains, however, to visualize how a subcloned A or B domain would be able to transiently associate with an A or B domain on a enzyme II^{Mtl} dimer at rates high enough to be comparable with wild-type enzyme.

Equilibrium Binding to Enzyme II

Enzymes II in the unphosphorylated state bind their substrates with extremely high affinity. Reported affinity constants are in the sub µM range. Purified *E. coli* II^{Mtl} solubilized in the detergent decylPEG binds mannitol with K_D = 100 nM [13]. Approximately the same value was reported when cytoplasmic membranes containing II^{Mtl} were solubilized directly in decylPEG. However, when the detergent was omitted in the latter case the affinity was even higher, K_D = 35 nM [10]. The availability of purified enzyme in large amounts allows for the assessment of the binding site stoichiometry. The high affinity binding with K_D = 100 nM amounted to 1 binding site per 2 molecules of enzyme. However, at higher mannitol concentrations a second site became apparent with an estimated affinity constant K_D = 10 µM. The data indicate that each II^{Mtl} monomer possesses a mannitol binding site and that dimer formation is accompanied by functional interaction between these binding sites.

Chemical modifications like alkylation with NEM or oxidation with diamide that inhibit the phosphorylation activity of the enzyme did not seem to have any significant effect on the high affinity binding site when the enzyme was solubilized in the detergent decylPEG [13, 17]. However, in the intact membrane these treatments reduced the affinity by a factor of 2-3. The reduction of the affinity was exclusively due to modification of the cysteine residue at position 384 in the B domain [17]. Apparently, the detergent effects the interaction between the B and C domains.

Mannitol binding was measured to II^{Mtl} embedded in cytoplasmic membrane vesicles with either an inside-out (ISO) or right-side-out (RSO) orientation [10]. The cytoplasmic domains of II^{Mtl} are at the interior face in the case of the RSO vesicles and at the exterior face in the case of the ISO vesicles. With the binding site fixed at the cytoplasmic side of the membrane, binding would only be detected to the ISO membranes and not to the RSO membranes. This would just be the other way around with the site fixed at the periplasmic side of the membrane. A third option is that the site is not fixed at either site of the membrane but can change its orientation. Binding would then be observed both to the RSO and ISO membrane vesicles. The experiments agreed with the prediction of the latter option. Scatchard analysis of the binding to both RSO and ISO membranes demonstrated a high affinity bindings with about equal K_Ds.

Kinetics of Binding

Conformational transitions of the translocator domain of II^{Mtl} have been detected by following the binding events in time [10, 18]. The technique used was flow dialysis which has a low time resolution. Our system has been optimized to a response time of $t_{1/2}$ = 10 sec. Nevertheless, interesting data could be collected with this technique. The time course of the binding of mannitol to RSO membranes in the backward and forward direction was too fast to be measured either at room temperature or at 4°. Surprisingly, the binding to- and release from the ISO membranes was slow enough to follow. At room temperature it took about one minute for the binding to equilibrate after the addition of mannitol to the vesicles. The release of bound mannitol, measured as the rate of exchange of bound ^{3}H-mannitol with excess ^{1}H-mannitol was even slower with a half time of about 1 minute. By itself, this asymmetry in the time course of the backward and forward processes indicated that they were enzyme catalyzed processes and that passive diffusion over the membrane was not significantly involved. The data has been taken as evidence that, at equilibrium, bound mannitol accumulated in state E_{per}:Mtl and therefore was bound to the periplasmic side of the membrane. The rapid binding to and dissociation from the RSO membranes would reflect the equilibrium between E_{per} and E_{per}:Mtl. The slow exchange between mannitol bound to ISO membranes and excess unlabeled mannitol would demonstrate the transition between E_{per}:Mtl and E_{cyt}:Mtl, followed by dissociation of mannitol at the cytoplasmic side (E_{cyt}:Mtl --> E_{cyt}). The reverse pathway would be measured with the binding to the ISO membranes. The small rate constants for the latter two processes indicate that turnover is very slow and has to be accelerated considerably when transport is coupled to phosphorylation.

The binding kinetics to the ISO membranes were studied in more detail by lowering the temperature to 4 °C [18]. This slowed down the transitions, making it possible to measure them more accurately. In fact, two time phases could be discriminated in the binding event. An initial rapid phase that relaxed within a few minutes was followed by a much slower phase that required more than an hour to equilibrate. It was shown that the slow phase was caused by a fraction of the sites that initially were in a state not accessible to the substrate. These empty sites would be slowly recruited to a state were binding could take place. The most straight-forward interpretation of this recruitment of sites would be translocation of a fraction of unloaded sites from the internal to the external phase of the ISO membrane (E_{per} --> E_{cyt}). The faster phase was interpreted as a combination of two steps i. the slow association step to the cytoplasmic facing binding site (E_{cyt} --> E_{cyt}:Mtl) and ii. a conformational change of E_{cyt}:Mtl to a state with a higher affinity, presumably E_{per}:Mtl. Therefore, both phases appear to measure conformational changes of the translocator domain.

The interpretation of much of the binding data given so far is based upon the assumption that the high affinity binding sites represent a population of independent sites. In the unphosphorylated II^{Mtl} these sites would open up either to the periplasmic or cytoplasmic side of the membrane independently of each other. The assumption ignores the evidence that the enzyme, in fact, is multimeric and that the data does suggest interaction between the binding sites on the monomeric units. To what extent this will effect the present interpretations will have to follow from future experimentation. In particular, these studies will focus on the identification of the two 'hidden' states detected in the binding kinetics to the ISO membranes at 4 °C described in the last paragraph. They were interpreted to be those states of the enzyme with the binding sites opened up to the internal phase of the vesicle (E_{per} and E_{per}:Mtl). However, in a more complex situation it could be possible that they actually are occluded states of the enzyme.

The Coupling Between Transport And Phosphorylation

The mechanistic coupling of transport and phosphorylation has been investigated by following the fate of mannitol bound to inside-out membrane vesicles upon phosphorylation of II^{mtl} [19]. The study resulted in the model presented in figure 3.

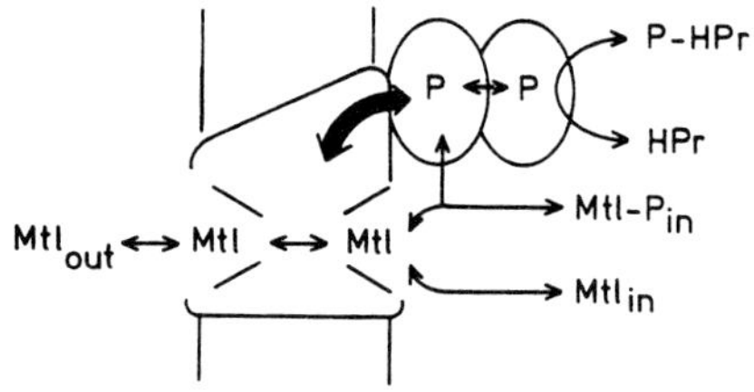

Figure 3

At equilibrium, bound mannitol is either at the periplasmic-facing binding site or in some occluded state. Transfer of mannitol from this state to the cytoplasmic volume involves at least part of the physical pathway for transport of mannitol from the outside to the inside of the cell. In the absence of the phosphoryl group donor, P-HPr, this transfer is very slow due to the slow isomerization step. It is clear that in the presence of P-HPr, when vectorial phosphorylation is catalyzed at a much higher rate, this part of the transport route has to be accelerated significantly. Therefore, phosphorylation of II^{Mtl} drastically reduces the activation energy for the translocation step. It was demonstrated that, once mannitol had arrived at state $E\text{-}P_{cyt}$:Mtl, less than half of the mannitol molecules were phosphorylated by the enzyme; more than half dissociated as unmodified mannitol into the cytoplasmic volume. These results were placed in the context of the domain structure of II^{Mtl} resulting in the model in Figure 3. The C domain constitutes a mannitol translocator. Domain A accepts the phosphoryl group from HPr and transfers it to domain B. The latter takes care of the mannitol kinase activity together with the internally oriented binding site on the translocator. In the absence of the phosphoryl group donor, P-HPr, the translocator would be able to catalyze at least part of the translocation of mannitol across the membrane, but at a very low rate. The process is accelerated 2-3 orders of magnitude when domain B is phosphorylated. Therefore, the state of phosphorylation of domain B modulates the activity of domain C. The interaction between the two domains is indicated by the wide arrow in Figure 3. Within this model, the following events take place when mannitol is added to the outside of the cells. It is assumed that the cells have a high phosphorylation potential, that is, II^{Mtl} will be phosphorylated. Initially, only half of the molecules that are transported into the cell become phosphorylated. The other half is released as free mannitol inside the cell. Mechanistically, the coupling between transport and phosphorylation is less than 50 %. External and internal free mannitol will rapidly equilibrate. Once this physiological steady-state is reached, cytoplasmic and periplasmic carbohydrate serve equally well as substrate. As a result, the phenomenological coupling between transport and phosphorylation under these conditions reaches 100 %. Every mannitol molecule that is phosphorylated by II^{Mtl} has also been transported by the enzyme, but it may have been in a previous turnover. The uncoupled reactions, cytoplasmic phosphorylation and facilitated diffusion, are readily recognized in the model and, in fact, it is the interplay between these two activities that is responsible for the vectorial phosphorylation of mannitol in the physiological steady-state. II^{Mtl} in the model of Figure 3 is a facilitated diffusion enzyme with a built in sugar trap. The most surprising conclusion of our studies is that the coupling between phosphorylation and transport is not at the level of the phosphorylation of the sugar, but at the level of the transport of the sugar. The translocator is activated by phosphorylation of the enzyme.

REFERENCES

1. Parker, L. L. and Hall, B. G. Genetics 1990; 124: 455-471
2. Reizer, J., Reizer, A., and Saier, Jr., M. H. Res. Microbiol. 1990; 141: 1061-1067
3. Erni, B and Zanolari, B. J. Biol. Chem 1986; 261: 16398-16401
4. Lee, C. A. and Saier, Jr., M. H. J. Biol. Chem. 1983; 258: 10761-10767
5. Erni, B., Zanolari, B, and Kocher, H. P. J. Biol. Chem. 1987; 262: 5238-5247
6. Erni. B., Zanolari, B., Graff, P. and Kocher, H. P. J. Biol. Chem. 1989; 264: 18733-18741
7. Reiche, B., Frank, R., Deutscher, J., Meyer, N. and Hengstenberg, W. Biochemistry 1988; 27: 6512-6516
8. Pas. H. H. and Robillard, G. T. Biochemistry 1988; 27: 5835-5839
9. Dorschug, M., Frank, R., Kalbitzer, H.R., Hengstenberg, W. and Deutscher, J. Eur. J. Biochem. 1984; 144: 113-119
10. Lolkema, J. S., Swaving-Dijkstra, D., ten Hoeve-Duurkens, R. H. and Robillard, G. T. Biochemistry 1990; 29: 10659-10663
11. Karplus, P. A. and Schmidt, G. E. Naturwissenschafter 1985; 72: 212-217
12. Van Weeghel, R. P., Van der Hoek, Y. Y., Pas, H. H., Elferink, M. G. L., Keck, W. and Robillard, G. T. Biochemistry 1991; 30: 1768-1773
13. Pas, H. H., Ten Hoeve-Duurkens, R. H. and Robillard, G. T. Biochemistry 1988: 27; 5520-5525
14. Van Weeghel, R. P., Meyer, G., Pas, H. H., Keck, W. and Robillard, G. T. Biochemistry 1991; in press
15. Grisafi, P. L., Scholle, A., Sugayama, J., Briggs, L., Jacobson, G. R. and Lengeler, J. W. J. Bact. 1989; 171: 2719-2727
16. Van Weeghel, R. P., Meyer, G. H., Keck, W. and Robillard, G. T. Biochemistry 1991; 30: 1774-1779
17. Lolkema, J. S., Swaving Dijkstra, D., ten Hoeve-Duurkens, R. H. and Robillard, G. T. Biochemistry 1991: 30: 6721-6726.
18. Lolkema, J. S., Swaving Dijkstra, D. and Robillard, G. T., manuscript submitted for publication
19. Lolkema, J. S., ten Hoeve-Duurkens, R. H., Swaving Dijkstra, D. and Robillard, G.T. Biochemistry, 1991; 30: 6716-6721.

SUGAR TRANSPORTERS

© 1992 Elsevier Science Publishers B.V. All rights reserved.
Molecular mechanisms of transport. E. Quagliariello, F. Palmieri, eds.

Substrate-binding Mechanism of Sugar Transporters Investigated by Time-resolved Photoaffinity Labelling

M. G. P. Page

Dept Microbiology, Biozentrum der Universitaet Basel. *Present address:* F.Hoffman-La Roche Ltd, Pharma Division Preclinical Research, 4002 Basel, Switzerland

INTRODUCTION

Transporters involved in the movement of small molecules across membranes carry out a complex cycle of reactions: recognition of the substrate in the external medium, its occlusion in a protein complex that will enable its passage across the hydrophobic core of the membrane, its release into the internal medium of the cell and, finally, the transporter must return to the state where external substrate can once more be recognized. Our understanding of the molecular events involved in any of these steps is sadly very limited, and will remain so until the high resolution structures of a transporter and some of the catalytic intermediates have been solved. An essential part of describing the mechanism of transport is the determination of the rates of interconversion of indentifiable intermediates in the catalytic cycle.

PHOTOAFFINITY LABELS FOR SUGAR TRANSPORTERS

Photoaffinity labelling is a well-established technique for identification of specific proteins and substrate binding sites within proteins [1,2]. Among sugar transporters, the technique has been applied to labelling of the erythrocyte glucose transporter [3-5], epithelial Na^+ - dependent glucose transporters [6,7], and the *E. coli* H^+-lactose transporter [8]. Labelling and competition experiments have indicated that certain inhibitors and alternative substrates are highly selective and even diagnostic for individual families of related transporters. For example, cytochalasin B binding has so far been found to be indicative of transporters related to the erythrocyte glucose transporter [9] and phloridzin binding has so far been found to be indicative of transporters related to the epithelial glucose transporter [10]. The galactoside analogues shown in Table 1 were recognized and reacted covalently after irradiation with wild type *E. coli* lactose transporter (Fig.1) and the sugar specificity mutants obtained by Brooker & Wilson, that have an increased affinity for glucosides [11]. They were also recognized by the melibiose transporter from *Escherichia coli* (Fig. 1). The glucoside analogues were not recognized by wild type lactose or melibiose transporters but the alpha-azidophloridzin bound to the two lactose tranporter mutants and azidophloridzin reacted with the sodium-dependent glucose transporters from rabbit intestinal epithelium and some *Archaeobacteria* (e.g

Halobacterium saccharovorum and *Natronobacterium gregoryi* , Fig.1) .
Thus the potentials exist for exploiting photolabelling to investigate mechanisms of substrate binding and to pursue evolutionary relationships between proteins in organisms where genetic techniques are as yet in a rather primitive stage.

Table 1
Photoaffinity labels for sugar transporters
LacY: lactose transporter in *E. coli* T206, MelB: melibiose transporter in *E. coli* RA11, RaSG: glucose transporter in rabbit intestinal epithelium, NaGreg: glucose transporter in *Natronobacterium gregoryi* .

Substrate	Dissociation constant in native membrane vesicles (μM)					
	LacY	LacY A177V	LacY Y326F	MelB	RaSG	NaGreg
I. 4-azido-2-nitrophenyl-α,D-galactopyranoside	10	30	400	5	>10^5	>10^5
II. 4'-methoxy-4-α,D-galactopryanosyloxybenzophenone	5	7	700	3	>10^5	>10^5
III. 4-(4(2,2,2-trifluoro-1-diazirinylethyl)benzamidophenyl-α,D-galactopyranoside	2	3	800	5	>10^5	>10^5
IV. Phenyl-α,D-2-deoxy-2-(3,3,3-trifluoro-2-diazo)propionamido-galactopyranoside	800	200	>1000	nd	nd	nd
V. 4-Azido-α-phloridzin	>10^5	700	300	>10^5	nd	500
VI. 4-Azidophloridzin	>10^5	500	600	>10^5	10	100

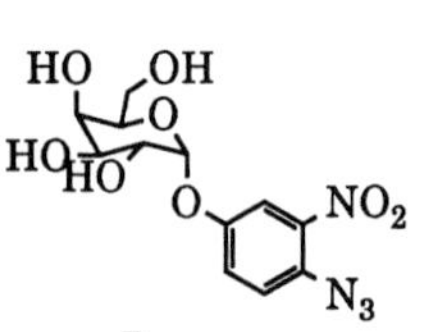

HO OH HO HO O O NO_2 N_3
I

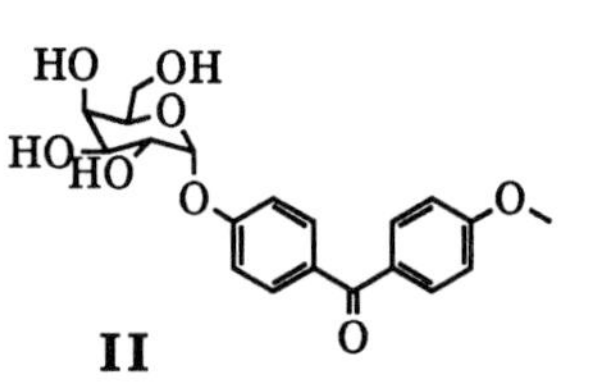

HO OH HO HO O O O O
II

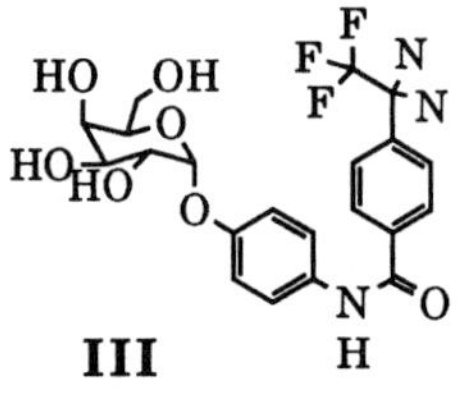

HO OH HO HO O O F F F N N N O H
III

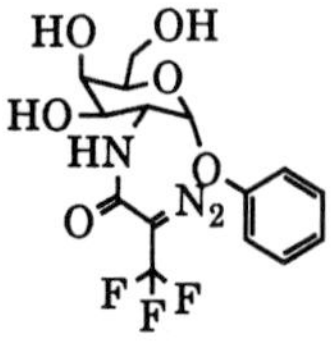

HO OH O HO HN O O N_2 F F F
IV

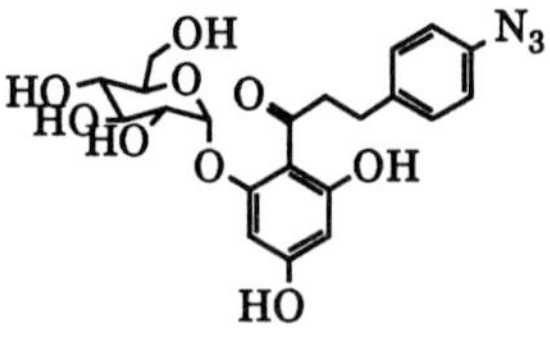

OH HO HO HO O O O OH HO N_3
V

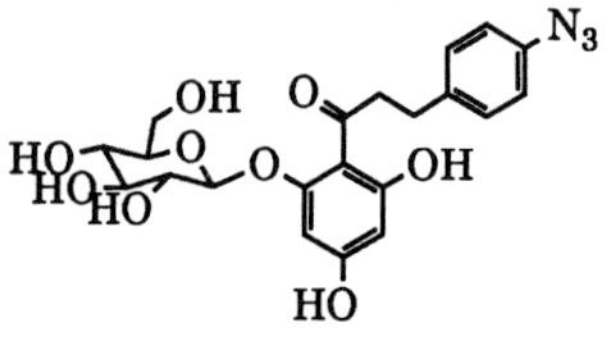

OH HO HO HO O O O OH HO N_3
VI

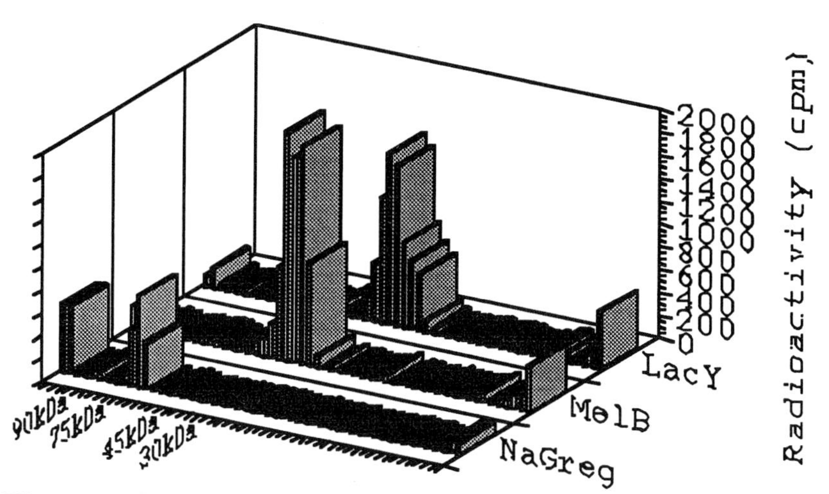

Figure 1. Photoaffinity labelling of sugar transporters
Native membranes derived from *Escherichia coli* T206 containing 2.2 nmol lactose transporter per mg membrane protein (LacY, labelled with [^{3}H]-*methyl* compound **II**), E.coli RA11 containing 0.78 nmol melibiose transporter per mg membrane protein (MelB, labelled with [^{3}H-*phenyl*] compound **I**) and *Natronobacterium gregoryi* containing 0.23 nmol glucose transporter per mg membrane protein (NaGreg, labelled with [^{3}H-*glucose*] compound **VI**) were dissolved in sample buffer for SDS-PAGE. After electrophoresis the gels were cut into slices, and the radioactivity in each slice measured by liquid scintillation counting after dissolution in 10% hydrogen peroxide.

When enough protein is available it is possible to go on to isolate peptides and characterize the position of labelling, shown in Fig. 2 for the lactose transporter from *Escherichia coli* . The labelling is concentrated in five of the recoverable peptides. Three (I,II & V) contain residues that are known to been exposed on the cytoplasmic face of the membrane [12] and could therefore comprise part of an internal binding site. Three of the peptides contain (peptide V) or lie close to (peptides II & IV) residues that are known to be exposed on the periplasmic face of the membrane and therefore might be part of an external binding site. Since peptide V falls into both categories, and appears to be labelled in several places along its length, it must be in contact with both sites or be a part of the sugar translocation route. There is some correspondence between the position of the labelling and the positions of residues either whose reactivity with electrophilic reagents is greatly altered by substrate binding (Fig. 2) or which are known from mutational studies to affect substrate binding (for example Ala177, Tyr236, Thr266, His 322 [11,13,14].

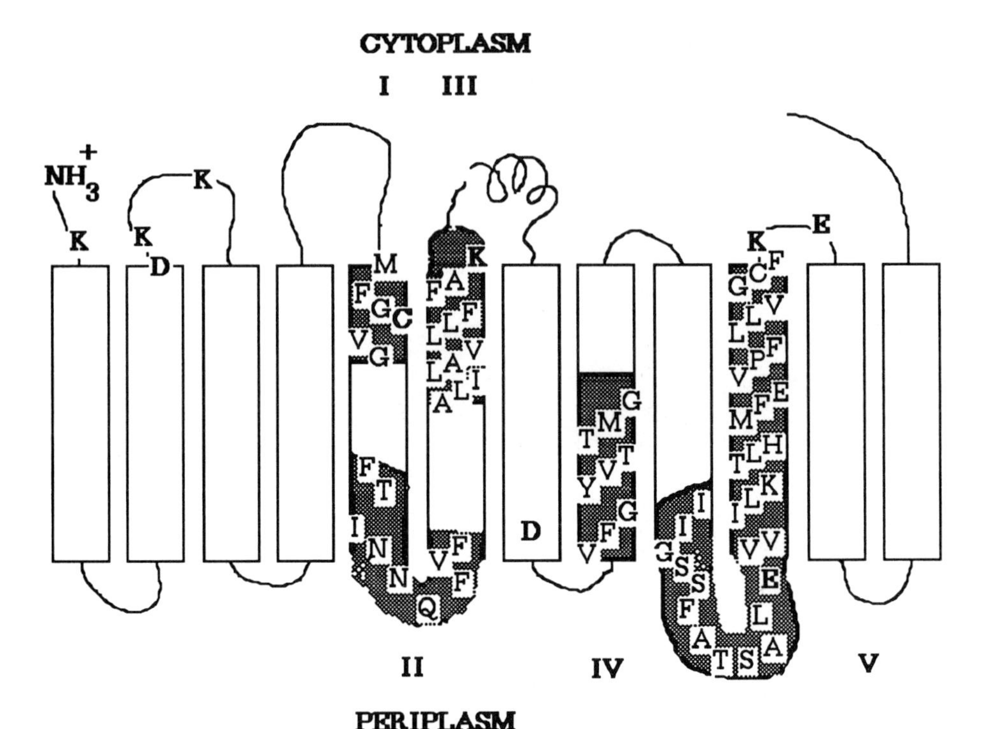

Figure 2. Sugar binding sites in the *E. coli* lactose transporter
Distribution of radioactivity in photoaffinity labelled lactose transporter from *Escherichia coli*. The stippled areas represent peptides labelled after photolysis of lactose permease equilibrated with 4'-methoxy-4-α,D-galacto-pyranosyloxybenzophenone (II in table 1). Residues are indicated by their single letter codes: those shown in bold face type are residues whose reactivity with small electrophilic reagents is affected by substrate binding [12].

TRANSIENT KINETICS OF H+-LACTOSE SYMPORT BY THE *ESCHERICIA COLI* LACTOSE TRANSPORTER

The transient kinetics of the lactose transporter have been investigated previously using radioactively labelled substrates [15,16]. In principle, intrinsic protein fluorescence, fluorescent sugars [17] and pH indicators [16] could also be used. The two intensely fluorescent galactosides (VII and VIII) and the pH indicator pyranine have been used to probe the mechanism of lactose transport by stopped-flow fluorimetry (Fig. 3 and 4). When the slowly transported substrate VII is mixed with an equimolar amount of active lactose transporter reconstituted into membrane vesicles, a rapid enhancement of its fluorescence occurs (Fig. 3, curve A): no significant change occurs if the transporter is inactivated or saturated with another substrate. The rapid increase in fluorescence contains at least two

VII VIII

exponential phases: one is complete within the mixing time while the second has a half-life of about 10ms. The rise in fluorescence is followed by a much slower decay with a half-life of about 70ms. When the bigger, and much more slowly transported, sugar VIII is used, only the first rise in fluorescence is seen (Fig. 4, curve A). Thus, this probably represents binding to externally exposed sites. The slower rise to a more fluorescent state that occurs with the transported sugar VII probably represents occlusion of the sugar and the third phase its release in the external compartment. Measurement of the amount of radioactively labelled sugar VII trapped inside the vesicles reveals that there is a lag phase corresponding to occlusion of the sugar in a form that cannot be released to the external compartment. The incorporation continues through a burst phase corresponding to 0.8mol sugar per mol of protein then continues at

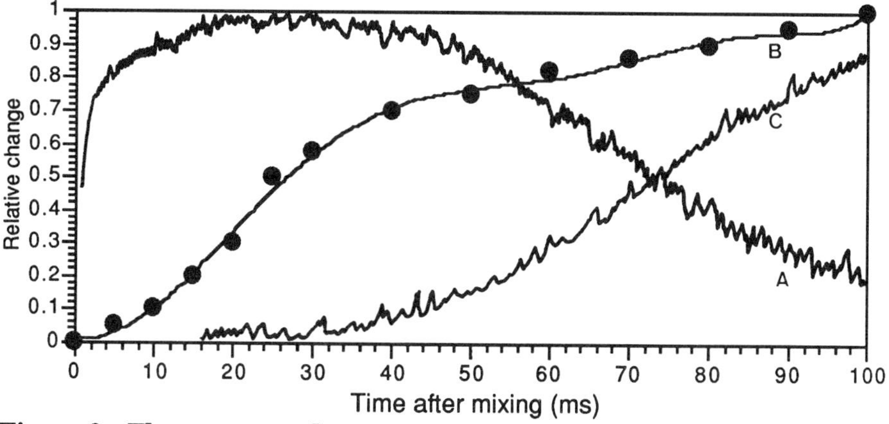

Figure 3. Fluorescence changes accompanying substrate binding to lactose transporter reconstituted into proteoliposomes. Curve A: changes produced when reconstituted vesicles equilibrated in 0.1M Tris-HCl, 10mM $MgCl_2$, 50mM KCl at pH 7.2 are mixed with sugar VII in the same buffer, the abscissa represents fluorescence increase in arbitary units. Curve B: the solid circles represent radioactivity from 6-[^{3}H-*galactose*]-sugar VII, the abscissa represents mol/mol protein. Curve C: changes produced when vesicles equilibrated in 2mM Tris-HCl, 10mM $MgCl_2$, 150mM KCl, 1mM Pyranine at pH7.2 are mixed with a 10mM solution of melibiose in the same buffer, the abscissa reprents fluorescence increase scaled to correspond to g-ions H^+ per mol protein.

a slow steady-state rate (Fig. 3, curve B). The accompanying pH change undergoes a much longer lag (50ms instead of 10ms) and then rises through a burst phase at a rate close to that of the decay in sugar fluorescence (Fig.3, curve C). This suggests that proton uptake is occurring only after a complete cycle of sugar uptake has occurred. Monitoring the changes in intrinsic protein fluorescence that occur during these reactions reveals that there are probably several protein conformational changes occurring. There is an initial quenching that accompanies the sugar binding reactions but it appears to contain several components, suggesting more than one intermediate on the pathway to the occluded form identified by sugar fluoresecence. Taken together, the various transients suggest the following reaction scheme for sugar VII:

$$CH_e + S_e \overset{25\mu M}{<==>} CHS_e \overset{70s^{-1}}{==>} [\,CHS^* \overset{18s^{-1}}{===>} CHS\,] \overset{>500s^{-1}}{==>} CH_i + S_i \overset{10s^{-1}}{==>} C_i + H_i \overset{5s^{-1}}{==>} C_e + H_e \overset{5nM}{<==>} CH_e$$

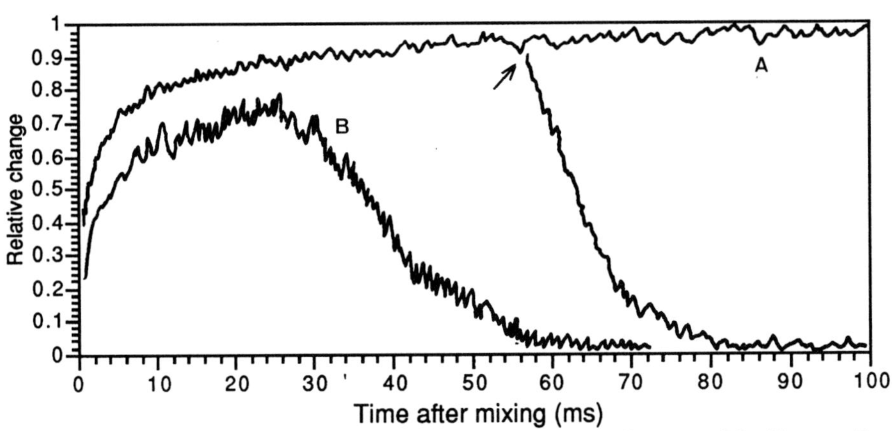

Figure 4 Fluorescence changes accompanying substrate binding to lactose transporter reconstituted into proteoliposomes. Curve A: changes in sugar fluorescence produced when vesicles equilibrated in 0.1M Tris-HCl, 10mM MgCl2, 50mM KCl at pH 7.2 are mixed with sugar VIII in the same buffer. The arrow indicates the ffect of mixing the suspension with 20mM melibiose. Curve B: the increase in quenching of intrinsic protein fluorescence produced when vesicles equilibrated in 0.1M Tris-HCl,10mM $MgCl_2$, 50mM KCl at pH 7.2 are mixed with melibiose in the same buffer.

TIME RESOLVED PHOTOAFFINITY LABELLING OF THE LACTOSE TRANSPORTER FROM *ESCHERICHIA COLI*

In order to build a a picture of how substrate is moved across the membrane we must find a way to identify the momentary interactions between transporter and substrate. The spectroscopic probes indicate a series of changes in the interactions, the photoaffinity probes indicate regions of the protein involved in these interactions. Can we put these together to follow the changing pattern that occurs during transport? It is clear that peptides II, IV and V, which were postulated to form part of an external binding site, are indeed labelled rapidly during the sugar binding phase indentified by stopped-flow fluorimetry (Fig. 5). Similarly peptides I and III, which were postulated to form part of an internal binding site, are labelled more slowly and first after a lag phase which corresponds to the sugar occlusion phase. The extent of labelling of the various peptides during the steady-state phase of transport (times >80ms) suggests that an internally directed site immeadiately preceeds the rate-limiting step but that it is possible to have both sites occupied. It is not clear whether this simply reflects partioning between the various intermediates of the transport cycle or it is because an individual transporter molecule is able to bind external sugar before the internal sugar has dissociated.

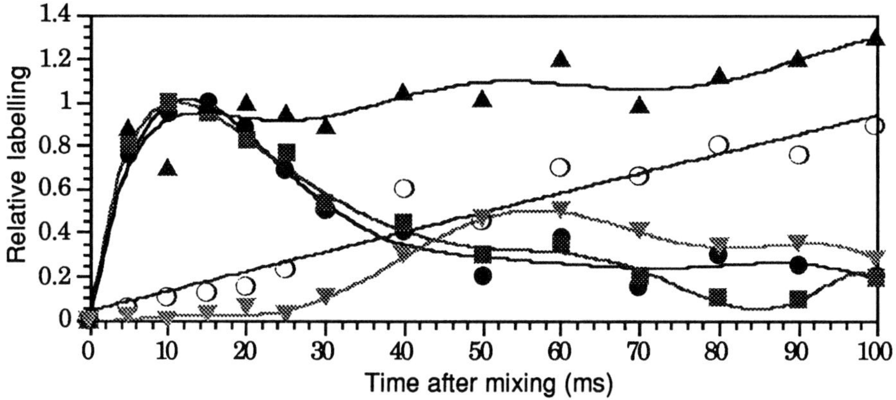

Figure 5. Time resolved photoaffinity labelling of the *Escherichia coli* lactose transporter. Intact cells of *E. coli* T206 were mixed rapidly with 10μM sugar II (table 1) and the mixture was passed down a delay tube before entering the irradiation cell. The length of the delay tube was adjusted to give the different times shown, and the irradiation time was about 2ms. The protein was isolated from the cells by octyl glucoside precipitation. The radioactivity recovered in the different fragments of the protein illustrated in Fig.2 is represented thus: peptide I by inverted triangles; peptide II by solid circles; peptide III by open circles; peptide IV by solid squares; peptide V by solid triangles. The radioactivity is scaled relative to the maximum extent of labelling of peptide II.

IMPLICATIONS OF TRANSIENT KINETICS FOR UNDERSTANDING THE MECHANISM OF SUBSTRATE BINDING TO THE *ESCHERICHIA COLI* LACTOSE TRANSPORTER

The transient kinetic studies have indicated the formation of an occluded ternary complex between the lactose transporter and its substrates. The photoaffinity labelling experiments strongly suggest two discreet sites involved in binding of substrate in, respectively, external and internal compartments and indicate that the path between the two sites is formed from the C-terminal part of the protein. There is some indication that, under certain circumstances, a quaternary complex involving two sugar molecules can also be formed, such that both internal and external sites are occupied simultaneously. Such a possibility has already been suggested from analysis of steady-state kinetics and binding data[18-20]. It remains to be seen whether it is merely an aberrant reaction occurring only with unnatural substrates, whether it is important in regulation of transport activity or whether this second mode of binding represents an essential part of the mechanism of transport.

REFERENCES

1. Bayley, H., Knowles, J. R. (1977) Methods in Enzymol. **46**, 69-114.
2. Chowdry, V., & Westheimer, F.H.(1979) Ann. Rev. Biochem. **48**, 293-325
3. Midgeley, P.J.W., Parker, B.A., & Holman, G.D. (1985) Biochem. Biophys. Acta **812**, 33-41.
4. Pessin, J.E., Tillerton, L.G., Isselbacher, K.T., Czech, M.F. (1984) Fed. Proc. **45**, 225-61.
5. Wadzinski, D.E., Shanahan, M.F., Clark, R.B., Ruoho, A.E. (1988) Biochem. J. **255**, 983-90.
6. Hosang, M., Gibbs, E.M., Diedrich, D.F., & Semenza, G. (1981) FEBS. Letts **130**, 244-8.
7. Hosang, M., Vasella, A., Semenza, G. (1981) Biochem. **20**, 5844-54.
8. Kaczorowski, G.J., Leblanc, G., & Kaback, H.R. (1980) Proc. Natl Acad. Sci. USA **77**, 6319-6323.
9. Cairns, M.T., McDonald, T.P., Horne, P., Henderson, P.J.F., & Baldwin, S.A. (1991) J. Biol. Chem. **266**, 8176-83
10. Semenza, G., Kessler, M., Hosang, M., Weber, J., & Schmidt, U. (1984) Biochim. Biophys. Acta **779**, 343-79
11. Brooker, R.J., & Wilson, T.H. (1985) J. Biol. Chem. **260**, 16181-6
12. Page, M.G.P., & Rosenbusch, J.P. (1988) J. Biol. Chem. **263**, 15906-14
13. Markgraf, M., Bocklage, H., & Mueller-Hill, B. (1985) Mol. Gen. Genet. **198**, 473-475
14. Kaback, H.R., Bibi, E., & Roepe, P.D. (1990) TIBS **15**, 309-314
15. Page, M.G.P., & West, I.C. (1984) Biochem. J. **223**, 723-31
16. Page, M.G.P., Rosenbusch, J.P., & Yamato, I. (1988) J. Biol. Chem. **263**, 15897-905
17. Schuldiner, S., & Kaback, H.R. (1977) Biochim. Biophys. Acta **472**, 399-418
18. Page, M.G.P. (1980) Biochem.Soc.Trans. **8**, 704
19. Page, M.G.P. (1987) Biochim. Biophys. Acta **897**, 112-26
20. Lolkema, J.S., & Walz, D. (1990) Biochemistry **29**, 10120-8

© 1992 Elsevier Science Publishers B.V. All rights reserved.
Molecular mechanisms of transport. E. Quagliariello, F. Palmieri, eds.

STUDIES OF MEMBRANE TOPOLOGY AND SUBSTRATE BINDING-SITE LOCATION IN THE HUMAN ERYTHROCYTE GLUCOSE TRANSPORT PROTEIN

Angela F. Davies[a], Michael T. Cairns[a], Anthony Davies[a], Richard A.J. Preston[a], Avril Clark[b], Geoffrey Holman[b] and Stephen A. Baldwin[a]

[a]Departments of Biochemistry & Chemistry, and of Protein & Molecular Biology, Royal Free Hospital School of Medicine (University of London), Rowland Hill Street, London NW3 2PF, U.K.

[b]Department of Biochemistry, University of Bath, Claverton Down, Bath BA2 7AY, U.K.

INTRODUCTION

Recent DNA cloning studies have revealed the existence of a large family of homologous sugar transporters, both active and passive, in organisms as diverse as mammals, plants, yeasts, blue-green algae and bacteria [1]. In mammals, five members of the family (GLUT1 - 5 in the terminology of Fukumoto *et al.* [2]) have so far been identified in various tissues [3]. The best characterised of these mammalian transporters, and indeed of the entire family, is the facilitated diffusion transporter from human erythrocyte membranes (GLUT1) [4]. This transporter is widely distributed in mammalian tissues, but is most abundant in those with a barrier function, such as the blood-brain barrier [5] and placental syncytiotrophoblast [6]. It remains the only member of the family to have been purified in functional form [7]. Biophysical investigations have revealed that this 492-residue integral membrane protein is highly α-helical [8-10], and hydropathic analysis suggests that it spans the lipid bilayer 12 times [4]. Studies using vectorial proteolytic digestion [11] and site-directed anti-peptide antibodies as topological probes [12,13] have demonstrated that a large hydrophilic domain near the centre of the sequence, and the hydrophilic *C*-terminal region, are located at the cytoplasmic surface of the membrane (Fig. 1). The extracellular site of glycosylation has also been located, on the loop connecting putative transmembrane helices 1 and 2 [4]. However, there is as yet little experimental evidence to support the predicted topology of other regions of the protein. In the present communication we present additional evidence for this topology, obtained from vectorial labelling experiments with the membrane-impermeant imidoester, isethionyl acetimidate.

Our knowledge of the regions of the protein that are involved in substrate translocation is also sparse. A widely favoured model for the mechanism of sugar translocation is that the transporter alternates between two conformations, in which a substrate-binding site (or sites) is exposed to either the extracellular or cytoplasmic face of the membrane [14]. There is now considerable kinetic evidence to support such a model, in which only a single binding site is accessible to substrate at any one time [15]. However, direct information about the location of the substrate-binding site(s), and about the nature of the conformational change, is minimal. In the present communication we describe the identification of substrate-binding sites in the transporter by photolabelling with two inhibitors of transport, ATB-BMPA (2N-4(1-azi-2,2,2-trifluoroethyl)benzoyl-1,3-bis(D-mannos-4-yloxy)-2-propylamine) and cytochalasin B. These photoactivable molecules are known to inhibit transport by binding at the exofacial and endofacial surfaces of the membrane respectively [16,17]. Identification of the sequence location of radioactive fragments produced by specific cleavage of the labelled protein has been facilitated by examining their reactivity towards a panel of site-directed anti-peptide antibodies raised against most of the hydrophilic regions of the transporter [13].

MATERIALS AND METHODS

Ethyl acetimidate was purchased from Aldrich, isethionyl acetimidate from Pierce and isethionyl [1-^{14}C]acetimidate from Amersham. New England Nuclear supplied [4-^{3}H]cytochalasin B. Glucose transporter was purified from human erythrocytes by the method of Baldwin and Lienhard [7], (scaled down appropriately for the purification of protein from small quantities of amidinated erythrocytes). Exhaustive amidination of the transporter (0.7mg/ml) was performed in 100mM tris(hydroxymethyl)methylaminopropane sulphonic acid (TAPS)/HCl, 1mM EDTA, pH 9.0 at 25°C. Ethyl acetimidate (100mM) was added initially and then again after 1h and 2h. Amidination was terminated after 3h by addition of a 5-fold excess (v/v) of 0.4M ammonium acetate, pH 5.0. Samples were then taken for determination of acetimidyl lysine by amino acid analysis, and for measurement of cytochalasin B binding activity by the method of Zoccoli *et al.* [18]. Amidination of intact erythrocytes (50% haematocrit) was performed for 1.5h in the TAPS buffer containing 80mM NaCl to maintain isotonicity, using 6.6mM isethionyl [1-^{14}C]acetimidate (17mCi/mmol). Membranes and purified glucose transporter were then prepared by standard procedures [7].

Photoaffinity labelling of the isolated transporter with tritiated ATB-BMPA or cytochalasin B was as previously described [16,19]. Cleavage of the labelled protein at lysine residues using endoproteinase Lys-C, and at both lysine and arginine residues using trypsin, was performed in 50mM sodium phosphate, 100mM NaCl, 1mM EDTA, pH 7.4 for 24h at 25°C in the presence or absence of 0.1% SDS. Fresh additions of protease (5%, w/w) were made initially, and again after 3 and 6h. Digestions were terminated by addition of a two-fold excess of aprotinin by weight over protease. Cleavage of the transporter at tryptophan residues was performed by treatment of the protein (150µg/ml) with 75µg/ml N-bromosuccinimide essentially as described by Holman & Rees [20]. Fragments produced by enzymic or chemical cleavage were then separated by SDS/polyacrylamide gel electrophoresis using the method of Schägger and von Jagow [21]. Radiolabelled fragments were identified by cutting the gel into 2mm slices for solubilization and liquid scintillation counting as previously described [19]. The M_r 18000 fragment produced by tryptic digestion of the transporter in the absence of SDS was isolated by preparative gel electrophoresis as described by Cairns *et al.* [11]. Western blotting procedures using anti-peptide antibodies were as described in Davies *et al.* [13].

RESULTS AND DISCUSSION

Chemical modification of lysine residues was chosen as a means of probing the membrane topology of the glucose transporter both because of the predicted accessibility of such residues at the extracellular surface of the erythrocyte, and because specific modification of amino groups can be performed under mild conditions using imidoesters. Acetimidyl lysine retains the positive charge found on the side chain of lysine at physiological pH, and is only slightly bulkier than lysine itself. Amidination of proteins therefore frequently results in little disturbance of protein function. Exhaustive modification of the purified glucose transporter with the membrane-permeant imidoester ethyl acetimidate was found to result in the amidination of between 14 and 15 lysine residues, out of a total of 16 lysine residues present in the protein. Scatchard plot analysis of cytochalasin B binding to the amidinated protein showed no loss of binding sites, and only a 2.5-fold increase in the dissociation constant for binding compared to a control sample. Retention of binding activity, and thus of tertiary structure, indicated that amidination of the transporter would be unlikely to affect the arrangement of the protein in the membrane. The imidoester isethionyl acetimidate was therefore chosen as a topological probe. This reagent also converts lysine residues to acetimidyl lysine, but its charged leaving group renders it membrane-impermeant. Membrane impermeability was confirmed by SDS/polyacrylamide gel electrophoretic analysis of membranes isolated from intact erythrocytes labelled with radioactive isethionyl acetimidate. This analysis showed that whereas both the anion transporter (band 3) and the glucose transporter (band 4.5) were radiolabelled, the cytoskeletal proteins spectrin and actin, which are located at the cytoplasmic surface of the membrane, were not labelled. It follows that the labelling of the glucose transporter must have been confined to the exofacial surface of the membrane.

The sequence location of the exofacial sites of labelling was investigated by cleavage of the purified transporter with trypsin. This procedure yielded the two large membrane-embedded fragments which we have previously shown to comprise residues 1-212 and 270-456 [11,13]. Both of these fragments were found to be radiolabelled, with the *N*-terminal fragment containing about twice the radioactivity of the *C*-terminal fragment. Together these two fragments accounted for all the radioactivity present in the undigested transporter, indicating lack of labelling of the seven cytoplasmic lysine residues known to be released from the protein in the form of small, water-soluble peptides upon tryptic digestion [11], and confirming the membrane impermeability of the imidoester used for labelling. Two of the three lysine residues in the *C*-terminal tryptic fragment, Lys_{451} and Lys_{456}, are known from vectorial proteolytic digestion experiments to be located on the cytoplasmic surface of the membrane [11]. The remaining residue, Lys_{300}, is therefore the probable site of labelling, although this remains to be confirmed by direct sequencing. This residue was predicted in our original model for the transporter topology to be exofacial [4]. The labelling of the *N*-terminal tryptic fragment by the impermeant imidoester is also consistent with the prediction of the model that four lysine residues within this part of the protein (residues 38, 114, 117 and 183) are exofacial or close to the extracellular surface of the bilayer, although once again direct sequencing will be required to establish which of these residues is actually labelled.

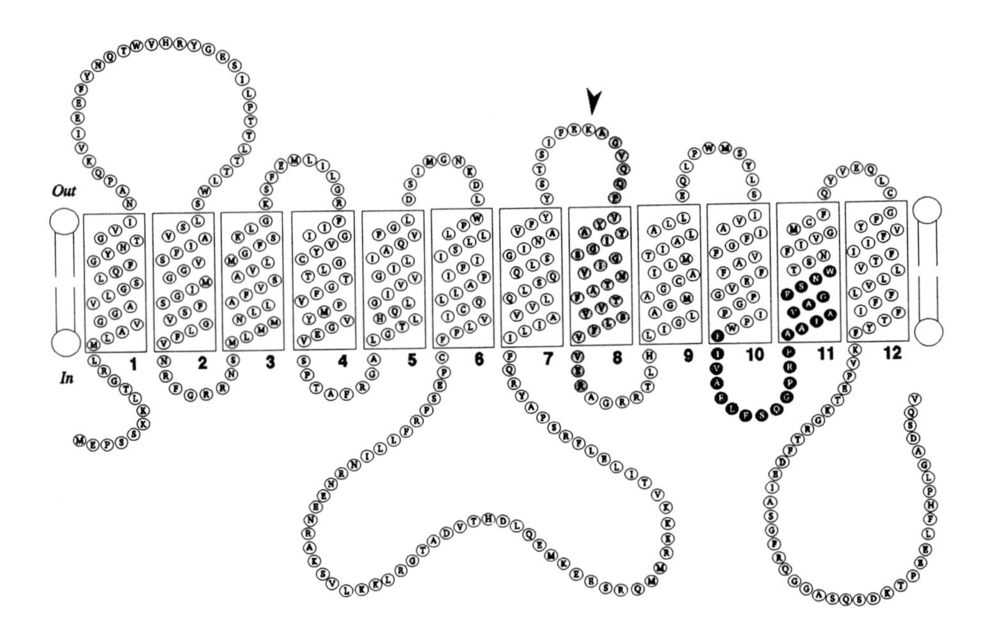

Figure 1. Model for the 2-dimensional arrangement of the human erythrocyte glucose transporter in the membrane.
Amino acid residues are identified by their single letter code. Lys_{300} is arrowed. Shaded and filled-in segments indicate the regions containing the sites photolabelled by ATB-BMPA and cytochalasin B respectively.

Previous photolabelling and enzymic cleavage experiments have shown that the *C*-terminal fragment (residues 270-456) produced by tryptic digestion of the native glucose transporter bears the sites of labelling by both ATB-BMPA and cytochalasin B [16,19]. In order to locate the sites more precisely within this region we digested the labelled transporter with specific proteases after denaturation in 0.1% SDS to expose more sites for cleavage. Cleavage at lysine residues by digestion with endoproteinase Lys-C yielded a fragment of apparent M_r 16500 which was labelled by both cytochalasin B and ATB-BMPA. This fragment was recognised on Western blots by anti-peptide antibodies raised against residues 326-340, 389-403 and 450-467, but not by antibodies against residues 293-306. This pattern of recognition, coupled with the location of potential cleavage sites, indicated that the fragment comprised residues 301-456 (predicted M_r 17027). Such a fragment would contain putative transmembrane helices 8-12 (Fig. 1). In contrast, digestion of the labelled transporter with trypsin in 0.1% SDS yielded a fragment of apparent M_r 11500 which was labelled by cytochalasin B but not by ATB-BMPA. This fragment was recognised by antibodies against residues 326-340 and 389-403, but not by antibodies against residues 293-306 or 450-467. The pattern of recognition indicated that the fragment comprised residues 331-451 (predicted M_r 12927) and so contained putative transmembrane helices 9-12. The observation that the endoproteinase Lys-C fragment (helices 8-12) but not the tryptic fragment (helices 9-12) was labelled by ATB-BMPA suggests that the site of labelling by this inhibitor lies within putative transmembrane helix 8, between residues Ala_{301} and Arg_{330} (Fig. 1). This conclusion was supported by studies on an ATB-BMPA-labelled M_r 18000 fragment of the transporter (residues 270-456) containing helices 7-12, isolated by preparative gel electrophoresis following digestion of the protein with trypsin under non-denaturing conditions. Digestion of this fragment with endoproteinase Lys-C in 0.1% SDS yielded a labelled fragment of apparent M_r 16500, identified by its reactivity towards anti-peptide antibodies as comprising helices 8-12, and an unlabelled fragment of apparent M_r 3000 comprising helix 7. These findings confirm that the site of labelling by this exofacial reagent lies within helix 8, and that neither helix 7 nor helices 9-12 are labelled.

The tryptic digestion experiments outlined above had demonstrated that the site of labelling by the endofacial ligand cytochalasin B lies within the region of the protein containing putative transmembrane helices 9-12. In order to locate the site of labelling within this region more precisely, N-bromosuccinimide was used to cleave the cytochalasin B-labelled protein at its tryptophan residues. This procedure yielded a number of different labelled fragments, the most abundant of which were two of apparent M_r 11500 and 14200. These fragments had identical mobility to bands on a Western blot of the digest that were recognised by antibodies raised against residues 450-467 of the transporter. A third fragment of apparent M_r 9000 recognised by these antibodies was not labelled by cytochalasin B. Only the largest of the three fragments was recognised by antibodies against residues 389-403. From their sizes and pattern of recognition by antibodies the three fragments appear to correspond to Met_{364}-Val_{492} (predicted M_r 13911), Phe_{389}-Val_{492} (predicted M_r 11378) and Thr_{413}-Val_{492} (predicted M_r 8829) respectively. Because the M_r 11500 fragment was labelled by cytochalasin B whereas the M_r 9000 fragment was not, it appears that the region Phe_{389} - Trp_{412} contains the cytochalasin B labelling site. However, direct sequencing of the labelled fragments will be required to confirm this location.

CONCLUSIONS

The results of the present study support the topological model previously proposed for the arrangement of the transporter in the membrane [4], although much further work will be required to confirm all the predictions of this model. Because of the duplication of sequences in the *N*- and *C*-terminal halves of the protein, we have previously proposed that the 12 helices are arranged in three-dimensions as two domains, each containing a bundle of 6 helices [1]. The results of the present study confirm previous observations that ATB-BMPA and cytochalasin B label sites in the *C*-terminal domain of the transporter (helices 7-12). The inhibitor forskolin and its derivatives are also known to photolabel the *C*-terminal half of the protein, and the site(s) of labelling have tentatively been identified as lying within helix 10 [22]. It therefore seems likely that this domain of the protein contains the substrate-translocation channel, probably involving helix 10, the amphipathic helices 8 and 11 identified as sites of inhibitor labelling in the present study, and helix 7 which we have previously shown to be highly conserved in sequence and amphipathic [1,4]. The function of the *N*-terminal domain is unclear.

ATB-BMPA and cytochalasin B are known to inhibit sugar transport by binding at the exofacial and endofacial surfaces of the membrane respectively [16,17]. The distinct and structurally separate locations of the outward-facing and inward-facing binding sites for these ligands identified in the present study suggest a mechanism for transport in which a conformational change, possibly involving helices 7, 8, 10 and 11, alternately exposes these outer and inner sites to a transported sugar. In the near future it should be possible to establish the validity of such a model by identifying more precisely the location of sites labelled by these inhibitors and other substrate analogues.

ACKNOWLEDGEMENTS

We are grateful to the SERC, MRC and Wellcome Trust for supporting this research, and to Professor G.E. Lienhard for the antibodies against residues 293-306 of the glucose transporter.

REFERENCES

1 Baldwin SA, Henderson PJF. Annu Rev Physiol 1989; 51: 459-471.
2 Fukumoto H, Kayano T, Buse JB, Edwards Y, et al. J Biol Chem 1989; 264: 7776-7779.

3 Bell GI, Kayano T, Buse JB, Burant CF, et al. Diabetes Care 1990; 13: 198-208.
4 Mueckler M, Caruso C, Baldwin SA, Panico M, et al. Science 1985; 229: 941-945.
5 Bagley PR, Tucker SP, Nolan C, Lindsay JG, et al. Brain Res 1989; 499: 214-224.
6 Barros LF, Baldwin SA, Jarvis SM, Cowen T, et al. J Physiol (Lond) 1991; (in press)
7 Baldwin SA, Lienhard GE. Methods Enzymol 1989; 174: 39-50.
8 Alvarez J, Lee DC, Baldwin SA, Chapman D. J Biol Chem 1987; 262: 3502-3509.
9 Chin JJ, Jung EKY, Jung CY. J Biol Chem 1986; 261: 7101-7104.
10 Chin JJ, Jung EKY, Chen V, Jung CY. Proc Natl Acad Sci U S A 1987; 84: 4113-4116.
11 Cairns MT, Alvarez J, Panico M, Gibbs AF, et al. Biochim Biophys Acta 1987; 905: 295-310.
12 Davies A, Meeran K, Cairns MT, Baldwin SA. J Biol Chem 1987; 262: 9347-9352.
13 Davies A, Ciardelli TL, Lienhard GE, Boyle JM, et al. Biochem J 1990; 266: 799-808.
14 Barnett JEG, Holman GD, Chalkley RA, Munday KA. Biochem J 1975; 145: 417-429.
15 Lowe AG, Walmsley AR. Biochim Biophys Acta 1986; 857: 146-154.
16 Clark AE, Holman GD. Biochem J 1990; 269: 615-622.
17 Devés R, Krupka RM. Biochim Biophys Acta 1978; 510: 339-348.
18 Zoccoli MA, Baldwin SA, Lienhard GE. J Biol Chem 1978; 253: 6923-6930.
19 Cairns MT, Elliot DA, Scudder PR, Baldwin SA. Biochem J 1984; 221: 179-188.
20 Holman GD, Rees WD. Biochim Biophys Acta 1987; 897: 395-405.
21 Schägger H, von Jagow G. Anal Biochem 1987; 166: 368-379.
22 Wadzinski BE, Shanahan MF, Seamon KB, Ruoho AE. Biochem J 1990; 272: 151-158.

TRANSPORT SYSTEMS IN MITOCHONDRIA

© 1992 Elsevier Science Publishers B.V. All rights reserved.
Molecular mechanisms of transport. E. Quagliariello, F. Palmieri, eds.

Topographical changes of the ADP/ATP carrier in the mitochondrial membrane related to conformational changes

G. Brandolin, I. Marty and P.V. Vignais

Laboratoire de Biochimie, Département de Biologie Moléculaire et Structurale, Centre d'Etudes Nucléaires, 85X, 38041 Grenoble cedex, France

INTRODUCTION

The paucity of actual 3D structures of membrane transport proteins that would be necessary to understand their structure-function relationships has led to the development of different methods to ascertain the topography of the polypeptide chains of the carriers within the membrane. Such approaches include the use of structure prediction algorithms and site-directed mutagenesis, as well as the determination of the sidedness of reactions with specific or unspecific chemical reagents, monoclonal or site-directed antibodies, endo- and exo-proteases.

Some of these methods have been applied in combination in our laboratory to study the topography of the membrane-bound ADP/ATP carrier in beef heart mitochondria, namely (i) specific photochemical labeling with modified substrates or inhibitors, (ii) immunochemical characterisation and (iii) controlled proteolytic cleavage.

The ADP/ATP carrier is a protein of the inner mitochondrial membrane that catalyses the import of cytosolic ADP into the matrix space of mitochondria in exchange for the export of matricial ATP generated by oxidative phosphorylation. The ADP/ATP transport system can be blocked by very specific inhibitors belonging to two different families : atractyloside (ATR) and carboxyatractyloside (CATR) on one hand, and bongkrekic acid (BA) and isobongkrekic acid (isoBA) on the other. Atractylosides and bongkrekic acids have been shown to bind to the ADP/ATP carrier from the cytosolic and from the matricial face of the membrane, respectively (for review see 1).

THE CONFORMATIONAL STATES OF THE ADP/ATP CARRIER

It is well established that the atractylosides and the bongkrekic acids recognize two different conformations of the carrier protein, referred to as the CATR and the BA conformations, respectively. The reversible transition of the ADP/ATP carrier between the CATR and the BA conformations is specifically induced by transportable nucleotides and for this reason is thought to be involved in the transport process itself.

The existence of two different conformations of the ADP/ATP carrier that could be trapped either by CATR or by BA has been demonstrated by several types of experiments carried out with the carrier either in the membrane-bound state or isolated in the presence of detergent. Depending on its conformational state, the carrier exhibits different chemical and immunochemical reactivities, as well as different susceptibility to proteases, corresponding to the modified accessibility of restricted domains of the polypeptide chain. This was illustrated, for instance, by immunological studies that have shown that antibodies directed to the CATR-carrier complex were unable to recognize the BA-carrier complex (2) and by the fact that tyrosine residues of the carrier protein were iodinated to a larger extent in mitochondria in the presence of CATR than in the presence of BA (3). One of the residues probing the conformational changes of the ADP/ATP carrier protein was identified as Cys 56 (4). This residue was of particular interest since it was the only cysteinyl residue out of four which was reactive to the alkylating reagent NEM. Moreover, its accessibility to

NEM was directly dependent on the conformation of the carrier since Cys 56 was alkylable only in the presence of ADP or ATP. This labeling was enhanced by BA and abolished by CATR, indicating that unmasking of Cys 56 occured only in the BA conformation of the carrier and in the BA-carrier complex.

As discussed later, the high mobility of the N-ter region of the ADP/ATP carrier, evidenced by the conformation-sensitive alkylation of Cys 56, has been confirmed by immunochemical approaches and by limited proteolysis experiments performed with the membrane-bound carrier (5).

Fluorescence techniques applied to the study of the ADP/ATP carrier protein allowed further caracterization of the BA and CATR conformational states. It was shown that the intrinsic fluorescence of the isolated carrier in detergent solution was significantly modified during the ADP or ATP-induced transition between the CATR and the BA conformations (6). Under our experimental conditions, the fluorescence of the carrier was increased upon addition of transportable nucleotides and further enhanced in the presence of BA. Detailed analysis of the kinetics of the transition observed at low temperature allowed the resolution of two sequential processes, namely the rapid binding of ADP or ATP to the carrier and the subsequent slower conformation change of the carrier protein. Non-transportable nucleotides were inefficient to trigger this transition.

Examination of the ADP or ATP-induced fluorescence changes of the ADP/ATP carrier protein isolated in different detergents revealed that the basal conformational state of the carrier in the absence of added nucleotide depended on the nature of the detergent used during the purification process. Thus, it appeared that lauryl amido dimethylpropylaminoxide (LAPAO) (7), a non-ionic detergent, stabilized the carrier in the CATR conformation whereas cholamidopropyl-dimethylammoniopropane sulfonic acid (CHAPS), which is zwitterionic, favored the BA conformation (8). This may be attributed to differences in constraints imposed to the carrier protein and possibly related to the different chemical nature of the detergents and to different micelle structures.

The transition of the membrane-bound ADP/ATP carrier between the CATR and the BA conformations have also been investigated by use of an extrinsic fluorescence technique, following the variations of fluorescence of naphthoyl-ADP (N-ADP) and naphthoyl-ATP (N-ATP), two fluorescent derivatives of ADP and ATP that bind with high affinity to the ADP/ATP carrier but are not transported (9,10). Release of N-ADP (or N-ATP) specifically bound to the carrier, induced by addition of CATR or BA, results in an increase of its fluorescence intensity and can be easily followed in a spectrofluorimeter. Analysis of the kinetics of release of bound N-ADP at low temperature permitted the discrimination between the CATR and the BA conformations of the carrier. It has been shown that the two conformers of the carrier co-exist in equilibrium in intact mitochondria; both bind N-ADP, but in the absence of any transportable nucleotide, only N-ADP bound to the CATR conformer was rapidly released by CATR and the remaining bound N-ADP, fixed to the BA conformer, was rapidly released by BA. Under these experimental conditions the spontaneous transition of the carrier between the two conformations was very slow. However, it could be considerably accelerated upon addition of ADP or ATP, resulting in the complete release of all bound N-ADP. Here again, non transportable nucleotides were inefficient.

In order to caracterize the structural modifications of the ADP/ATP carrier related to its functionning we have investigated the topography of the membrane-bound carrier trapped in the two different conformations adopted in the course of the transport process, namely the CATR and the BA conformations. The sidedness of restricted segments of the polypeptide chain of the carrier has been determined in both conformations by use of non permeant probes such as proteolytic enzymes and sequence-directed antibodies generated from synthetic peptides (Figure 1). Our results indicate that several regions of the carrier polypeptide chain are mobile and able to probe the conformational changes undergone by the carrier.

N-terminal peptide

Ac-Ser-Asp-Gln-Ala-Leu-Ser-Phe-Leu-Lys-Asp-Tyr—OVA

40-50 peptide

Ala-Ser-Lys-Gln-Ile-Ser-Ala-Glu-Lys-Gln-Tyr—OVA

C-terminal peptide

OVA—Cys-Val-Leu-Val-Leu-Tyr-Asp-Glu-Ile-Lys-Lys-Phe-Val

Figure 1. Structure of the synthetic peptide conjugates used to generate sequence-specific antibodies for the immunochemical study of the ADP/ATP carrier topography.

TOPOLOGY AND FLEXIBILITY OF THE NH_2 TERMINAL DOMAIN OF THE ADP/ATP CARRIER

The first direct evidence that the N-terminal part of the ADP/ATP carrier is highly mobile was given by Boulay and Vignais (4), who identified Cys 56 as a probe of the conformational changes of the carrier. In the neighbourhood of Cys 56, lysine groups at positions 42 and 48 were reported to be labelled by pyridoxal phosphate in a conformational sensitive manner (11).

Recently, we have explored in detail the arrangement of the N-terminal region of the ADP/ATP carrier protein and its topological modifications related to the conformational changes of the carrier using sequence directed antibodies and an Arg specific endoprotease (5). To probe the sidedness of the N-terminal region of the carrier, polyclonal antibodies were raised in rabbits immunized against an eleven residue synthetic peptide corresponding to the N-terminal sequence of the carrier. Specificity of antibodies was assessed by immunoblotting experiments performed with mitochondrial lysates. Reactivity of the N-terminal antibodies to the membrane-bound carrier in bovine mitoplasts was assayed by ELISA with mitoplasts coated on the wells of microtiter plates ; the immunotitrations indicated that the binding of antibodies was related to the amount of coated particles and that it was abolished in the presence of competing N-terminal peptide. These results demonstrated the cytosolic exposure of the N-terminal region of the ADP/ATP carrier.

The possibility of a disorganisation of the mitochondrial membrane upon binding to the plastic of microtiter plates resulting in the exposure of previously unaccessible epitopes led us to carry out back-titration experiments. In this procedure, antibodies were first allowed to react with mitoplasts in suspension and then the unreacted antibodies remaining in the supernatant were titrated by ELISA in microtiter plates coated with the N-terminal peptide. We found that increasing the amount of mitoplasts resulted in a decrease of titrable antibodies in the supernatant thus confirming the orientation of the N-terminal region of the carrier protein towards the cytosolic compartment.

We investigated the possible modulation exerted by the conformational states of the ADP/ATP carrier on the accessibility of the N-terminal region of the membrane-bound carrier towards anti N-terminal antibodies. The immunoreactivity of the carrier in mitoplasts was assayed, applying the ELISA techniques described above, in the presence of CATR or in the presence of BA. Immunotitrations indicated a much higher reactivity of the N-terminal region to anti N-terminal antibodies when the carrier was in the CATR conformation than in the BA conformation. This result corroborated previous observations that revealed the ability of the N-terminal region of the carrier to sense the conformational changes of the ADP/ATP carrier related to the transition between the CATR and the BA conformations and thus probably occuring during the ADP/ATP transport process.

For a further study of the topography and the conformation-sensitive arrangement of the N-terminal region of the ADP/ATP carrier, we combined immunological analysis with limited enzymatic digestion techniques. We investigated the protease susceptibility of the membrane-bound carrier detecting the generated proteolytic fragments by immunoblot analysis either with polyclonal antibodies raised against the SDS-denatured ADP/ATP carrier (12) or sequence-directed antibodies. To localize the enzymatic cleavage sites either on the matricial or on the cytosolic face of the membrane, proteolysis experiments were performed of the carrier in the membrane of mitoplasts or of inside-out submitochondrial particles (SMP).

As shown by Western blot analysis, digestion of the ADP/ATP carrier with an Arg-specific endoprotease was not possible when using mitoplasts, demonstrating that no Arg group in the carrier was accessible to the protease from the cytosolic face of the membrane. In contrast, cleavage of the carrier occurred when using SMP, generating only one fragment of about 25 kDa that was not reactive to anti N-terminal antibodies. From these results it was concluded that the cleavage site was exposed to the matrix compartment and located close to the N-terminal extremity of the polypeptide chain. Only two Arg residues, Arg 30 and Arg 59 are present in the sequence of the ADP/ATP carrier in this region and either of these residues could be candidate as the cleavage site, considering the size of the generated proteolytic fragment. Demonstrations of the cytosolic orientation of the N-terminal sequence of the carrier as well as the exposure of Arg 30/59 to matrix implies the presence of a transmembrane segment extending between residues 10 and 30 as predicted by the hydropathy analysis of the ADP/ATP carrier sequence. The considerably different susceptibility of the carrier polypeptide chain to the Arg endoprotease on the two faces of the mitochondrial membrane confirmed the overall asymmetry of the carrier and ruled out the possibility of an antiparallel orientation of monomers in the oligomeric assembly.

To explore the influence of the CATR and BA conformational states of the ADP/ATP carrier on the accessibility of the polypeptide chain to proteases, we investigated the enzymatic proteolysis of the membrane-bound carrier with trypsin and with a lysine endoprotease using four types of particles, namely mitoplasts-CATR, mitoplasts-BA, SMP-CATR and SMP-BA. The carrier was cleaved exclusively in SMP pretreated with BA. The generated fragments were identified following their immunoreactivities to antipeptide antibodies and by direct sequencing, allowing the localization of three cleavage sites at bonds : Lys 42-Glu43, Lys 146-Gly 147 and Lys 244-Gly 245. It could therefore be concluded that these residues are exposed to the matrix compartment when the ADP/ATP carrier is in the BA conformation. Further, they belong to polypeptide segments mobile enough to render them not accessible to proteases when the carrier is shifted to the CATR conformation. These results confirm the matrix exposure of the region of the carrier comprising Cys 56 as well as the mobility of this region depending on the conformational changes of the ADP/ATP carrier.

Mapping of the inhibitor and nucleotide binding sites of the ADP/ATP carrier with photoactive derivatives of atractyloside or of ADP (13-15) indicated that these binding sites are located in the C-terminal half of the polypeptide chain, quite remote from the N-terminus. Thus, the N-terminal region, although not being directly involved in the binding of specific ligands of the carrier, undergoes topological modifications. These might be propagated along the polypeptide chain, or result from interactions with other mobile regions, either of the same carrier molecule or more probably, of an adjacent monomer in a dimeric or tetrameric oligomer.

ARRANGEMENT OF THE PEPTIDE CHAIN OF THE ADP/ATP CARRIER IN THE MITOCHONDRIAL MEMBRANE

The ADP/ATP carrier has been sequenced and is a protein 297 amino acid long (16). As it is an integral membrane protein, the determination of its secondary structure is based largely on hydrophobicity distribution analysis along its sequence. Even though this approach gives variable predictions, depending on the algorithm used (17), it permitted the detection of five hydrophobic regions long enough to span the membrane as α-helices and separated by hydrophilic sequences (18, 19). Although it contains polar residues, the existence of a sixth possible transmembrane region has been postulated. Comparison of the sequence of the carrier with itself in a Diagon plot showed the presence of three internal repeats, approximately 20-40 amino acid long, in the hydrophilic segments, separated by 100 and 200 residues from the diagonal. On the basis of these data, an arrangement of the polypeptide chain of the carrier within the membrane was proposed, which consisted of three repeated units, each containing two transmembrane segments connected by an hydrophilic loop, and connected themselves one to each other by hydrophilic segments. Theoretical structure analysis of other mitochondrial carriers sequenced so far, including the uncoupling protein from brown adipose tissue, the phosphate carrier and the oxoglutarate carrier, yielded similar hydrophobicity profiles, suggesting similar transmembrane arrangements (19, 20).

Other models of folding of the polypeptide chain of the ADP/ATP carrier have been proposed essentially based on data obtained from labelling experiments with pyridoxal phosphate, a lysine-specific reagent (11, 21, 22). However, as discussed later, these results are open to criticism.

The threefold symetry model of the ADP/ATP carrier would fit with the presence of three equidistant lysine groups, Lys 42, Lys 146 and Lys 244, located on the same face of the membrane (the matrix face) when the carrier is in the BA conformation. The even number of transmembrane segments implies that the NH_2- and COOH-terminal ends of the carrier are oriented towards the same side of the membrane, i.e. the cytosolic side referring to the orientation of the N-terminal region (5). To check the validity of this model, we investigated the orientation of the C-terminal region of the membrane-bound ADP/ATP carrier, using immunological approaches and ELISA techniques, as described for the probing of the N-terminal segment topography. Antibodies directed to a thirteen aminoacid long peptide corresponding to the C-terminal sequence of the carrier were shown to react with the carrier in mitoplasts coated onto microtiter plates. However, this result was not corroborated in back-titration experiments that indicated a greater immunoreactivity of the carrier in the membrane of SMP. Since back-titration assays should be considered as more reliable because they allow antibodies to react with particles in suspension, they would indicate the matrix exposure of the C-terminus of the carrier but these experiments could not lead to clear-cut conclusions, in contrast to the determinations of the orientation of the N-terminus of the ADP/ATP carrier (5) or of the N-terminal and C-terminal ends of the phosphate carrier (23), that were based on the same experimental approaches.The matrix orientation of the C-terminal region of the ADP/ATP carrier would be conflicting with the cytosolic exposures of the C-terminus established for the UCP (24) and for the phosphate carrier (23) if one assumes that the three carriers share similar membrane foldings on the basis of their overall sequence homologies.

The experimental data that have allowed insights into the topography of the ADP/ATP carrier are summarized on Table 1. They were obtained through the use of probes shown to react with the membrane-bound carrier from only one side of the mitochondrial membrane.

Pyridoxal phosphate (PLP) has been used to probe the lysine group distribution of the ADP/ATP carrier on each side of the mitochondrial membrane (11, 21, 22). Although PLP is considered to be membrane impermeable, confusing labeling results were reported, inconsistent with other topological data. For instance, Lys 9, which is obviously oriented towards the cytosol was not labeled. Lysine residues at positions 22, 42, 48, 106, 162, were assigned to translocation paths whereas, as shown on Table 1, only the regions spanning the residues 153-200 and 250-281 in the bovine ADP/ATP carrier sequence were involved in the binding of photoactivable specific ligands. Photolabeling experiments performed with 2 azido ADP on the yeast ADP/ATP carrier by the Klingenberg's group resulted in the mapping of the region 160-170 (15), thus corroborating our findings; no labeling of a region corresponding to the segment 250-281 in the bovine ADP/ATP carrier was detected, probably due to improper experimental conditions. To accomodate the Lys labeling data with the threefold symetry of the carrier and, for the sake of homology with the UCP, Klingenberg proposed a model of the transmembrane arrangement of the ADP/ATP carrier that included three hydrophilic paths (16). Two of them carried binding sites, in accordance with our photolabeling results but in contradiction with the reorienting gated pore concept that postulates a single binding center for substrates and inhibitors proposed by the same author.

Table 1
Summary of topological data concerning the membrane-bound ADP/ATP

Probing technique	Cytosolic accessibility	Matricial accessibility	References
(^{3}H) NAP$_4$ ATR labeling	Segment Cys159-Met200		13
(^{3}H) NEM binding		Cys 56	4
2 azido ADP binding	Segments Phe153-Met200 and Tyr250-Met281		14
anti N-ter antibodies binding	N-terminal segment (in the presence of CATR)		5
Arg endoprotease cleavage on SMP		Arg 30 and/or Arg59	5
Trypsin and Lys endoprotease cleavage on SMP		Lys 42-Gln 43 Lys146-Gly147 Lys 244-Gly 245	25

A DYNAMIC MODEL FOR THE ADP/ATP CARRIER

Examination of the topological data shown on Table 1 indicates that the regions of the carrier that contain the protease-sensitive bonds Lys 146-Gly 147 and Lys 244-Gly 245 and that are able to move during the conformational changes of the ADP/ATP carrier between the CATR and the BA conformations, are both located adjacent to the two segments photolabeled with 2 azido ADP. These observations made very probable the involvement of such domains of the peptide chain in the translocation path of the carrier. However, the cytosolic exposure of the photolabeled regions, demonstrated from the photolabeling of the carrier in mitochondria with non-permeant probes, was not compatible with the matricial orientation of bonds Lys 146-Gly 147 and Lys 244-Gly 245 that were enzymatically cleaved in the carrier in the BA conformation. These conflicting data may be reconcilied in a dynamic model of the arrangement of the peptide chain of the ADP/ATP carrier in which the two regions of the polypeptide chain suggested to belong to the translocation path are differently accessible to proteases from the matrix side of the membrane depending on the conformation of the carrier. Such loops are amphipathic and clearly too polar to cross the lipid bilayer, but they may be inserted in the membrane, either as β-sheets or as short α-helices, in a way that hydrophilic residues pack together, as would be the case, for instance, if they faced one another in dimers of the carrier. The existence of short α-helices in the transmembrane arrangements of membrane carrier proteins has been postulated by Lodish (26). We found that four segments extending from residues 140 to 151 and 153 to 167 on one hand and 237 to 249 and 253 to 267 on the other hand, if organized in α-helices, show clustering of polar amino acids on one face of the helice. This is clearly shown in the Edmunson-wheel representation of their sequences (Figure 2). Thus, polar faces of helices might be juxtaposed in carrier dimers allowing the formation of an hydrophilic translocation channel whereas the hydrophobic periphery of helices interacts with the apolar lipid phase of the membrane. Similar arrangements of amphipathic helices have been recently proposed by Jähnig (27) for two *E. coli* membrane proteins : the lactose permease and the OmpA protein.

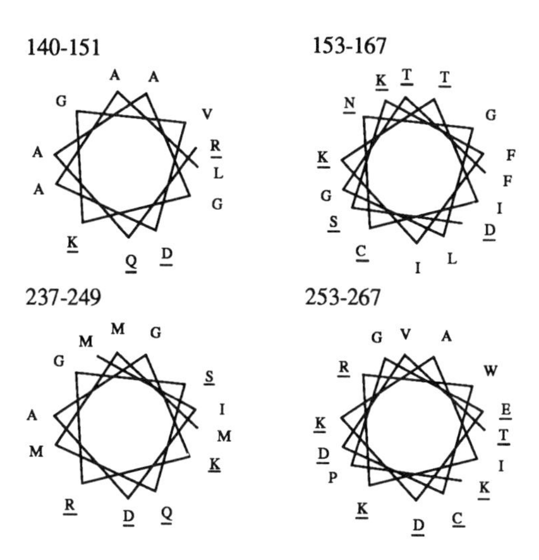

Figure 2. The Edmunson-wheel representation of the amphiphilic sequences located close to the nucleotide binding region. Polar residues are underlined.

It should be added that the existence of a hypothetical translocation path made by the spatial juxtaposition of amphipathic peptide segments in carrier dimers is not dependent on the structure of the carrier monomers. Such an arrangement of the carrier would obviously imply a twofold symetry and, as the orientation of the C-terminal end of the ADP/ATP carrier relative to the membrane plane is still to be determined, models of the arrangement of the carrier monomer either with five or with six of transmembrane segments would both fit with our proposal. As an example the scheme in Figure 3 illustrates a possible arrangement of a carrier dimer showing the translocation path formed by the association of amphiphilic helices.

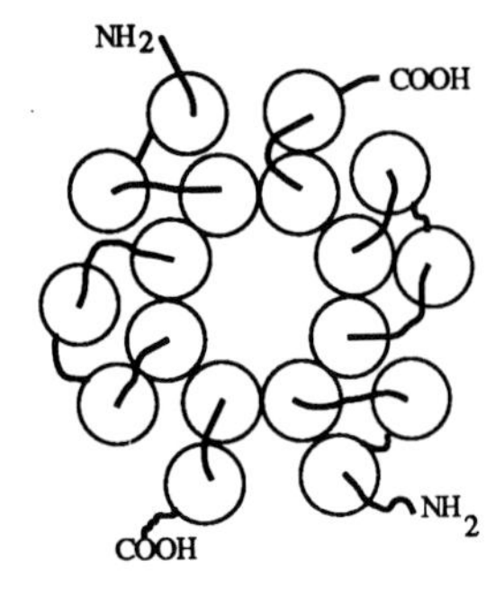

Figure 3. Suggested arrangement of the polypeptide chain of the carrier in the membrane (Top view from the cytosol) showing the formation of an hydrophilic channel by the juxtaposition of amphiphilic short helices.

REFERENCES

1 Vignais PV, Block MR, Boulay F., Brandolin G, Lauquin GJM. In: Bengha G, ed. Structure and Properties of Cell Membranes. CRC: Boca Raton FL, 1985; II: 139-179.
2 Buchanan BB, Eiermann W, Riccio P, Aquila H, Klingenberg M. Proc Natl Sci USA 1976; 73: 2280-2284.
3 Brdiczka D, Schumacher D. Biochem Biophys Res Commun 1976; 73: 823-832.
4 Boulay F, Vignais PV. Biochemistry 1984; 23: 4807-4812.
5 Brandolin G, Boulay F, Dalbon P, Vignais PV. Biochemistry 1989; 28: 1093-1100.
6 Brandolin G, Dupont Y, Vignais PV.Biochemistry 1985; 24:1991-1997.
7 Brandolin G, Doussière J, Gulik A, Gulik-Kzywicki T, Lauquin GJM, Vignais PV. Biochim Biophys Acta 1980; 592: 592-614.
8 Block MR, Vignais PV. Biochemistry 1986; 25: 374-379.
9 Block MR, Lauquin GJM, Vignais PV. Biochemistry 1982; 21: 5451-5447.
10 Block MR, Lauquin GJM, Vignais PV. Biochemistry 1983; 22: 2202-2208.
11 Bogner W, Aquila H, Klingenberg M. Eur J Biochem 1986; 165: 611-620.
12 Boulay F, Lauquin GJM, Vignais PV. Biochemistry 1986; 25: 7567-7571.
13 Boulay F, Lauquin GJM, Tsugita A, Vignais PV. Biochemistry 1983; 22: 477-484.
14 Dalbon P, Brandolin G, Boulay F, Hoppe J, Vignais PV. Biochemistry 1988; 27: 5141-5149.
15 Mayinger P., Winkler E, Klingenberg M. FEBS Lett 1989; 244: 421-426.
16 Aquila H, Misra D, Eulitz M, Klingenberg M. Hoppe-Seyler's Z Physiol Chem 1982; 363: 345-349.
17 Fasman GD, Gilbert WA. Trends Biochem Sci 1990; 15: 89-92.
18 Saraste M, Walker JE. FEBS Lett 1982; 144: 250-254.
19 Runswick MJ, Pwell SJ, Nyren P, Walker JE. EMBO J 1987; 6: 1367-1373.
20 Runswick MJ, Walker JE, Bisaccia F, Iacobazzi V, Palmieri F. Biochemistry 1990; 29: 11033-11040.
21 Bogner W, Aquila H, Klingenberg M. FEBS Lett 1982; 146:259-261.
22 Klingenberg M. Arch Biochem Biophys 1989; 270: 1-14.
23 Capobianco L, Brandolin G, Palmieri F. Biochemistry 1991; 39: 4963-4969.
24 Eckerskorn C, Klingenberg M. FEBS Letters 1987; 226: 166-170.
25 Marty I, Brandolin G, Vignais PV. submitted 1991.
26 Lodish HF. Trends Biochem Sci 1988; 13: 332-334.
27 Jähnig F. In: Fasman GD, Ed. Prediction of Protein Structure and the Principles of Protein Conformation. Plenum Press: New York, 1989; 707-717.

© 1992 Elsevier Science Publishers B.V. All rights reserved.
Molecular mechanisms of transport. E. Quagliariello, F. Palmieri, eds.

GLUTAMINE TRANSPORT IN RAT KIDNEY MITOCHONDRIA

S. Passarella[a], A. Atlante[b] and E. Quagliariello[c].

[a]Dipartimento di Scienze Animali, Vegetali e dell'Ambiente, Università del Molise, Campobasso, Italy.

[b]Centro di Studio sui Mitocondri e Metabolismo Energetico, Consiglio Nazionale delle Ricerche, Bari, Italy.

[c]Dipartimento di Biochimica e Biologia Molecolare, Università di Bari, Bari, Italy.

Mitochondrial metabolite transport has been extensively studied since the mid 1960's, and the specificity and kinetic properties of many anion-transporting systems have been established (1). More recently a number of new carriers for metabolites has been discovered in different mitochondria (2-8). Nonetheless, in spite of the major role played by the transport processes in the whole cellular metabolism, how certain metabolites enter mitochondria has not yet been elucidated. This applies to glutamine.

Enzymes responsible for glutamine metabolism are located in the mitochondria; thus glutamine must necessarily cross the mitochondrial membranes. Glutamine transport has mainly been studied in kidney mitochondria, but, as stressed in (9,10), elucidation of the properties of the glutamine carrier/s has proved difficult.

In Scheme 1 the mitochondrial glutamine metabolism is reported: glutaminase activity causes ammonia and glutamate formation in the matrix. In addition oxoglutaramate can be formed as a result of transamination with 2-oxoglutarate (11). Removal of glutamate, an inhibitor of glutaminase (12), via glutamate dehydrogenase and aspartate aminotransferase fills up the citric cycle metabolite pool.

Glutamine uptake by mitochondria was first shown by Kovacevic and co-workers (13) who demonstrated that isolated rat kidney mitochondria (RKM) swell in an isotonic solution of glutamine when incubated in the presence of a respiratory inhibitor; moreover the existence of an electrogenic glutamine/glutamate antiporter was proposed in pig kidney mitochondria (14).

The study of mitochondrial glutamine transport was intensified when investigators recognized that this process might be important in the regulation of renal ammoniagenesis in metabolic acidosis. It should be noted that in the kidney ammonia formation derives from glutamine metabolism, even though the occurrence of the purine nucleotide cycle in this organ has

recently been shown (15). Interestingly an increased rate of glutamine transport into kidney mitochondria during acidosis was reported (16-18).
Nonetheless at present the mechanism of glutamine transport into kidney mitochondria remains unresolved. As shown in Table 1 no general consensus has been reached with respect to glutamine transport.

TABLE 1
THE STATE OF ART FOR GLUTAMINE TRANSPORT IN KIDNEY MITOCHONDRIA

	Normal RKM		Acidotic RKM	
UNIDENTIFIED GLN CARRIER	Goldstein	(19)	Adam	(17)
			Goldstein	(19)
			Shapiro	(20)
GLUTAMINE UNIPORTER	Kovacevic	(13)	Kovacevic	(13)
	Cheema-Dhadli	(21)	Brosnan	(22)
			Tannen	(23)
			Cheema-Dhadli	(21)
GLUTAMINE/ GLUTAMATE ANTIPORTER	Crompton	(14)		

It should be noted that the existence of glutamine uniporter and glutamine/glutamate antiporter has also been excluded by certain laboratories (14,23).
To gain further insight into glutamine uptake by RKM, previously developed assay systems (2,3) were used, which allow for a measurement of metabolite efflux from mitochondria under conditions in which the intramitochondrial reactions can either freely occur or be selectively blocked by externally added specific inhibitors (see Scheme 1).
The capability of externally added glutamine to cause the efflux from mitochondria of either endogenous substrates or metabolites formed during glutamine metabolism was tested.
Glutamate and malate were chosen as possible counteranions for incoming glutamine. This choice was made in the light of (14) and because the appearance of malate in the extramitochondrial phase had already been reported by Cheema-Dhadli and Halperin (21), with the dicarboxylate carrier proposed to mediate malate/phosphate exchange.

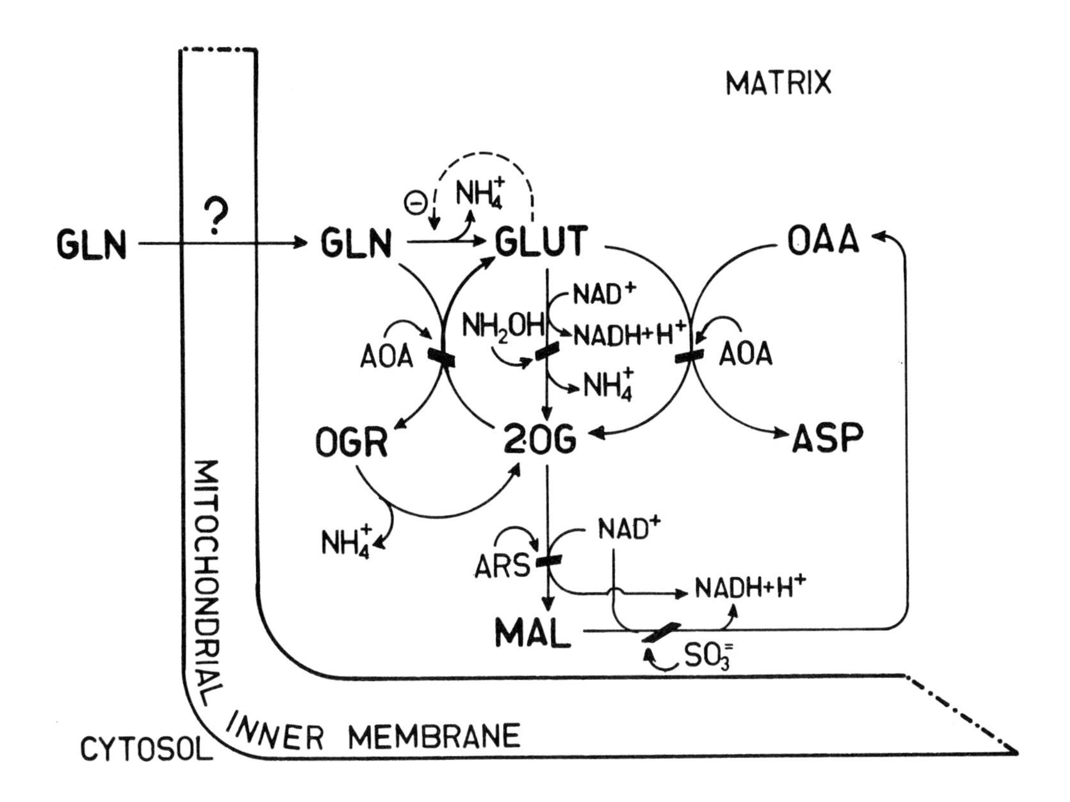

SCHEME 1

Scheme 1. GLUTAMINE METABOLISM. Abbreviations: AOA, aminooxyacetate; ARS, arsenite; ASP, aspartate; GLN, glutamine; GLUT, glutamate; MAL, malate; NH_2OH, hydroxylamine; OAA, oxaloacetate; 2-OG, 2-oxoglutarate; OGR, oxoglutaramate; $SO_3^{=}$, sulphite.

In the phase outside mitochondria glutamate (Fig.1A) or malate (Fig.1B) concentration is negligible, since no change of absorbance occurs following the addition of either glutamate (A) or malate (B) detecting system, to RKM. Further addition of

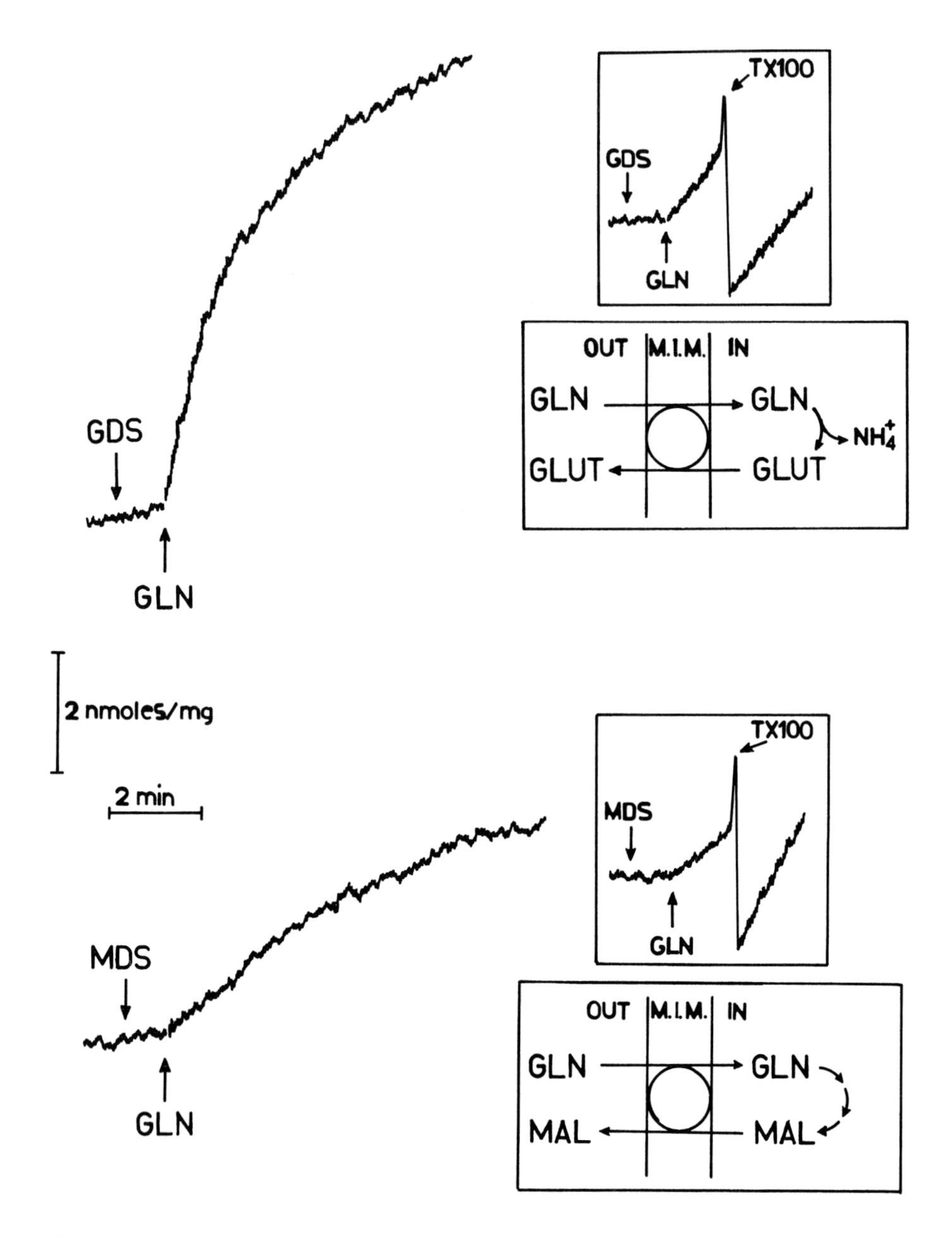

Fig. 1 Appearance of both glutamate and malate in the extramitochondrial phase caused by the addition of glutamine to rat kidney mitochondria. RKM (1.5 mg protein) were incubated at 20°C for 1 min in 2.0 ml of standard medium containing 0.2 M sucrose, 10 mM KCl, 20 mM Hepes-Tris pH 7.2, 1 mM $MgCl_2$ plus 1.25 µM FCCP and 2 µg rotenone. Where indicated, additions

were as follows: glutamate detecting system (GDS, consisting of 1 mM NAD^+ plus 5 e.u. glutamate dehydrogenase) (in A), malate detecting system (MDS, consisting of 0.2 mM $NADP^+$ plus 0.2 e.u. malic enzyme) (in B), 1.5 mM glutamine (GLN), 0.5 % Triton-X-100 (TX100). In both cases NAD/NADP reduction was followed by measuring the increase in A_{340}.
In the insets the traces of Triton experiments are reported.

---

glutamine (1.5 mM in both cases), which has no effect when added alone, causes an increase of absorbance, which shows the appearance of glutamate (A) or malate (B) outside mitochondria.
A possible explanation of these findings could be that glutamine enters mitochondria either via uniporter or in exchange with endogenous glutamate or malate. Inside the matrix glutamine is deaminated to glutamate which either in turn is exchanged for further glutamine (A) or gives malate which could exit from mitochondria in a carrier mediated process in exchange with further glutamine (B) (see insets Fig.1).
Both glutamate and malate efflux were also found in acidotic RKM as obtained with one week 0.28 M NH_4Cl treatment as in (24) (not shown).
These findings raise the question as to whether the rate of glutamine transport across the mitochondrial membrane limits glutaminase activity, i.e. whether the rate of substrate appearance in the extramitochondrial phase is the rate of glutamine transport.
Triton-X-100 (insets of Fig.1, A and B) was used to investigate this point, according to (6). In the case of glutamine/glutamate exchange in normal mitochondria the rate of absorbance increase proved to be the rate of glutaminase, whereas under acidotic conditions glutamine transport was found to limit the glutaminase activity.
Both in normal and in acidotic rats, glutamine/malate exchange was found to be the rate limiting step of the investigated processes.
To further substantiate the theory that substrate efflux is a carrier mediated process, the dependence of the rate of $NAD^+/NADP^+$ reduction on increasing glutamine concentration was investigated for glutamine/glutamate (for acidotic RKM only) and glutamine/ malate exchanges.
Saturation kinetics were found. In acidotic RKM Km values, i.e. the substrate concentration which gives half maximum $NAD^+/NADP^+$ reduction rate, were equal to 1.2 mM and 0.6 mM and Vmax values were 3.1 and 2.5 nmoles/min x mg protein for glutamine/glutamate and glutamine/malate exchanges, respectively.

TABLE 2
[^{14}C]-GLUTAMINE UPTAKE BY GLUTAMATE-MALATE LOADED RKM.

t(s)	a) [^{14}C]GLN uptake	b) GLN/GLUT exchange	c) GLN/MAL exchange	ΔGLN uniport [a-(b+c)]
	(nmoles/mg)			
		NORMAL RATS		
20	3.7	2.9	0.8	0.0
45	7.0	5.6	1.2	0.2
60	7.6	5.9	1.4	0.3
300	9.1	7.3	2.0	-0.2
		ACIDOTIC RATS		
20	1.2	0.6	0.5	0.1
45	3.8	1.4	0.9	1.5
60	4.7	1.7	1.1	1.9
300	5.8	2.7	2.3	0.8

Glutamate-malate-loaded RKM (1.5 mg protein) obtained essentially according to (25) were preincubated at 2°C in 1 ml of the standard medium in the presence of 2 µg rotenone, 1 mM arsenite, 10 mM aminooxyacetate, 10 mM hydroxylamine. After a 1 min incubation, the reaction was started by addition of 3 and 1.5 mM [^{14}C]glutamine (in normal and acidotic RKM respectively) and stopped by rapid addition of an inhibitor mixture (10 mM benzylmalonate, 5 mM phenylsuccinate and 0.2 mM mersalyl) at time t indicated, followed by the rapid centrifugation of the mitochondria.
In control experiments the intramitochondrial glutamate concentration proved to not significantly vary during the exchange measurements.
[^{14}C]Glutamine (GLN) uptake (**a**), glutamine/glutamate (GLN/GLUT) exchange (**b**) and glutamine/malate (GLN/MAL) (**c**) exchange were measured in the same postmitochondrial extract as described in (26-28). The glutamine uniporter activity is measured as [**a**-(**b**+**c**)].

Determination of Km and Vmax values for glutamine/glutamate exchange in normal mitochondria merits further discussion. In this case the impermeable inhibitor benzylmalonate which was found to inhibit this exchange. According to Halestrap (see 1), Km and Vmax values (7 mM and 34.5 nmoles/min x mg protein, respectively) were obtained by plotting the reciprocal of the rate of change in absorbance as a function on benzymalonate concentration. The resulting Dixon plot extrapolated at zero inhibition provides a measure of transport in the absence of inhibition.

In normal RKM, Km and Vmax values for glutamine/malate exchange were 3 mM and 1.5 nmoles/min x mg protein, respectively.

As expected, in acidotic RKM both glutamine and glutamate metabolism is increased as suggested by the 80% decrease of the rate of glutamine/glutamate carrier at 2 mM glutamine and by the increase (250% at 2 mM glutamine) of the rate of glutamine/malate carrier.

The capability of many metabolites to impair glutamine/ glutamate and glutamine/malate translocators was also tested: no effect was observed in the presence of fumarate, aspartate, oxoglutarate, ornithine, phosphate, pyruvate, cis-aconitate, istidine, arginine, lysine, valine etc.

Contrarily certain inhibitors of mitochondrial translocators such as benzylmalonate, phenylsuccinate, mersalyl, methyl-glutamate, glutamate hydroxamate, block both the processes. A further distinction between these carriers is given by the specificity of glutamate dimethylesther and thiomalate in completely blocking glutamine/glutamate without significantly affecting the glutamine/malate exchange.

The reported results are consistent with the existence of two translocators in RKM.

To further substantiate this conclusion as well as to ascertain the existence of the glutamine uniporter, use was made of mitochondria loaded with both glutamate and malate and the uptake of ^{14}C-glutamine measured as a function of time (Table 2).

The reported experimental findings strongly suggest that in RKM glutamine uptake occur by means of three different carriers: the glutamine/glutamate and glutamine/malate antiporter and the glutamine uniporter active mainly in acidotic RKM.

REFERENCES

1 LaNoue KF, Schoolwerth AC. In: Ernster L, ed. New Comprehensive Biochem: Bioenergetics. New York: Elsevier, 1984; 9: 221-268.

2 Passarella S, Barile M, Atlante A, Quagliariello E. Biochem

Biophys Res Commun 1984; 119: 1039-1046.
3 Passarella S, Atlante A, Barile M, Quagliariello E. Biochem Biophys Res Commun 1984; 121: 770-778.
4 Passarella S, Atlante A, Quagliariello E. Biochem Biophys Res Commun 1985; 129: 1-10.
5 Atlante A, Passarella S, Giannattasio S, Quagliariello E. Biochem Biophys Res Commun 1985; 132: 8-18.
6 Passarella S, Atlante A, Barile M, Quagliariello E. Neurochem Research 1987; 12: 255-264.
7 Passarella S, Atlante A, Quagliariello E. Biochem Biophys Res Commun 1989; 158: 870-879.
8 Passarella S, Atlante A, Quagliariello E. Eur J Biochem 1990; 193: 221-227.
9 Kovacevic Z, McGivan JD. Physiol Reviews 1983; 63: 547-605.
10 Schoolwerth AC, LaNoue KF. Annu Rev Physiol 1985; 47: 143-171.
11 Cooper AJL, Meister A. J Biol Chem 1974; 249: 2554-2561.
12 Kovacevic Z. Biochim Biophys Acta 1975; 396: 325-334.
13 Kovacevic Z, McGivan JD, Chappel JB. Biochem J 1970; 118: 265-274.
14 Crompton M, Chappel JB. Biochem J 1973; 132: 35-46.
15 Bogusky RT, Lowenstein LM, Lowenstein JM. J Clin Invest 1976; 58: 326-335.
16 Pitts RF. In: Orloff J, Berliner RW, eds. Handbook of Physiol. Renal Physiology. Washington, DC: Am Physiol Soc, 1973; 15: 455-496.
17 Adam W, Simpson DP. J Clin Invest 1974; 54: 165-174.
18 Tager JM, Slater EC. Biochim Biophys Acta 1963; 77: 227 -245.
19 Goldstein L. Am J Physiol 1975; 229: 1027-1033.
20 Shapiro RA, Curthoys NP. FEBS lett 1978; 91: 49-52.
21 Cheema-Dhadli S, Halperin ML. Can J Biochem 1978; 56: 23-28.
22 Brosnan JT, Hall B. Biochem J 1977; 164: 334-337.
23 Tannen R. Am J Physiol 1978; 235: F265-F277.
24 Kunin AS, Tannen RL. Am J Physiol 1979; 237: F55-F62.
25 Schoolwerth AC, LaNoue KF, Hoover WJ. J Biol Chem 1983; 258: 1735-1739.
26 Palmieri F, Quagliariello E. Methods Enzymol 1979; 56: 279-301.
27 Bernt E, Bergmeyer HU. In: Bergmeyer HU, ed. Methods of Enzymatic Analysis. Academic Press, New York and London 1963; 384-388.
28 Hohorst HJ. In: Bergmeyer HU, ed. Methods of Enzymatic Analysis. Academic Press, New York and London, 1963; 328-332.

© 1992 Elsevier Science Publishers B.V. All rights reserved.
Molecular mechanisms of transport. E. Quagliariello, F. Palmieri, eds.

STRUCTURAL AND FUNCTIONAL PROPERTIES OF TWO MITOCHONDRIAL TRANSPORT PROTEINS: THE PHOSPHATE CARRIER AND THE OXOGLUTARATE CARRIER

F. Palmieri, F. Bisaccia, L. Capobianco, V. Dolce, V. Iacobazzi, C. Indiveri and V. Zara

Department of Pharmaco-Biology, Laboratory of Biochemistry and Molecular Biology, University of Bari, Bari (Italy)

The inner mitochondrial membrane contains several transport systems for metabolites, which are necessary for oxidative phosphorylation, for the transfer of reducing equivalents and for important metabolic pathways. The transport systems (carriers) which have been most extensively studied are: the ADP/ATP carrier, the phosphate carrier (PiC), the uncoupling protein (UCP), the oxoglutarate carrier (OGC), the dicarboxylate carrier (DIC), the aspartate/glutamate carrier (AGC), the pyruvate carrier, the citrate (tricarboxylate) carrier (CIC) and the carnitine carrier. Although these carriers have been identified in intact mitochondria, the state of their characterization is very different. All the carriers indicated above have been purified and functionally reconstituted into liposomes (1-3). The primary structure of the first three carriers has been determined by amino acid analysis and/or by DNA sequencing. These proteins are formed by three segments of about 100 amino acids, which are homologous one to another. The clear homology between these carriers led to the development of a concept of a carrier family, which is supposed to have originated from a common ancestor gene (4-5). It is proposed that also the other mitochondrial carriers fall into the same family, an idea which is supported by the very similar Mr of around 30 kDa for most of the carriers so far identified. With respect to elucidation of transport mechanism and regulation the ADP/ATP carrier, the PiC and the AGC are the best characterized. In this paper several structural and functional properties of previously purified carriers are reported.

cDNA of human heart PiC. We have isolated a full length cDNA clone encoding the precursor of the human heart mitochondrial PiC protein (6). The entire clone is 1330 bp in length with 5' and 3'-untranslated regions of 48 and 184 bp, respectively. The open reading frame encodes the mature protein consisting of 312 amino acids, preceded by a presequence of 49 amino acids. The mature human PiC sequence (6) differs in 21 amino acids from that of bovine heart (5) and in 18 amino acids from that of rat liver (7). Out of these differences 6 between man and beef and 2 between man and rat are non conservatives. The PiC from man is much less homologous with that from yeast (8) (33%), although the homology increases to 49% if only the hydrophobic regions are considered.

Transmembrane topology of the PiC. It is likely that all the mitochondrial metabolite carriers so far sequenced share a similar arrangement in the membrane, because their hydrophobic profiles are very similar. The precise transmembrane organization of the PiC as well as of any other mitochondrial carriers remains to be determined. In order to investigate the topology of the PiC in the mitochondrial membrane, in collaboration with G. Brandolin we have raised antibodies against synthetic peptides corresponding to the following regions of the bovine heart PiC : the N-terminal region (A1 - Y10), the C-terminal region (S304 - Q313), R140 - A155, and A195 - C208. In preliminary experiments the specificity of the generated antisera was checked by ELISA and Western blot. All four antibodies reacted specifically with the corresponding peptide. However, only two antisera, i.e. the anti N-terminal and the anti C-terminal antibodies, reacted with the PiC protein in mitochondrial lysates of bovine heart. We then examined whether the four peptide-specific antisera were able to react with the membrane-bound carrier. ELISA were performed with coated freeze-thawed mitochondria because in these particles the inner mitochondrial membrane is made accessible to antibodies by damage of the outer membrane (9). It has been found that the binding of the anti N-terminal antibodies as well as the binding of the anti C-terminal antibodies to freeze-thawed mitochondria increases with the concentration of the antibodies and with the amount of mitochondria (10). The identity of the N-terminal and the C-terminal regions of the membrane-bound carrier as the sites of reaction of the two antibodies with the mitochondria is supported by the observation that when the N-terminal and C-terminal peptides were added together with the respective antisera to the mitochondria, the reaction of the antibodies with the mitochondria was drastically reduced (10). Similar results were obtained with freshly prepared mitoplasts. In contrast to the anti N-terminal and the anti C-terminal antibodies, the anti R140 - A155 and the anti A195 - C208 antisera did not bind to mitochondria or mitoplasts at all. With submitochondrial particles (SMP), in which the membrane is thought to be at least predominantly inverted, the intensity of the reaction of the anti N-terminal and the anti C-terminal antibodies was much lower (about 40% of that observed with mitochondria). These results indicate that both the N-terminal and the C-terminal ends of the membrane-bound PiC are exposed at the cytoplasmic side of the inner mitochondrial membrane (10). A cytosolic location of the C-terminus of the PiC has also been proposed by Ferreira et al. (11).

For a further insight into the transmembrane arrangement of the PiC in the inner mitochondrial membrane, enzymatic digestion of the PiC in freeze-thawed mitochondria and SMP was combined with the immunological approach. Carboxypeptidase A markedly decreased the binding of anti C-terminal antibodies (but not of anti N-terminal antibodies) to the PiC in freeze-thawed mitochondria (10). Likewise, the reaction of the mitochondrial PiC with the anti C-terminal serum was decreased by the treatment with trypsin. The latter observation has also been made by Ferreira et al. (11). In addition to this, we have found that trypsin also decreased the binding of anti N-terminal antibodies to the PiC in freeze-thawed mitochondria. Also from these experiments the conclusion emerges that both termini of the PiC are located on the external side of the inner mitochondrial membrane, where they can be degrated by the action of specific enzymes. In further experiments the membrane-bound PiC was cleaved by other two proteases, Arg-endoprotease and Lys-endoprotease. Arg-endoprotease cleaved the PiC in SMP, but not in right-side-out particles.

Upon treatment of SMP with this enzyme, the anti N-terminal serum revealed the appearance of two peptides with a Mr about 14.5 K and the anti C-terminal serum identified a peptide with a Mr of about 17 K. These results indicate that Arg-endoprotease cleavage sites of the PiC are present only at the matrix side of the inner mitochondrial membrane at Arg-140 and/or Arg-152 (10). This means that the PiC is asymmetric and, if it is dimeric (or multimeric), an antiparallel orientation of the monomers can be excluded. Lys-endoprotease cleaved the PiC in freeze-thawed mitochondria yielding three peptides of Mr about 29 K, 20 K and 10 K which were immunodetected only by the anti N-terminal serum, and two peptides of Mr about 11 K and 24K which were immunodetected only by the anti C-terminal serum. On the basis of these results we conclude that three cleavage sites are available for Lys-endoprotease on the cytosolic side of the membrane corresponding to Lys 96, Lys 198 (or Lys 203) and Lys 288.

In summary, our results on the topology of the membrane-bound PiC provide evidence that the two terminal ends face the cytosol, the loop containing Arg 140 (or Arg 152) is exposed towards the matrix and the two loops containing Lys 96 and Lys 198 (or Lys 203) are exposed towards the cytosol.

cDNA of bovine heart OGC. Recently, in collaboration with M.J. Runswick and J.E. Walker, we have determined the primary structure of the OGC (12), which we had isolated from bovine heart mitochondria in 1985. First, several protein sequences were obtained from cyanogen bromide and elastase fragments of the OGC isolated from bovine heart. On the basis of these partial sequences, several oligonucleotides were synthesized and employed as hybridization probes in attempts to isolate clones encoding the OGC from a bovine heart cDNA library. All of these experiments, however, were unsuccessful, due to the low level of expression of the OGC. Therefore, we have adopted a new strategy based upon the polymerase chain reaction (PRC) (12). In a first set of PCRs, degenerated oligonucleotides corresponding to the N- and C-terminal ends of available protein sequences were used as forward and reverse primers to amplify specific segments of a cDNA made from bovine heart poly(A^+) mRNA. The products of the reactions were cloned and the clones of interest were recognized with hybridization probes based upon the central regions of the available protein sequences. By this procedure, short cDNAs encoding the central part of the available protein sequences were obtained. These central sequences were then used to synthesize specific primers for further PCR experiments. From the sequences of several overlapping cDNA clones the complete cDNA sequence was obtained (12). It is 1217 nucleotides in length and is terminated by a poly(A) tail separated by 22 nucleotides from the preceding sequence AATAAA, a typical signal for polyadenylation of RNA. This cDNA sequence encodes a protein of 314 amino acids including the initiator methionine (12), with a calculated Mr of 34 K which is in good agreement with the value estimated by gel electrophoresis for the OGC purified from bovine heart. The polypeptide sequence predicted from the DNA sequence agrees with the available amino acid sequences of the purified OGC. One of these sequences corresponds to residues 5-19 of the entire protein. Therefore, the OGC has no presequence as the AAC and the UCP, but in contrast to the PiC.

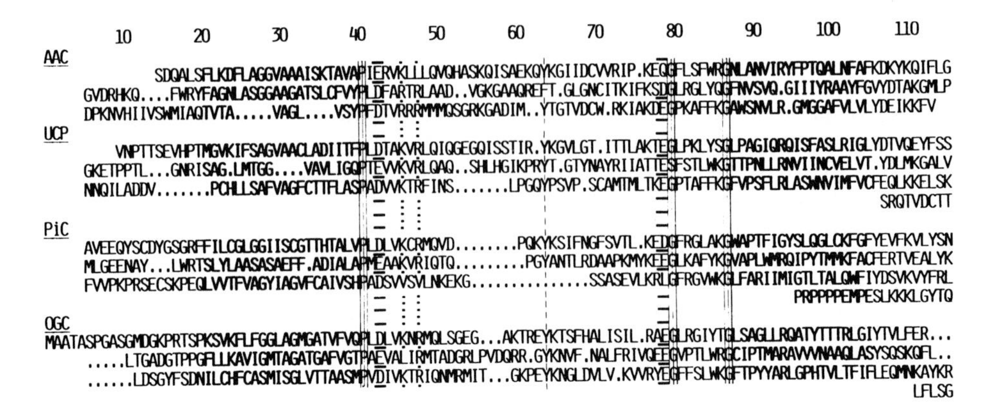

Fig. 1. Alignment of the amino acid sequences of four mitochondrial carrier proteins. The alignments are based upon the dot plots. The hydrophobic regions are shaded.

Structural relationship of the mitochondrial carrier protein family. Comparison of the protein sequence with itself reveals that the OGC has a tripartite structure made of related sequences about 100 amino acid long (12). Moreover, the repetitive elements of the OGC are related to those found in the AAC, the PiC and the UCP (Fig. 1), showing that the OGC is another member of the mitochondrial carrier protein family. The hydrophobic profiles (12) and the alignment of the 12 related sequences (Fig. 1) clearly indicate that these four proteins have a similar structure with six hydrophobic domains, probably folded in the membrane as α-helices, linked by extramembranous loops. The proposed extramembranous regions, which are hydrophilic, are among the most variable parts of the proteins, and presumably they are the sites of the substrate binding centers and the gates. In contrast, the junctions between the hydrophobic and the hydrophilic regions are the most highly conserved. A distinctive feature is the full conservation of prolin in position 41, of an acidic residue in position 43, and of glycine in position 87 throughout the 12 elements. Glycine in position 80 and an aromatic residue in position 85 are conserved in 11 out of 12 elements. In 10 out of 12 elements are conserved two basic residues in positions 46 and 48, a tyrosine residue in position 64, and an acidic residue in position 79. In order to check the proposed transmembrane arrangement of the OGC, we have raised antibodies against the N- and C-terminal ends of the bovine heart carrier protein. However, these antibodies did not react with the membrane-bound OGC both in bovine heart mitochondria and in SMP.

Structure of the human and bovine OGC genes. Since for the AAC multiple genes have been detected (13), we have investigated the possibility that more than one gene exists also for the OGC by using the bovine cDNA as a hybridization probe. By Southern blot analysis of digests of bovine and

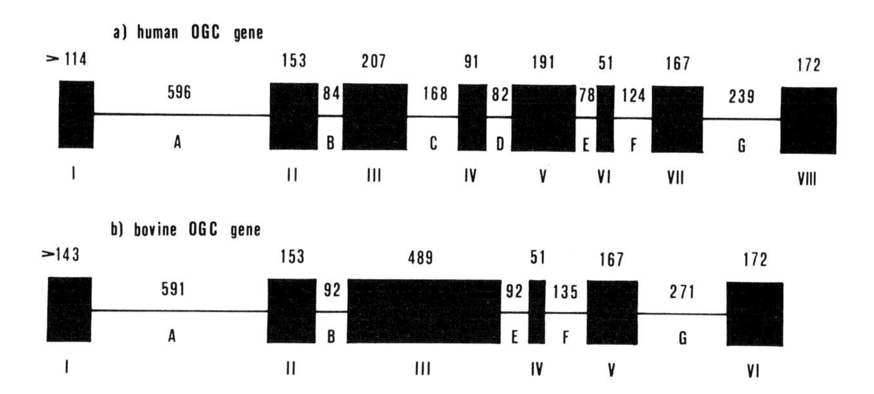

Fig. 2. Structure of the human and bovine OGC genes. In each gene, the exons and the introns are shown as black boxes and solid lines, respectively.

human DNA, single hybridizing bands were detected. These experiments, and also PCR's performed on bovine and human genomic DNA templates, suggest that these two genomes contain a single gene for the OGC (12). We have now sequenced the human and bovine OGC gene. As shown in Fig. 2, the human OGC gene contains 8 exons separated by 7 introns. The bovine gene has the same organization except that introns C and D are absent. The 5 introns in common interrupt the coding sequences at exactly the same positions in the two genes. When considering the location of the introns relative to the proposed transmembrane arrangement of the OGC, it appears that there is a tendency of all the introns to interrupt the coding sequence at the level, or near, the extramembranous loops. This finding has already been observed for the AAC and the UCP, although its significance, if any, remains to be established.

In Fig.3 the amino acid sequences of the human and the bovine OGC are compared. A partial sequence of the rat OGC is also shown for comparison. It is clear that the homology between the OGC from these three species is very high. The human OGC sequence differs only in 12 amino

```
Bovine  MAATASPGASGMDGKPRTSPKSVKFLFGGLAGMGATVFVQPLDLVKNRMQLSGEGAKTREYKTSFHALISILRAEGLRGIYTGLSAGLLRQATYTTTRLGIYTVL
Human   MAATASAGAGGMDGKPRTSPKSVKFLFGGLAGMGATVFVQPLDLVKNRMQLSGEGAKTREYKTSFHALTSILKAEGLRGIYTGLSAGLLRQATYTTTRLGIYTVL
Rat     ........................................................................................................

Bovine  FERLTGADGTPPGFLLKAVIGMTAGATGAFVGTPAEVALIRMTADGRLPVDQRRGYKNVFNALFRIVQEEGVPTLWRGCIPTMARAVVVNAAQLASYSQSKQFLL
Human   FERLTGADGTPPGFLLKAVIGMTAGATGAFVGTPAEVALIRMTADGRLPADQRRGYKNVFNALIRITREEGVLTLWRGCIPTMARAVVVNAAQLASYSQSKQFLL
Rat     ......................................................................................................LL

Bovine  DSGYFSDNILCHFCASMISGLVTTAASMPVDIVKTRIQNMRMIDGKPEYKNGLDVLVKVVRYEGFFSLWKGFTPYYARLGPHTVLTFIFLEQMNKAYKRLFLSG
Human   DSGYFSDNILCHFCASMISGLVTTAASMPVDIAKTRIQNMRMIDGKPEYKNGLDVLFKVVRYEGFFSLWKGFTPYYARLGPHTVLTFIFLEQMNKAYKRLFLSG
Rat     DSGYFSDNFLCHFCASMISGLVTTAASMPVYIVKTRIQNMRMIDGKPEYKNGLDVL[illegible]KVVRYEGFFSLWKGFTPYYARLGPHTVLTFIFLEQMNKAYKRLFLSG
```

Fig. 3. Amino acid sequences of the OGC from man, beef and rat.

acids from that of beef. The corresponding human cDNA differs in 67 nucleotides out of 942 from that of beef.
Dimeric structure of the OGC. During the isolation of the OGC it was frequently observed a co-purification of a protein with a Mr twice that of the OGC polypeptide. Since one possibility was that the OGC was cross-linked by oxidation of SH groups, we have investigated this phenomenon by using reagents which catalyze the formation of disulfide bridges. The isolated OGC was incubated with 0.5 mM Cu^{2+}-phenanthroline and 30% acetone or propanol for 30 min at 0°C. After denaturation of the samples at 37°C for 30 min, the degree of cross-linking was determined by SDS-gel electrophoresis. Upon treatment with Cu^{2+}-phenanthroline and the above mentioned solvents there was a clear-cut increase in the intensity of a protein band with an apparent Mr of 66 K, which was immunodecorated by antibodies raised against the isolated OGC. The solvents were probably required to cause a small perturbation of the structure and/or a local dehydration of the carrier which permit juxtapositioning of the SH groups involved in the cross-linking. Several observations give evidence for a S-S bridge-mediated cross-linking. First, the cross-linking was prevented by the SH reagents mersalyl and eosin-5-maleimide added before Cu^{2+}-phenanthroline; second, the cross-linking was not reversed by the same SH reagents added after Cu^{2+}-phenanthroline; and third, the S-S reducing reagent dithioerythrol completely reconverted the cross-linked product into the monomeric OGC. The cross-linking may depend on the dimeric state of the isolated OGC or on the collision of separate OGC molecules. Since the extent of cross-linking did not change on increasing the concentration of the OGC 7 fold, we conclude that disulfide bridge(s) are formed between the two subunits of the preexisting OGC dimer. There are only 3 cysteines in the OGC polypeptide chain: Cys 184, Cys 221 and Cys 224. In order to localize the cysteines involved in the disulfide-bridge, cross-linked OGC was cleaved at pH 9 after cyanylation of the free cysteines by 2-nitro-5-thiocyanobenzoate. On cleavage of the cross-linked OGC, a fragment with a Mr slighly lower than 30 K appeared. On the other hand, a peptide of 18 K Mr was present in both cross-linked and control samples. These results give evidence, although not conclusive, that two S-S bridges are formed at the level of the Cys 221 and 224 of the two monomers.
Reaction mechanism of mitochondrial carriers. In mitochondria a very complex kinetic behaviour was found in the case of the OGC (14). Therefore, in collaboration with T. Dierks and R. Krämer, we have investigated the transport mechanism of the isolated OGC (15). The OGC, which catalyzes a 1:1 exchange of oxoglutarate against malate, has been functionally incorporated into liposomes. Since the external affinities for oxoglutarate and malate both were different from the internal affinities, the oxoglutarate carrier had been inserted unidirectionally into the liposomal membrane. On the other hand, the exchange reaction itself was perfectly symmetrical. Thus both the Km values and the maximum exchange rates for oxoglutarate and malate were independent of the nature of the counter-substrate present at the other side of the membrane. Under these defined conditions we have analyzed the antiport mechanism in two-reactant initial velocity studies varying both the internal and the external substrate concentrations within

the same experiment. When the kinetic data were plotted according to Lineweaver-Burk plots, a pattern of straight lines was obtained intersecting at a common point on the abscissa. This intersecting pattern, opposed by the parallel pattern of ping-pong reactions, is indicative of a sequential type of mechanism, which is observed whenever the two substrate molecules bind before catalysis occurs. Thus, the oxoglutarate carrier had to form a ternary complex with the two counter-substrates prior to translocation. The intersection point on the abscissa indicates that the affinity of the carrier for one substrate bound to one side of the membrane was not influenced by the counter-substrate bound to the other side. Only the transport rate was stimulated with increasing concentration of the counter-substrate, as was apparent from the different ordinate intercepts and consequently from the different slopes. This mutual dependence could be quantitatively analyzed in replots of the slopes and the ordinate intercepts of the primary curves vs. the reciprocal concentration of the so-called "non-varied" substrate. The concentration-independent K_m values, obtained by extrapolation from secondary plots, agreed well with the apparent K_m values at finite substrate concentration. Furthermore, the K_i values, representing substrate interaction with the unloaded carrier, more or less coincided with the corresponding K_m values. Thus, the affinity was almost identical whether binding occurs to the free or to the single-substrate occupied carrier. From these kinetic results we can conclude first, that we have one binding site inside and one binding site outside and second, that the OGC functions according to a sequential type of mechanism including a ternary complex between the carrier and one internal and one external substrate molecule. This catalytic complex is formed after rapid and independent binding of the two counter-substrates, which then triggers a concerted transport into opposite directions.

It has been shown by Dierks et al. (16) that the reconstituted AGC functions according to a sequential type of mechanism and we have evidence that also in the case of the purified CIC (17) and of the purified DIC (18) a sequential mechanism is operating. Thus, four mitochondrial carriers have been proven to function according to this type of mechanism. We would like therefore to propose that the close relationship of the mitochondrial "carrier family" is expressed not only in the similarity of the primary structure and conformation but also in a common type of kinetic mechanism underlying antiport function.

ACKNOWLEDGEMENTS

This work was supported by the C.N.R. Target Project "Ingegneria genetica".

REFERENCES

1. Krämer, R. and Palmieri, F. (1989) Biochim. Biophys. Acta 974 : 1-23

2. Bisaccia, F., De Palma, A. and Palmieri, F. (1989) Biochim. Biophys. Acta 977 : 171-176

3. Indiveri, C., Tonazzi, A. and Palmieri, F. (1990) Biochim. Biophys. Acta 1020 : 81-86

4. Aquila, H., Link, T.A. and Klingenberg, M. (1987) FEBS Lett. 212 : 1-9

5. Runswick, M.J., Powell, S.J., Nyren, P. and Walker, J.E. (1987) EMBO J. 6 : 1367-1373

6. Dolce, V., Fiermonte, G., Messina, A. and Palmieri, F. (1991) DNA Sequence - J. DNA Sequencing and Mapping, 2 : 131-134

7. Ferreira, G.C., Pratt, R.D. and Pedersen, P.L. (1989) J. Biol. Chem. 264 : 15628-15633

8. Phelps, A., Schobert, C.T. and Wohlrab, H (1991) Biochemistry 30 : 248-252

9. Brandolin, G., Boulay, F., Dalbon, P. and Vignais, P.V. (1989) Biochemistry 28 : 1093-1100

10. Capobianco, L., Brandolin, G. and Palmieri, F. (1991) Biochemistry 30 : 4963-4969

11. Ferreira, G.C., Pratt, R.D. and Pedersen, P.L. (1990) J. Biol. Chem. 265 : 21202-21206

12. Runswick, M.J., Walker, J.E., Bisaccia, F., Iacobazzi, V. and Palmieri, F. (1990) Biochemistry 29 : 11033-11040

13. Cozens, A.L., Runswick, M.J. and Walker, J.E. (1989) J. Mol. Biol. 206 : 261-280

14. Sluse, F.E., Duyckaerts, C., Liébecq, C. and Sluse-Goffart, C.M. (1979) Eur. J. Biochem. 100 : 3-17

15. Indiveri, C., Dierks, T., Krämer, R. and Palmieri, F. (1991) Eur. J. Biochem. 198 : 339-347

16. Dierks, T., Riemer, E. and Krämer, R. (1988) Biochim. Biophys. Acta 943 : 231-244

17. Bisaccia, F., De Palma, A., Dierks, T., Krämer, R. and Palmieri, F., in preparation

18. Indiveri, C., Marinò, F., Dierks, T., Krämer, R. and Palmieri, F., in preparation

© 1992 Elsevier Science Publishers B.V. All rights reserved.
Molecular mechanisms of transport. E. Quagliariello, F. Palmieri, eds.

Recombinant expression of the mitochondrial uncoupling protein (UCP)

B. Miroux, L. Casteilla, S. Raimbault, S. Klaus, Bouillaud F. and D. Ricquier

Centre de Recherche sur l'Endocrinologie Moléculaire et le Développement, Centre National de la Recherche Scientifique, 9 rue Jules Hetzel, 92190 Meudon, France

INTRODUCTION

In most cells mitochondria function as ATP synthesizer. A unique situation is presented by the brown adipocytes which are specialized mammalian adipocytes involved in heat production and regulation of body temperature. In such cells, the Uncoupling Protein (UCP), located in the inner mitochondrial membrane, short-circuits the tight coupling between respiration and ADP phosphorylation. Its activation results in elevated respiration rate and energy dissipation as heat (1). In fact the UCP is a proton and anion translocator of which the activity is strongly inhibited by purine nucleotides whereas free fatty acids activate it (2, 3). The activity of the purified UCP has been reconstituted in liposomes by several groups (4-6). The primary structure of the protein was determined from purified hamster UCP (7) and cDNAs or genes cloned in rat, mouse, calf and man (8). UCP was the second mitochondrial carrier to be sequenced after the ADP/ATP carrier (7). A significant homology between these 2 mitochondrial transporters was noticed (1, 7, 8). Later this homology was extended to the mitochondrial Phosphate carrier (9) and the Oxoglutarate carrier (10). Clearly, these different mitochondrial transporters belong to the same family, among which UCP is a recently evolved and a simple ion carrier (Fig. 1). It is assumed that any progress in the functional organization of one given mitochondrial transporter will be relevant to the other members of the family. Expression systems are necessary in order to address questions about the structure-function relationships of mitochondrial carriers using site-directed mutagenesis. The unique expression of UCP in brown adipocytes make it a good candidate for such a program.

METHODS

Several methods were used in order to obtain the heterologous expression of UCP. We first expressed UCP in *Xenopus* oocytes (11). This was achieved via the injection of UCP mRNA (transcribed from a pTZ plasmid in which the complete UCP cDNA had been inserted) into the vegetative pole of mature oocytes. UCP was synthesized and correctly transported into mitochondria of oocytes. However the non-possibility to demonstrate its ability to bind GDP and uncouple the respiration led us to consider other expression systems : mammalian cells in culture, *Escherichia coli*, yeasts. We describe here the expression of partial or entire UCP in mammalian cells and bacteria. An expression vector pECE-UCP was constructed and used to transfect CHO and COS cells (12). The expression of a fragment of UCP in bacteria was achieved through the construction of DNA encoding bacterial malE protein fused to a piece of UCP cDNA.

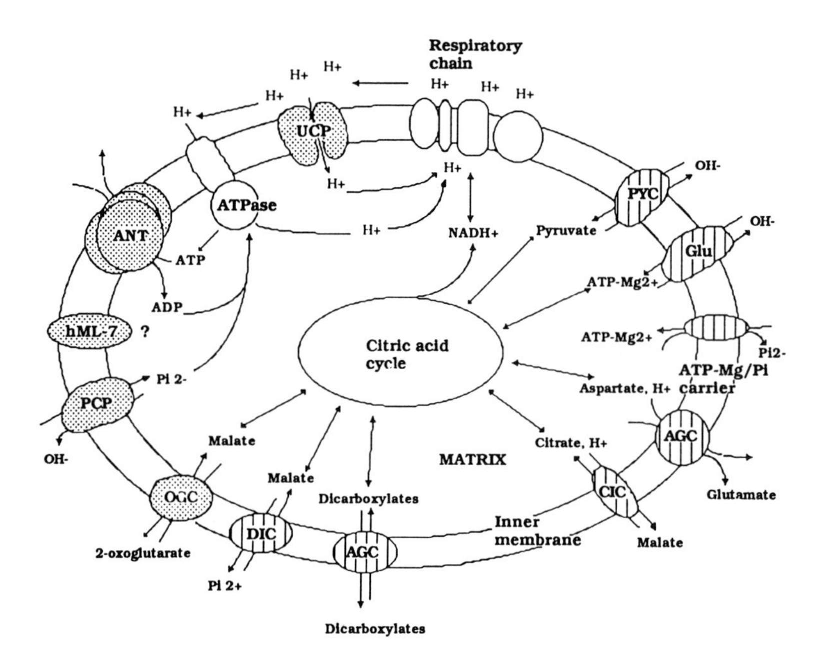

Figure 1. The transport systems of the inner membrane of mitochondria. The dotted areas correspond to carriers of which the primary structure is known.

RESULTS

Expression of UCP in CHO and COS cells

The expression of rat UCP cDNA was tested in transiently transfected CHO and COS cells. In both types of cells UCP was detected into mitochondria using antibodies. Permanent expression of UCP was achieved in stable transformed CHO cell lines co-transfected with pSV2neo plasmid (12). The protein expressed in the recombinant CHO-B5 cell line displayed the functional characteristics of brown fat UCP : GDP-sensitive uncoupling of respiration of isolated mitochondria, lowering of the mitochondrial membrane potential.

As it is shown in Fig. 2, the B5 cells were more resistant to acidic pH. Such a behaviour could be related to diminished mitochondrial ATP synthesis, increased glycolysis and activation of lactate production. The B5 cells expressing UCP were also shown to be able to grow in minimum medium containing neither proline nor glutamine whereas in these conditions control CHO cells do not grow. Such data suggest that expression of UCP led to phenotypic alterations in CHO cells. However such putative effects of UCP should be checked is several other CHO cell lines expressing this protein.

EFFECT OF MEDIUM pH ON
THE SURVIVAL OF CHO CELLS

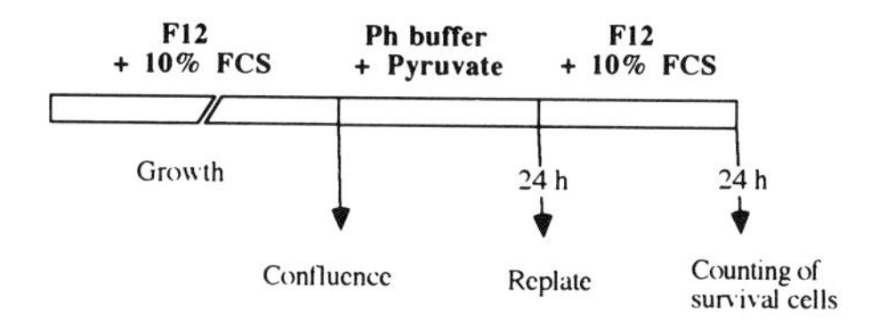

% F12	ph 6,5	ph 8,0
K1	14	47
K1 UCP-	26	62
9B5 UCP+	54	27

Figure 2. Effect of medium pH on growth of CHO cells. K1 : wild CHO cells ; K1-UCP- : transformed cells, neomycin-resistant, non-expressing UCP ; 9 B5 UCP+ : recombinant cells expressing UCP.

Expression of mutants of UCP in mammalian cells

The homology between UCP and the Adenine Nucleotide Translocator is particularly high at the C-terminal end (13) corresponding to the putative terminal alpha-helix (7). Interestingly, photolabeling of both proteins with 2-Azido-alpha-^{32}P-adenosine Diphosphate demonstrated that this domain was involved in nucleotide binding (14, 15). To address the question of the nucleotide binding site in UCP, we constructed several mutants in which aminoacids presumably located in the nucleotide binding domain were deleted. Transient expression revealed that these mutated forms of UCP were synthesized and incorporated into mitochondria of CHO and COS cells. However we were unable to clone recombinant cell lines expressing altered UCP. One explanation could be that the expression of UCP no more inhibitable by nucleotides, induced the death of cells.

Expression of a fragment of UCP in E. coli

In previous experiments the expression of UCP in E.coli revealed to be unsuitable since most UCP molecules were embedded in inclusion bodies (Bouillaud and Raimbault, unpublished data). In order to circumvent such a difficulty, another strategy was developed based on the expression of hybrid protein. Hybrids between the maltose-binding protein (MalE) of *Escherichia coli* and central or C-terminal domains of rat UCP were constructed at the DNA level (Fig. 3). Then the fusion proteins were expressed in *E. coli*. The opportunity to purify MalE by affinity chromatography was taken to purify hybrid proteins. All the procedures used were according to Bedouelle and Duplay (16). Following their purification, the fusion proteins were used to screen polyclonal antibodies against UCP and select antibodies that specifically react with the corresponding regions of UCP. These antibodies are presently used to study the transmembrane topology of the transporter.

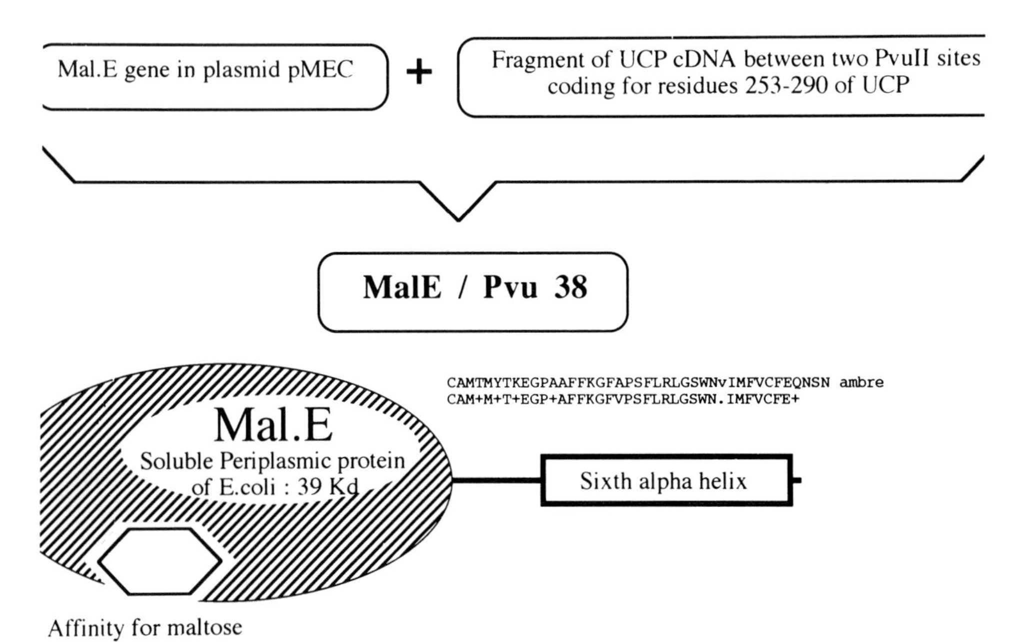

Figure 3. Expression of a fragment of UCP as a hybrid protein. The gene encoding bacterial MalE protein was linked to the DNA corresponding to the putative C-terminal helical domain of UCP.

PERSPECTIVES

It may be anticipated that site-directed mutagenesis of UCP expressed in CHO cells will contribute to the identification of aminoacids involved in nucleotide binding. However it is obvious that the reconstitution of the activity of mutated UCP will be required. Such a requirement implies the development of high level expression systems such as expression in yeasts. Interestingly 2 groups recently reported the expression of rat UCP in *Saccharomyces cerevisiae* (17, 18). In other respects ,the selection of antibodies made with fusion proteins seems to be fruitful to study the folding of UCP in the membrane. Recently several groups have generated antibodies against synthetic peptides corresponding to short domains of the ADP/ATP Carrier (19) and the mitochondrial Phosphate Carrier (20, 21). The hypothesis of a similar membranous folding for these 3 mitochondrial transporters will be confirmed or rejected soon.

REFERENCES

1 Klaus S, Casteilla L, Bouillaud F, Ricquier D. Int J Biochem 1991; 23: 791-801.
2 Nicholls DG, Heaton R. Physiol Rev 1984; 64: 1-64
3 Locke RM, Rial E, Scott I, Nicholls DG. Eur J Biochem 1982; 129: 373-380.
4 Klingenberg M, Winkler E. EMBO J 1985; 4: 3087-3092.
5 Strielman PJ, Schalinske KL, Shrago E. J Biol Chem 1985; 260: 13402-13405.
6 Jezek P, Orosz DE, Garlid KD. J Biol Chem 1990; 265: 19296-19302.
7 Klingenberg M. TIBS 1990; 15: 108-112.
8 Ricquier D, Casteilla L, Bouillaud F. FASEB J 1991; 5: 2237-2242.
9 Runswick MJ, Powell SJ, Nyren P, Walker JE. EMBO J 1987; 6: 1367-1379.
10 Runswick MJ, Walker JE, Bisassia F, Iacobazzi V, Palmieri F. Biochemistry 1990; 29: 11033-11040.
11 Klaus S, Casteilla L, Bouillaud F, Raimbault S, Ricquier D. Biochem Biophys Res Commun 1990; 167: 784-789.
12 Casteilla L, Blondel O, Klaus S, Raimbault S, et al. Proc Natl Acad Sci USA 1990; 87: 5124-5128.
13 Bouillaud F, Weissenbach J, Ricquier D. J Biol Chem 1986; 261: 1487-1490.
14 Dalbon P, Brandolin G, Boulay F, Hoppe J, Vignais PV Biochemistry 1988; 27: 5141-5149.
15 Mayinger P, Klingenberg M. EBEC Reports 1990; 6 : 64.
16 Bedouelle H, Duplay P. Eur J Biochem 1988; 171: 541-549.
17 Prieto S, Bouillaud F, Raimbault S, Rial E, et al. EBEC Reports 1990; 6: 32.

18 Murdza-Ingliss DL, Patel HV, Freeman KB, Jezek P, et al. J Biol Chem 1991; 260: 11871-11875.
19 Brandolin G, Boulay F, Dalbon P,Vignais P. Biochemistry 1989; 28: 1093-1100.
20 Ferreira GC, Pratt RD, Pedersen PL. J Biol Chem 1990; 265: 21202-21206.
21 Capobianco L, Brandolin G, Palmieri F. Biochemistry 1991; 30: 4963-4969.

© 1992 Elsevier Science Publishers B.V. All rights reserved.
Molecular mechanisms of transport. E. Quagliariello, F. Palmieri, eds.

TOPOLOGY OF THE MITOCHONDRIAL PORIN.

Vito De Pinto and Ferdinando Palmieri

Department of Pharmaco-Biology, Laboratory of Biochemistry and Molecular Biology, University of Bari and CNR Unit for the Study of Mitochondria, Trav. 200 v. Re David, 4, I-70125 Bari (ITALY)

INTRODUCTION

Porin or voltage dependent anion selective channel (VDAC) is the protein responsible for the high permeability of the outer mitochondrial membrane. Mitochondrial porins are polypeptides of Mr ranging between 30 and 37 kDa. They show low anion selectivity, as demonstrated in lipid bilayer reconstitution studies, and molecules as large as 6-8 kDa have been shown to pass through the pore at applied voltages of 5-10 mV (1-4). The pore has been shown to be voltage-dependent, with the extent of this dependence varying among different tissues (3). In yeast deletion of the gene encoding this polypeptide leads, after some adaptation, to the overexpression of other pore components with different functional features (5). Furthermore it has been shown that the hexokinase-binding protein present in the outer mitochondrial membrane is identical with the mitochondrial porin (6-7). Recent papers have provided valuable information about the structure of porin. The primary structures of porins from *Saccharomyces cerevisiae* (8), *Neurospora crassa* (9) and human B-lymphocytes (10) have been determined. A great deal of interest is nowaday focussed in the structural arrangement of the polypeptide chain(s) of porin and its/their role in the properties of the channel. In this paper we show a model of the transmembrane topology of the mitochondrial porin based on experiments with proteolytic enzymes and an antiserum against the N-terminal region of porin. This model is compared to other mitochondrial porin models and to bacterial porin structures determined by crystal analysis.

METHODS

Immunological techniques. The antisera employed were raised in rabbit against the purified bovine heart porin and against the first 19 residues of porin purified from human B-lymphocytes with the amino terminal acetylated, as found in the purified protein (10). Western blot experiments were performed as described in (3). The ability of antisera to react with membrane-bound porin was tested by ELISA (enzyme-linked immunosorbent assay) using microtitration polyvinylchloride plates (Titertek). Freshly prepared intact or broken bovine heart mitochondria, diluted at the appropriate concentrations, were coated on plate wells overnight at 4°C under gentle shaking (100 µl per microtiter plate's well). After coating, the plates were incubated with 1% BSA in order to saturate the unspecific sites. 100 µl of primary antisera were used and the incubations lasted 2 h at room temperature. A peroxidase-conjugated anti-rabbit Ig, diluted 2000-fold, was employed as secondary antibody with 0.4 mg/ml o-phenylenediamine and 0.03% H_2O_2 as substrates. The developed absorbance was determined at 450 nm.

Proteolytic digestions Intact and broken mitochondria were used for proteolytic digestion of membrane-bound porin. 1 mg samples were incubated with the proteolytic enzymes for 30 minutes and at a temperature of 37°C. At the end of the incubation, the reactions were stopped and aliquots of 100 µg mitochondrial proteins were loaded on slab SDS-PAGE and analyzed by immunoblotting.

Structure prediction methods. The analysis of amphipathic secondary structures was performed on the basis of the Kyte and Doolittle work (11). The sided hydropathy values for sided α-helices and β-sheets and the corresponding averaged hydropathy values were obtained by summing up the hydropathy indices of the respective residues, using corrected weights (11).

RESULTS

The properties of many different purified mitochondrial porins have been compared and are summarized in Table 1. In all mitochondria studied so far purified porins showed on SDS-PAGE a single polypeptide of Mr ranging between 29 and 37 kDa in different species (3, 4, 14-16). These polypeptides have been identified as porin on the basis of their functional properties, of their peptide maps and of their immunological cross-reactivity with antisera raised against other porins (3). When reconstituted in black lipid membranes the porins tested showed a deduced diameter of 1.7 nm with the only exceptions of the *Paramecium tetraaurea* porin (12) and of the socalled emergency porin shown in yeast (5).

The availability of an antiserum directed against the human porin N-terminal end allowed us to investigate this interesting domain of the protein. In previous papers the N-terminal region of porin was suggested to form an amphipathic α-helix (9). Although the primary sequence is not very similar, this putative amphipathic α-helix seems preserved in organisms as distantly related as human and *S.cerevisiae* or *N.crassa* (10). Interestingly, fish, echinoderma and insect porins, for which sequence data are not available, were detected by this antiserum indicating structural similarities in this region (Table 1). Two hypothesis presently exist about the role of the N-terminal sequence. Since porin has no cleavable pre-sequence, the N-terminal α-helix could be responsible for the import pathway of the protein (9). Alternatively, the N-terminal α-helix could be involved in the structural changes that porin undergoes during its partial closure (13).

Table 1
A COMPARISON OF EUKARYOTIC PORINS

Organism	Channel diameter[a]	Subunit M.W.[b]	Cross-reactivity with antisera against				Ref.
			BHP[c]	N-term	yeast	*N.crassa*	
Neurospora crassa	1.7	31000	-	-	+	+++	14
S. cerevisiae (2nd porin)	1.2	20000?					5
Paramecium tetraaurea	1.3	37000	-	-	+	++	12
Drosophila melanogaster	1.7	31000	-	+	++	-	15
Paracentrotus lividus	1.7	34000	++	++	-	N.D.	d
Anguilla anguilla	1.7,1.2	32000	++	++	-	-	16
Rattus norvegicus	1.7	35550	+++	+++	-	-	3,4

[a]Determined in 1 M KCl.[b]Determined by SDS-PAGE. [c]Abbreviations used: BHP: bovine heart porin; N-term: antiserum to the 19 N-terminal amino acids of human porin; yeast: *S. cerevisiae.* [d]De Pinto et al., unpublished results. The sign - indicates absence of cross-reactivity. The sign + indicates presence of cross-reactivity and the number of + indicates the intensity of the immunological reaction.

The antisera against the whole protein and against the N-terminus region of porin and specific proteases were used in this paper to study the accessibility of hydrophilic domains of membrane-bound porin.

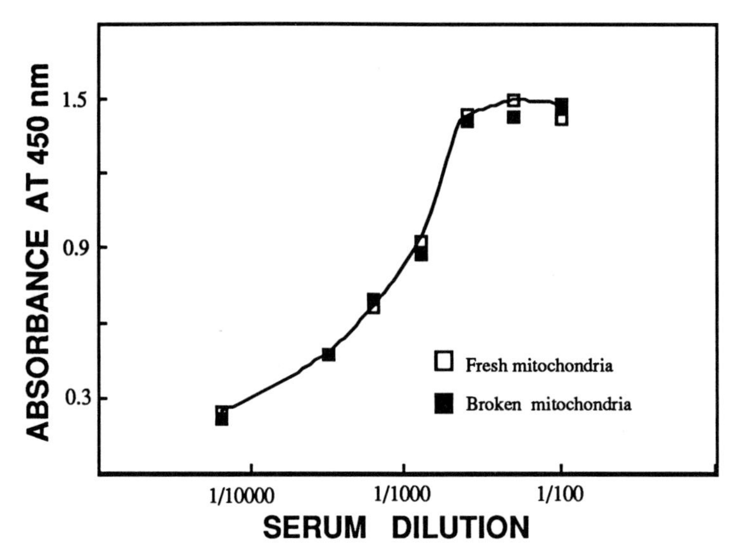

Figure 1. Reactivity of the anti-N-terminal antiserum toward the membrane-bound porin in intact and broken mitochondria, assayed by ELISA.

The binding of these antibodies to mitochondria increased with the amount of mitochondria to a plateau value. This means that some water-exposed epitopes were recognized by the antisera in the physiological transmembrane conformation of porin.

The immunotitrations obtained with anti-N-terminal antibodies using intact and broken mitochondria are presented in Fig. 1. In these titrations, osmotic shock and freeze/thawing were used to break the mitochondrial membranes, making the inner as well as the outer side of the outer membrane accessible to antibodies. Both types of particles yielded the same results in ELISA assays (Fig. 1), suggesting that the N-terminal region of porin is located on the external and not on the internal side of the outer mitochondrial membrane.

In order to identify other extramembranous segments of porin and their orientation, the access of different proteases to the peptide chain of the membrane-bound protein was investigated using intact and broken bovine heart mitochondria. The fragments generated upon proteolysis were detected by immunoblotting using the antisera raised against the bovin heart porin and the N-terminus of the human porin.

In experiments with carboxypeptidase A it was checked the possibility that the C-terminus is water-exposed. The treatment with carboxypeptidase A of intact and broken mitochondria did not increase either the electrophoretic migration of porin or the intensity of the immunodetected band (17), indicating the lack of porin cleavage. It is likely, therefore, that the C-terminus is not exposed to the water-phase since it is not accessible to the enzyme.

In other experiments intact and broken mitochondria were subjected to endoproteolysis, looking for cleavage sites located in water-exposed domains of porin. Only a few of the proteases assayed were able to cleave porin, and the cleavage occurs in a very specific way, usually at only one site. Based on the number of the fragments generated upon enzymatic digestion of porin and their molecular weight, we propose that there is one cleavage site for trypsin, located between residues 108 and 119 (where 5 lysines and 1 arginine are present), two cleavage sites for chymotrypsin at positions 117 and 172, and one for protease V8 at D_{227} or D_{229} (17). It should be stressed that porin was completely cleaved by protease V8, trypsin and chymotrypsin in intact mitochondria. This suggests that all the porin molecules have the same orientation in the outer mitochondrial membrane, since in the case of antiparallel orientation only part of the molecules would have been cleaved.

On the basis of the experiments described above and thanks to a refined computer-based analysis of the mammalian porin sequence we have drawn the model of the transmembrane arrangement of porin shown in Fig. 2A (17). Most of the protein is embedded in the phospholipid bilayer. The amphipathic N-terminal α-helix is accessible to the water-phase. Two major loops protrude into the external water-phase. All large hydrophilic domains and all the cleavage sites are exposed to the outside of the mitochondrial outer membrane. The most important antigenic epitopes should be localized in these regions, since the C-terminal part of the protein seems poorly detectable by polyclonal antisera. All these results are compatible with a model where most if not all the extramembranous domains of porin are exposed to the outside and all the porin molecules have the same orientation.

DISCUSSION

We have drawn a suggestive model of the transmembrane arrangement of the mitochondrial porin, the protein responsible of the permeability of the mitochondrial outer membrane. Our topological model is based on the above reported experimental results. Furthermore the computer analysis predicted the presence of 16 amphipathic β-strands which can cross the membrane.

In previous papers 12 (18), 15 (9) or 19 (19) β-strands were predicted from the analysis of the yeast porin sequence. In particular,

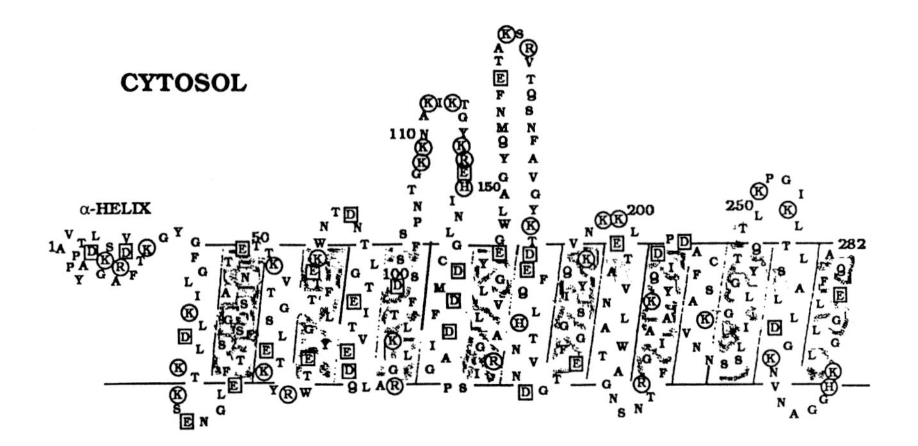

INTERMEMBRANE SPACE

Figure 2 A. Model of transmembrane arrangement of the polypeptide chain of human porin (see also ref. 17). 16 amphipathic β-strands cross the membrane. Positively charged amino acids are indicated by circles, negatively charged by squares.

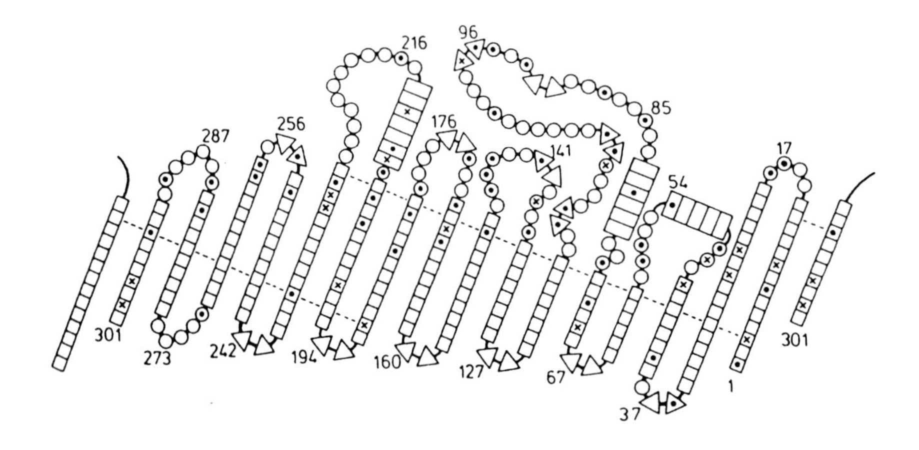

Figure 2 B. Topology of porin from *Rhodobacter capsulatus* determined from crystals at 1.8 Å resolution. The transmembrane arrangement of the 16-stranded β-sheet barrel is viewed from the barrell outside. Appropriate marks indicate the secondary structure assignments (quadrangle= β; rectangle= α; triangle= hydrogen bond reverse turn). Asp and Glu residues are marked by dots (●) and His, Lys and Arg by crosses (×). Reproduced with permission from (20).

Blachly-Dyson et al. (18) after site-directed mutagenesis experiments proposed a model of the transmembrane arrangement of *S. cerevisiae* porin in which 12 β-strands and the amino-terminal α-helix cross the membrane. This 12-strand model omits large part of the sequence from the membrane since mutations in that range did not alter the channel behaviour of the corresponding mutant-proteins. Comparison with our model showed that important residues are also within the membrane in human porin, but most of them are not conserved in the human porin although the channel properties seem to be identical.

It is interesting to compare our model with the crystal structure of porin from *Rhodobacter capsulatus* which was determined at 1.8 Å resolution (20) (Fig. 2 B). Weiss et al. also found that 16 β-strands cross the membrane to form a "β-barrel" structure, which is large enough to form a channel. Two major loops are water-exposed protruding from the same side of the membrane. There is no trace, however, of an N-terminal amphipathic α-helix. The distribution of charged residues in *R. capsulatus* porin is remarkable. In one half of the structure, divided at the barrel equator, there are as many as 44 negative and 12 positive charges, whereas in the other half there are only 7 positive and 7 negative charges. This electrostatic asymmetry should be relevant for the functional properties of the pore (20). Also in our model an asymmetric distribution of charged residues can be observed. 18 positively and 11-12 negatively charged residues (including those of the amphipathic N-terminal α-helix) are exposed to the external water-phase, while only 6-8 positive and 3 negative are exposed to the internal water-phase.

While the basic orientation of mitochondrial porin in the membrane seems to be quite clear, it is not yet clear how many polypeptides are involved in the formation of a single channel-unit. Recent results from electron microscopy of frozen-hydrated *N. crassa* outer mitochondrial membrane crystals favour a monomer channel. Mannella et al. (21) showed that a β-barrel consisting of a 3.8 nm-diameter alpha-carbon cylinder, with a 0.5 nm-thick "shell" of amino acid residues on either side, would be consistent with most of the functional features of porin (1-4). The inner diameter of such a channel would be 2.8 nm, a value which is in agreement with a "β-barrel" type structure formed by 15-19 transmembrane β-segments.

ACKNOWLEDGMENTS

The authors thank F. Thinnes (Göttingen, Germany) for supplying the antiserum against the porin N-terminus and T. Link (Frankfurt/Main) for helping them in the computer-based analysis. This work was supported by the C.N.R. Target Project "Biotechnology and Bioinstrumentation" and by the Ministero dell'Università e della Ricerca Scientifica e Tecnologica (MURST).

REFERENCES

1. Colombini, M. (1979) Nature 279, 643-645
2. Benz, R. (1985) Crit. Rev. Biochem. 19, 145-190
3. De Pinto, V., Ludwig, O., Krause, J. Benz, R. and Palmieri, F. (1987) Biochim. Biophys. Acta 894, 109-119
4. Colombini, M. (1989) J. Membrane Biol. 111,103-111
5. Dihanich, M., Schmid, A. and Benz, R. (1989) Eur. J. Biochem. 181, 703-710
6. Linden, M., Gelefors, P. and Nelson, B. (1982) FEBS Lett. 141, 189-192
7. Fiek, C., Benz, R., Roos, N. and Brdiczka, D. (1982) Biochim. Biophys. Acta 688, 429-437
8. Mihara, K. and Sato, R. (1985) EMBO J. 4, 769-774
9. Kleene, R., Pfanner, N., Pfaller, R., Link, T.A., Sebald, W. Neupert, W. and Tropschung, M. (1987) EMBO J. 6, 2627-2633
10. Kayser, H., Kratzin, H.D., Thinnes, F.P., Götz, H., Schmidt, W.E., Eckart, K. and Hilschmann, N. (1989) Hoppe-Seyler 370, 1265-1278
11. Link, T.A., Schägger, H. and von Jagow, G. (1987) in: Cytochrome systems: Molecular Biology and Bioenergetics (Papa, S. et al eds.) pp. 289-300 Plenum press, New York
12. Ludwig, O., Benz, R. and Schultz, J.E. (1989) Biochim. Biophys. Acta 978, 319-327
13. Mannella, C. (1990) Experientia, 46, 137-145
14. Freitag, H., Neupert, W. and Benz, R. (1982) Eur. J. Biochem. 123, 629-639
15. De Pinto, V., Benz, R., Caggese, C. and Palmieri, F. (1989) Biochim. Biophys. Acta 987, 1-7
16. De Pinto, V., Zara, V., Benz, R., Gnoni, V. and Palmieri, F. (1991) Biochim. Biophys. Acta 1061, 279-286
17. De Pinto, V., Prezioso, G., Thinnes, F., Link, T.A. and Palmieri, F. (1991) Biochemistry, in press
18. Blachly-Dyson, E., Peng, S., Colombini, M. and Forte, M. (1990) Science, 247, 1233-1236
19. Forte, M., Guy, H.R. and Mannella, C. (1987) J. Bioenerg. Biomembr. 19, 341-350
20. Weiss, M.S., Kreusch, A., Schiltz, E., Nestel, U., Welte, W. Weckesser, J. and Schulz, G.E. (1991) FEBS Lett. 280, 379-382
21. Mannella, C., Guo, X. and Cognon, B. (1989) FEBS Lett. 253, 231-234

MUTAGENESIS OF TRANSPORTERS

© 1992 Elsevier Science Publishers B.V. All rights reserved.
Molecular mechanisms of transport. E. Quagliariello, F. Palmieri, eds.

Recent Studies on the Lactose Permease of *Escherichia coli*

E. Bibi and H.R. Kaback

Howard Hughes Medical Institute, Department of Physiology, Molecular Biology Institute, University of California Los Angeles, Los Angeles, CA 90024-1574

The lactose (lac) permease of *Escherichia coli* is a polytopic cytoplasmic membrane protein that catalyzes the coupled transport of β-galactosides and H^+ with a stoichiometry of 1 (i.e., β-galactoside/H^+ symport or cotransport; cf. ref's 1,2 for reviews). The permease has been solubilized from the membrane, purified to homogeneity and reconstituted into phospholipid vesicles (3,4) and is functional as a monomer (5). The *lacY* gene has been cloned and sequenced, and the amino-acid sequence of the permease has been deduced from the DNA sequence (6). Based on circular dichroism and hydropathy analysis of the primary sequence, a secondary-structure model was proposed (7) in which the polypeptide has twelve hydrophobic domains in α-helical conformation that traverse the membrane in zig-zag fashion connected by hydrophilic loops with the N- and C-termini on the cytoplasmic face of the membrane (Fig. 1). The model is consistent with other spectroscopic measurements (8), chemical modification (9), limited proteolysis (10,11) and immunological studies (12-18), but none of these approaches differentiates between the 12-helix structure and other models (cf. ref. 8). However, recent analyses of *lacY-phoA* fusions (19) have resolved the alternatives by providing strong, exclusive support for the 12-helix motif. In addition, it was demonstrated (19) that approximately half of a transmembrane domain is needed to translocate alkaline phosphatase to the external surface of the membrane. Thus, the alkaline phosphatase activity of fusions engineered at every third amino-acid residue in putative helices III and V increases as a step function as the fusion junction proceeds from the 8th to the 11th residue of each of these transmembrane domains. In order to study the location of specific residues in putative helix X of the permease which contains 3 amino-acid residues (K319, H322, E325) that are important for substrate recognition and/or H^+ translocation, we have constructed a set of chimeric proteins with alkaline phosphatase devoid of the leader peptide fused to each amino-acid residue in helix X. A sharp discontinuity in alkaline phosphatase activity at H322-M323 is observed, implying that these residues are located in the middle of the membrane (20). Furthermore, the alkaline phosphatase activity observed with each of the fusion proteins is correlated with the amount of mature phosphatase detected in the periplasm.

In the remainder of this article, three types of studies with deletion mutants in lac permease are described with respect to stability, topology in the membrane, activity and ability to exhibit *in vivo* complementation.

Expression of "split" lac permeases

Many proteins maintain tertiary structure when the peptide backbone is cleaved, and substrate binding and/or catalytic activity may be retained. Among many examples, bacteriodopsin can be split into two fragments that reconstitute to form an active complex (21). In a similar vein, lac permease binds ligand after proteolysis, although transport activity is abolished (10). Thus, forces between different domains within a protein are able to maintain three-dimensional structure when the peptide backbone is not intact. Furthermore, studies with the β-adrenergic receptor (22) and the sodium channel (23) indicate that functional complexes are formed even when the mRNAs encoding these proteins are expressed as discontinuous fragments.

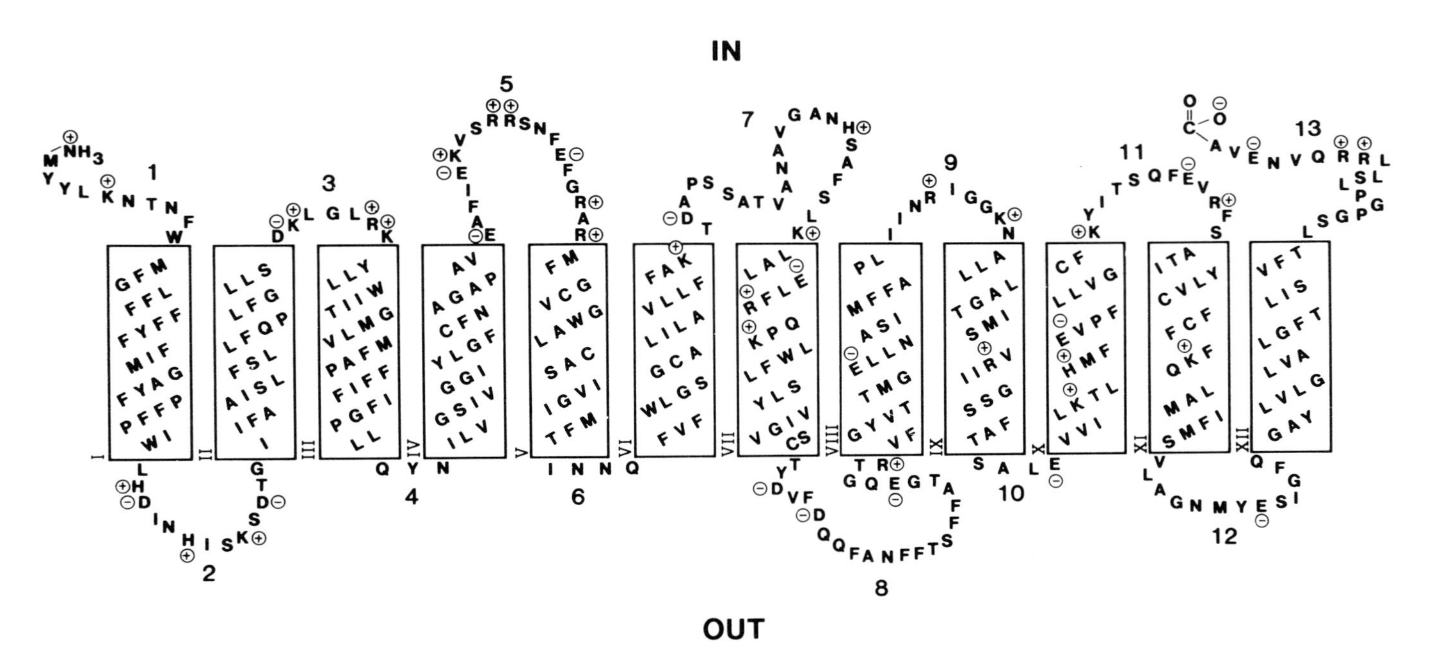

FIGURE LEGENDS

Figure 1. Secondary-structure of lac permease. The single amino-acid code is used, hydrophobic transmembrane helices are shown in boxes (cf. ref. 7).

In order to develop an *in vivo* system to study the assembly of integral membrane proteins, we showed that synthesis of two polypeptide fragments from independently cloned portions of the *lacY* gene leads to functional association in the membrane (24). The *lacY* gene was restricted into two approximately equal-size fragments and subcloned individually or together under separate *lac* operator/promoters in plasmid pT7-5. Under these conditions, lac permease is expressed in two portions: (i) the N-terminus, the first six putative helices and most of putative loop 7; and (ii) the last six putative helices and the C-terminus (i.e. N_6/C_6). As judged by [^{35}S] methionine pulse-chase experiments, immunoblotting studies and transport assays, the N- and the C-terminal portions of *lac* permease are proteolyzed when expressed independently. Dramatically, however, when the complementing polypeptides are expressed together, an association between the two polypeptides leads to a more stable, catalytically-active complex. More recently it has been shown that co-expression of independently cloned fragments of the *lacY* gene encoding N_2 and C_{10} (25), N_1 and C_{11} or $N_{6.5}$ and $C_{5.5}$ (26) form stable molecules in the membrane which interact to form functional permease. When the fragments are expressed by themselves, however, the polypeptides are relatively unstable and exhibit no transport activity. The observation that stable, functional permease can be formed when N_1 and C_{11} are co-expressed as individual polypeptides argues against the hypothesis that the N-terminus is inserted as a helical hairpin.

Topology and organization of lac permease

Secondary-structure models for most polytopic membrane proteins exhibit multiple hydrophobic domains in a linear array that traverse the membrane in zigzag fashion as α-helices connected by hydrophilic segments ("loops"). The putative transmembrane domains are generally identified as hydrophobic stretches in the primary sequence that are approximately 20 amino-acid residues in length, since a polypeptide of this length in α-helical conformation is sufficiently long to cross the hydrophobic core of the membrane, and the position of the loops is dictated by the orientation of the putative helices. Regarding the orientation of the proteins with respect to the plane of the membrane, one proposal (27) implies that the primary information for orientation of polytopic membrane proteins is determined by the oriented insertion of the first N-terminal transmembrane domain, followed by the passive, serpentine insertion of the remainder of the protein. The other hypothesis (28) suggests that topogenic information is spread throughout the protein.

Using the *lac* permease, we have studied the insertion and the stability of in-frame deletion mutants (29). So long as the first N-terminal and the last four C-terminal putative α-helical domains are retained, stable polypeptides are inserted into the membrane, even when an odd number of helical domains is deleted. Moreover, even when an odd number of helices is deleted, the C-terminus remains on the cytoplasmic surface of the membrane, as judged by lac permease-alkaline phosphatase (*lacY-phoA*) fusion analysis. In addition, permease molecules devoid of even or odd numbers of putative helices retain a specific pathway for downhill lactose translocation. The findings imply that relatively short C-terminal domains of the permease contain topological information sufficient for insertion in the native orientation regardless of the orientation of the N-terminus.

Functional complementation of lac permease deletion mutants

Although there is strong evidence that the lac permease is functional as a monomer (cf. ref. 5), recent experiments demonstrate that certain paired in-frame deletion constructs are able to complement functionally (30). For example, we have constructed lac permease molecules deleted of helices III and IV (N_2C_8) or helices IX and X (N_8C_2). Although cells expressing the deletion mutants individually are unable to catalyze active lactose accumulation, cells simultaneously expressing N_2C_8 and N_8C_2 catalyze transport up to 60% as well as cells expressing wild-type permease. Moreover, complementation does not occur at the level of DNA, but probably at the protein level. The findings with N_2C_8/N_8C_2 were expanded by examining additional pairs of deletion mutants: N_2C_6/N_8C_2 or N_4C_6/N_8C_2

exhibit diminished but significant transport activity, N_4C_6/N_6C_4 or N_2C_6/N_6C_2 exhibit only marginal activity, and the combinations N_4C_4/N_8C_2, N_2C_4/NN_8C_2 or N_6C_4/N_8C_2 exhibit no activity whatsoever. Therefore, the ability to complement functionally is a specific property of certain pairs of permease molecules containing relatively large deletions of transmembrane hydrophobic domains.

One possible interpretation of the results is that there are specific interactions between membrane spanning domains in wild-type permease and that disruption of these interactions by deletion leaves a "potential gap" in the structure that can be filled by interaction with another molecule containing the deleted segment. By this means, a permease molecule deleted of helices IX and X (N_8C_2) might "accept" these helices from a "donor" molecule deleted of helices III and IV (N_2C_8) and/or vice versa. In order to test this hypothesis, *E. coli* T184 was transformed with plasmids encoding P28S and N_8C_2 or H322K and N_2C_8 as potential donor/acceptor pairs, but transport assays demonstrate that the pairs do not complement functionally. Similarly, ΔH322 does not exhibit functional complementation with N_2C_8 nor does ΔW78 functionally complement with N_8C_2. Thus, the simplistic explanation does not appear to be the case.

Although nothing is known about the 3-dimensional structure of lac permease, it has been shown recently (see above and ref. 24-26)) that co-expression of independently cloned fragments of the *lacY* gene form stable molecules in the membrane which interact to form functional permease. When the fragments are expressed by themselves, however, the polypeptides are relatively unstable and exhibit no transport activity. Since these observations may be related to the phenomenon described here, we suggest that permease mutants containing missense mutations or point deletions, like the intact wild-type molecule, form relatively compact structures that are unable to form intermolecular complexes. On the other hand, molecules containing deletions in certain hydrophobic domains (for example, N_2C_8 and N_8C_2) may be in a more "relaxed" state and therefore able to interact to form functional oligomers.

REFERENCES

1 Kaback, HR. Physiol of Membr Disorders, ed. TE Andreoli, JF Hoffman, DD Fanestil, SG Schultz. Plenum Publishing Corp.; 1986: 387-408.
2 Kaback, HR. Harvey Lect Series 1989: 83: 77-103.
3 Newman, MJ, Foster, D, Wilson, TH, Kaback, HR. J Biol Chem 1981: 256: 11804-11808.
4 Viitanen, P, Newman, MJ, Foster, DL, Wilson, TH Kaback, HR. Methods in Enzymol 1985: 125: 370-377.
5 Costello, MJ, Escaig, J, Matsushita, K, Viitanen, PV, Menick, DR, Kaback, HR. J Biol Chem 1987: 262: 17072-17082.
6 Büchel, DE, Groneborn, B, Müller-Hill, B Nature (London) 1980: 283: 541-545.
7 Foster, DL, Boublik, M, Kaback, HR. J Biol Chem 1983: 258: 31-34.
8 Vogel, H, Wright, JK, Jähnig, F. EMBO J 1985: 4: 3625-3631.
9 Page, MGP, Rosenbusch, JP. J Biol Chem 1988: 263: 15906-15914.
10 Goldkorn, T, Rimon, G, Kaback, HR. Proc Natl Acad Sci USA 1983: 80: 3322-3326.
11 Stochaj, V, Bieseler, B, Ehring, R. Eur J Biochem 1986: 158: 423-428.
12 Carrasco, N, Tahara, SM, Patel, L, Goldkorn, T, Kaback, HR. Proc Natl Acad Sci USA 1982: 79: 6894-6898.
13 Seckler, R, Wright, J.K, Overath, P. J Biol Chem 1983: 258: 10817-10820.
14 Seckler, R, Wright, JK. Eur J Biochem 1984: 142: 269-279.
15 Carrasco, N, Viitanen, P, Herzlinger, D, Kaback, HR. Biochem 1984: 23: 3681-3687.
16 Herzlinger, D, Viitanen, P, Carrasco, N, Kaback, HR. Biochem 1984: 23: 3688-3693.
17 Carrasco, N, Herzlinger, D, Mitchell, R, DeChiara, S, Danho, W, Gabriel, TF, Kaback, HR. Proc Natl Acad Sci USA 1984: 81: 4672-4676.

18 Seckler, R, Möröy, T, Wright, JK, Overath, P. Biochem 1986: 25: 2403-2409.
19 Calamia, J, Manoil, C. Proc Natl Acad Sci USA 1990: 87: 4937-4941.
20 Ujwal, ML, Bibi, E, Chang, C-Y, Manoil, C., Kaback, HR.1990 (in preparation).
21 Liao, MJ, Huang, KS, Khorana, G. J Biol Chem 1984: 259: 4200-4204.
22 Kobilka, BK, Kobilka, TS, Daniel, K, Regan, JW, Caron, MG, Lefkowitz, RJ. Science 1988: 240: 1310-1316.
23 Stühmer, W, Conti, F, Suzuki, H, Wang, X, Noda, M, Yahagi, N, Kubo, H, Numa, S. Nature 1989: 339: 597-603.
24 Bibi, E, Kaback, HR. Proc Natl Acad Sci USA 1990: 87: 4325-4329.
25 Wrubel, W, Stochaj, U, Sonnewald, U, Theres, C, Ehring, R. J Bacteriol 1990: 172: 5374-5381.
26 McKenna, E, Bibi, E, Hardy, D, Pastore, JC, Kaback, HR. 1990 (in preparation).
27 Hartmann, E, Rapoport, TA, Lodish, HF. Proc Natl Acad Sci USA 1989: 86: 5786-5790.
28 Popot, JL, Engelman, DM. In: Protein Form and Function, eds. Bradshaw, R A, Purton, M, Elsevier Publishing Comp: Cambridge, 1990; 147.
29 Bibi, E, Verner, G, Chang, C-Y, Kaback, HR. Proc Natl Acad Sci USA 1991: 88: 7271-7275.
30 Bibi, E, Kaback, HR. 1991 (submitted).

© 1992 Elsevier Science Publishers B.V. All rights reserved.
Molecular mechanisms of transport. E. Quagliariello, F. Palmieri, eds.

Na^+-MELIBIOSE PERMEASE OF *Escherichia coli*: SITE-DIRECTED MUTAGENESIS OF ASPARTIC ACIDS LOCATED IN TRANSMEMBRANE SEGMENTS.

Gerard LEBLANC, Thierry POURCHER, Marcel DECKERT and Marie-Louise ZANI
Laboratoire J. Maetz, Département de Biologie Cellulaire et Moléculaire du Commissariat à l'Energie Atomique, Villefranche s/mer, FRANCE

Among bacterial secondary transport systems (or cation-coupled co-transport or symport), the melibiose (*mel*) permease of *Escherichia coli* is of particular interest as it displays the unusual property to accumulate α-galactosides using H^+, Na^+ or Li^+as coupling species depending on the substrate or ionic environment (1-4). These coupling properties of *mel* permease have been used for detailed analysis of *mel* permease function (2, 4-7). Despite intensive efforts over the past decade, the coupling process of this co-transport mechanism - including the mechanism of cation recognition and role of the cation in the coupling process - is not understood at the molecular level.

The cloning of *melB* gene, which codes for the melibiose transport protein (8,9), has opened the possibility of utilizing DNA recombinant technology and *in vitro* mutagenesis strategies for investigation of the co-transport mechanism at the molecular level. The purpose of this paper is to summarize recent site-directed mutagenesis studies on *mel* permease which suggest that four Aspartic acid residues, putatively clustered in the hydrophobic core of the NH_2 terminal domain of the permease (13), are involved in cation-permease interaction. For the sake of clarity, relevant catalytic properties and molecular characteristics of *mel* permease are summarized prior to description of the results.

FUNCTIONAL PROPERTIES

The α-disaccharide melibiose is actively accumulated by *mel* permease in *E. coli* RA11 (*mel A*$^-$, *melB*$^+$, *Δ lac Y*) resuspended in media devoided of Na^+ ions (for review see ref. 4). Addition of Na^+ ions at a concentration of about 10 mM produces maximal stimulation of sugar uptake, the effect resulting from reduction of the transport constant Kt. Proton and Na^+ have been identified as coupling species in Na^+-free and Na^+-containing media respectively, by means of pH or Na^+-sensitive electrodes measurements (1). Sugars with a β configuration (such as β-thio-methyl galactopyranoside, TMG) are co-transported with either Na^+ or Li^+ but not with H^+(2). Binding of the high affinity ligand α-nitrophenyl galactoside (NPG) on right-side-out membrane vesicles harboring *mel* permease is competitively inhibited by melibiose or TMG and is specifically enhanced by Na^+ or Li^+ (affinity increase), H^+ being much less efficient activator (10). Further binding analyses suggest that H^+, Na^+ or Li^+ compete for the same binding site and also that cation and sugar substrates bind in a stoichiometric ratio of 1/1 to the permease. Consistently, simultaneous measurement of Na^+ and sugar flows in de-energized cells also indicates a 1/1 stoichiometry (6).

Active transport of melibiose requires generating an electrochemical potential gradient of H^+ (or Na^+) across the membrane. Mechanistic studies on RSO membrane vesicles led to suggest a transport model in which binding of cation and sugar to the permease on the outer surface of the membrane and their release into the cytoplasm are ordered processes, giving rise to a model in which Na^+ binds first and is released last. Finally, the rate of co-substrate release into the cytoplasm, and hence permease cycling, varies according to the coupling cation ($H^+ > Na^+ > Li^+$) and is dependent on the magnitude of the electrical potential (3).

BIOCHEMICAL IDENTIFICATION AND SECONDARY STRUCTURE

Sequencing of the *melB* gene indicates that *mel* permease is a protein of 469 amino acid residues, i.e. with a molecular weight of 52,069 Da (9). The transport protein has been identified by two independent means. Cloning the *melB* gene into the T7 polymerase-promotor expression system and labeling its product *in vitro* enabled Pourcher et al. (13) to demonstrate that the product of *melB* is a cytoplasmic membrane protein which migrates with an apparent molecular mass of 38-40 kDa in SDS-polyacrylamide gels. Botfield and Wilson (12) showed that antipeptide polyclonal antibodies (MBct10) directed against the putative COOH terminal domain of mel permease react with the same polypeptide (see also ref 4). As the transport protein contains a large amount of non polar residues (70 %) and does not exhibit a typical signal sequence which makes unlikely post-translational processing, the lower apparent mass of the *mel* permease most likely reflects excess binding of negatively charged SDS molecules on this hydrophobic protein as previously proposed for other prokaryote permeases .

A secondary structure model for *mel* permease consisting of 12 hydrophobic segments in an α-conformation has been proposed on the basis of different criteria (Figure 1, (13)). Hydropathy profiling suggests that the polypeptide contains at least 10 hydrophobic stretches (mean length >20 residues) connected by hydrophilic charged segments (8). Moreover, the NH_2 and COOH termini seem to be co-localized on the cytoplasmic surface of the membrane (12). As for other bacterial membrane proteins (14), clusters of arginine and/or lysine residues are preferentially located in cytoplasmic loops connecting transmembrane segments . Finally, fusion studies of mel permease and alkaline phosphatase (phoA) which allows distinction between periplasmic and cytoplasmic domains of membrane proteins, provided support for the 12 transmembrane segment model (15).

significant active transport of melibiose in Na^+-free medium. This suggests that these mutated permeases retain the basic coupling properties of a H^+-coupled co-transport mechanism although Na^+-coupled transport capacity is lost. These findings support the conclusion that the major defect in D51C or D55C permease function results from a modification of the cationic selectivity profile of *mel* permease. Experiments are presently underway to examine if these results and conclusions apply to the other modified permeases.

Replacement of Asp 31, 51, 55 or 120 by neutral amino acids modifies the cationic dependence of *mel* permease and does not impair its ability to recognize the sugar. This suggests that these acidic residues contribute specifically to cation-permease interaction. Moreover, their location in putative hydrophobic segments of the N-terminal domain of the transport protein further suggests that this particular domain of *mel* permease is involved in this interaction. We recall here that aspartic or glutamic residues lying in the transmembrane domain of various cation or H^+ (or H_3O^+) -translocating devices appear to be essential for either recognition or translocation of the charged species across the membrane. This is the case for Asp 61 in the Fo sector of the *E. coli* H^+-translocating ATPase (22), Glu 325 for H^+-lactose symport by *lac* permease (23) or Asp 85, 96 and 212 for proton pumping by bacteriorhodopsin (24). In addition, one Asp and three Glu residues have recently been proposed as ligands for one or both of the two high affinity Ca^{2+} binding sites in the Ca^{2+}-ATPase of the sarcoplasmic reticulum (25). Whether the cationic site is restricted to the N-terminal domain or includes other domains of the permease (C-terminal membrane spanning segments ?) remains to be established.

The change in cationic properties of the mutated permeases is unlikely to result from a general structural defect of the transport protein as they all retain the ability to recognize the co-transported sugar. The location of D31, D51, D55 and D120 on adjacent membrane spanning segments suggests that these acidic residues contribute to cation-permease interaction by forming an acidic pocket in which the carboxylates take part in the coordination of the coupling ion. This interpretation would give support to Boyer's hypothesis which suggests that appropriately placed O (or N) atoms in symporters like *mel* permease could provide cation-binding domains akin to those in the crown ethers or cryptates, both of which form coordination complexes with Na^+ and H_3O^+ (26). One cannot exclude that one (or more) of these Asp residues might be structurally important for appropriate positioning of additional amino-acids interacting with the cation. Site-directed mutagenesis of other nearby polar residues lying in the NH_2 terminal domain of *mel* permease (7 Tyr, 2 Asn, 2 Ser and 4 Met) is presently being analysed in order to determine if these residues are important for cation recognition by *mel* permease.

In conclusion, the results presented above implicate the participation of the membrane-embedded NH_2 terminal domain of *mel* permease - and in particular the acidic residues of this domain - in the interaction of the transport protein with the coupling ion.

ACKNOWLEDGMENTS
We wish to thanks R. Lemonnier and P. Lahitette for excellent technical assistance This work was supported in part by the Centre National de la Recherche Scientifique, France (DO 638). M.Deckert's present address is Laboratoire d'Immunologie, INSERM U434, Hopital l'Archet Nice, France.

REFERENCES

1 Tsuchiya T. and Wilson T.H. (1978),*Membr. Biochem.* **2**, 63-79.
2 Wilson D.M. and Wilson T.H. (1987),*Biochem. Biophys. Acta* **904** , 191-200.
3 Leblanc G., Bassilana M. and Pourcher T. (1988) in :" *Molecular basis of biomembrane transport* " Eds Palmieri F. and Quagliariello E. . Elsevier Science Publishers pp 53-62.
4 Leblanc G., Pourcher T., Bassilana M. and Decker M. (1991), *Melibiose transport,* in *Ion-coupled sugar transport in microorganisms,* (Page M. and Henderson P.J.F.), CRC Press, Boca Raton, in press.
5 Bassilana M., Pourcher T. and Leblanc G. (1987), *J. Biol. Chem.* **262**, 16865-16870.
6 Niiya S., Yamasaki K., Wilson T.H. and Tsuchiya T. (1982), *J.Biol. Chem* **257**, 8902-8906.
7 Kawakami T., Akizawa Y., Ishikawa T., Shimamoto T., Tsuda M. and Tsuchiya T. (1988), *J. Biol. Chem.* **263**, 14276-14280.
8 Hanatani M., Yazyu H., Shiota-Niiya S., Moriyama Y, Kanazawa H., Futai M. and Tsuchiya T. (1984), *J. Biol. Chem.* **259**, 1807 -1812.
9 Yazyu H., Shiota-Niiya T., Shimamoto T., Kanazawa H., Futai M. and Tsuchiya T. (1984), *J. Biol.Chem.* **259**, 4320-4326.
10 Damiano-Forano E., Bassilana M. and Leblanc G. (1986),*J. Biol. Chem.* , **261** , 6893-6899.
11 Pourcher T., Bassilana M., Sarkar H.K., Kaback H.R. and Leblanc G. (1990), *Biochemistry* **29** , 690-696.
12 Botfield M.C. and Wilson T.H. (1989), *J. Biol.Chem.* **264** , 11649-11652.
13 Pourcher T., Sarkar H.K., Bassilana M., Kaback H.R. and Leblanc G. (1990), *Proc. Natl. Acad. Sci. (USA)*, **87**, 468-472.
14 Von Heijne G. (1986), *EMBO J.* **5** , 3021-3027.
15 Botfield M.C.(1989), Ph. D. Thesis: " Structure/ function of the melibiose carrier of E. coli", Harvard Medical University , Cambridge , Mass.
16 Henderson P.J.F. (1990), *J. Bioenerg. Biomemb.* **22**, 525-569.
17 Poolman B., Royer T.J., Mainzer S.E. and Schmidt B.F. (1989), *J. Bact.* **171**, 244-253.
18 Botfield M.C. and Wilson T.H. (1989), *J. Biol.Chem.* **263** , 12909-12915.
19 Pourcher T., Bassilana M., Sarkar H.R , Kaback H.R. and Leblanc G. (1990), *Phil. Trans. R. Soc. , Lond.* B **326**, 411-423.
20 Pourcher T., Bassilana M., Sarkar H.R., Kaback H.R. and Leblanc G. (1991), in preparation.
21 Pourcher T., Deckert M., Bassilana M. and Leblanc G. (1991), *Biochem. Piophys. Res. Comm.* **178**, 1176-1181.
22 Miller M.J., Oldenburg M. and Fillingame R.H. (1990), *Proc.Natl.Acad.Sci. USA* **87**, 4900-4904.
23 Carasco N., Püttner I.B., Antes L.M., Lee J.A., Larigan J.D., Loklema J.S., Roepe P.D. and Kaback H.R. (1989), *Biochemistry* **28**, 2533- 2539.
24 Stern L.J., Ahl P.L., Marti T., Mogi T., Dunach M., Berkowitz S., Rothschild K.J. and Khorana G. (1989), *Biochemistry* **28** , 10035-10042.
25 Clarke D.M., Loo T.W., Inesi G. and MacLennan D.H. (1989), *Nature* **339**, 476-478.
26. Boyer P.D. (1988), *Trends in Biochem Sci.* **13**, 5-7.

© 1992 Elsevier Science Publishers B.V. All rights reserved.
Molecular mechanisms of transport. E. Quagliariello, F. Palmieri, eds.

MUTAGENIZED ADP/ATP CARRIER FROM *Saccharomyces*

Martin Klingenberg*, Meinrad Gawaz, Michael G. Douglas, Janet E. Lawson

*Institute for Physical Biochemistry, University of Munich,
Goethestrasse 33, 8000 Munich 2, Fed. Rep. Germany

Introduction

The ADP/ATP carrier (AAC) was the paradigm of the mitochondrial carrier family. It was the first of the mitochondrial and also of any biological carriers to be isolated (1) and of which the primary structure became known (2). The similarities in the primary sequence of other mitochondrial carriers, such as the uncoupling protein (UCP) (3), phosphate carrier (4, 5, 6), the ketoglutarate carrier (7), provided evidence that the mitochondrial carriers form a family of proteins which is derived from a common ancestral source. The striking feature of the AAC carrier as well as of this family is an internal tripartite structure. Each element of about hundred residues is proposed to span twice the membrane resulting in a total of six membrane spanning sections with three large hydrophilic domains on the matrix site (8). As a result, the carrier molecules are constructed of three similar building blocks which are suggested to be arranged around a threefold pseudo symmetry axis traversing the membrane. Two of these monomers are arranged about a twofold axis forming the carrier dimers (9). It is still an open question whether the translocation channel is located within one monomer along the pseudo threefold axis or between the two monomers along the twofold axis.

An important working basis for the elucidation of the structure/function relationship in the AAC carrier is the single binding center gated pore mechanism (10, 11). This model, as based on a vast amount of experimental data, implies a single binding center within the translocation path for the solute with gates around the binding center. Also only one of the two gates can be opened at any time. By opening either the outer or the inner gate, the binding center is accessible to the inner or outer face. Thus, the carrier essentially exists in two different states according to the configuration of these gates. In the inner state the binding center can be accessed by the solute from the inside and in the external state from the outside.

The investigation of the structure/function relationship would imply a search for the binding center within the structure and for the two gates. It will be asked which residues within the primary structure are participating at those sites. Apart from being involved in the translocation, residues might be indirectly essential as they are important for maintaining the structure by being located in an access route to the binding center, or by providing the conformational flexibility required in the translocation. Furthermore, certain residues might be operative for sensing the membrane potential which drives the ADP/ATP exchange.

Isoforms of AAC

One of the obvious methods for elucidating the functional role of these residues is mutagenesis. In yeast there are excellent preconditions for mutations of the AAC either by site–directed mutagenesis or by spontaneous mutations. However, there are three genes for the AAC in yeast cells only one of which, the AAC2, is the isoform which is actively expressed under normal growth conditions (12, 13, 14).

For characterizing the isoform AAC1, the AAC1 gene was introduced on a plasmid into an AAC2–deleted yeast strain (15). Only with the multicopy vector it was possible to produce sufficient AAC1 for isolation and usage in reconstitution (16). The molecular activity of the reconstituted AAC1 and AAC2 were measured in proteoliposomes in which the number of AAC molecules incorporated were determined by ^{3}H–CAT. The molecular translocation activity of AAC1 is only 40% of AAC2 (16). There is no difference in KM for ADP or ATP between both isoforms. Evidently, the structural changes in the isoforms influence the catalytic efficiency which is reflected in the maximum rate rather than in the binding of the nucleotide. Since, differently from enzymes, the catalytic activity in carriers is conditioned by the activation energy of the protein conformation change, an influence primarily on the maximum rate should be expected (16a). With these results for the first time functional differences of isoforms of a solute carrier have been determined.

Site–directed mutagenesis

For elucidating the structure/function relationship in the ADP/ATP carrier site–directed mutagenesis were applied. For this purpose, into yeast strains from which both the AAC1 and AAC2 were deleted, point–mutated genes were inserted by a plasmid vector. The selection of these mutations was based partially on a folding model of the AAC as derived from the primary structure and on information about the sidedness of the C– and N–terminals.

The model, as shown in Figure 1, incorporates the tripartite structure of the AAC with three domains each containing two transhelical segments. Both the N– and C–terminals are assumed, as based on circumstantial evidence, to be located on the c–side (17, 18). A portion of these three matrix side–oriented hydrophilic segments is assumed to participate in the translocation channel (19). In each of the three domains, within the second helix, the only charged residue is an arginine, i.e. R96, R204 and R294. The mutants of each of these were generated by mutation R/L. In addition, R96 was mutated into H, D and P. Because of their intramembraneous location these arginine residues may prove to be essential for transporting the anionic substrates.

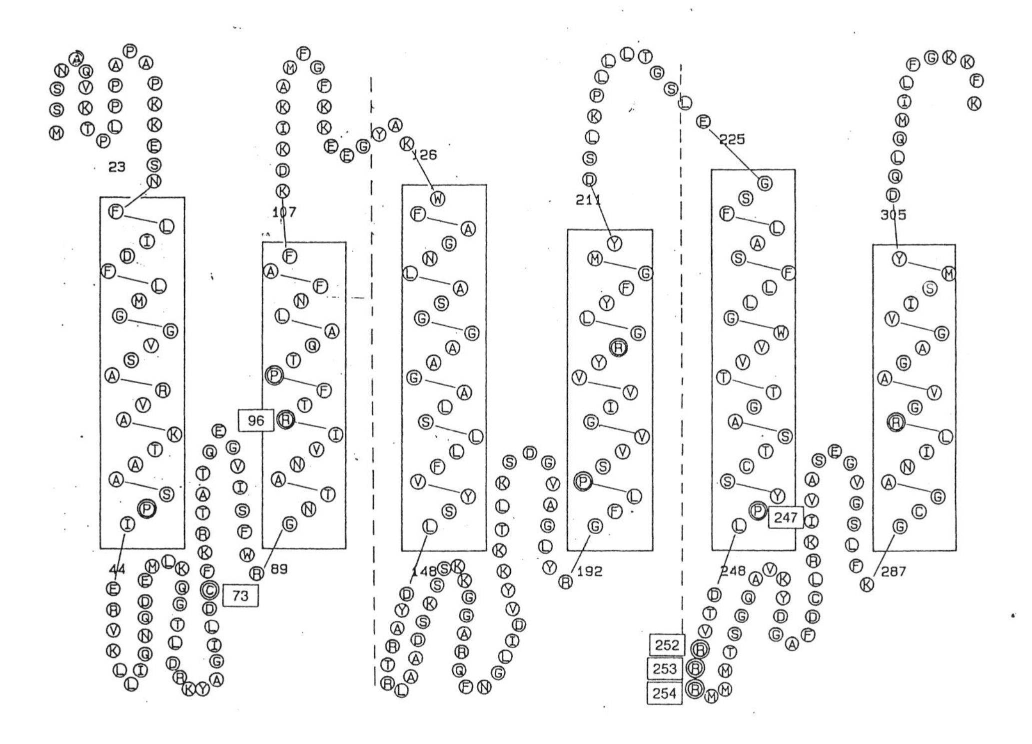

Figure 1. The suggested folding of the ADP/ATP carrier (AAC2) from *Saccharomyces*. The residues used for site–directed mutagenesis are designated by the boxed numbers. These residues and those which are noted in the text for the three intrahelical positions are emphasized by a second circle.

Another conspicuous feature are proline residues within the transmembrane α–helical segments. Although proline is generally assumed to break α–helices, its occurrence within transmembrane segments is now quite well accepted. It is assumed that proline is instrumental for the flexibility of these helices required for the conformational changes involved in transport. For this reason, proline 247 was modified to glycine.

Two other potentially important groups of residues are cysteine residues which occur three times within the carrier, and the RRR triplet 252 to 254 at the matrix side of the third repeat domain. Out of the three cysteines, C73 is analogous to C56 in bovine heart mitochondria which has been identified to be an "essential" SH–group (20). Alkylation of this SH–group inhibits transport. Most importantly, this alkylation is only possible in the "m–state" of the AAC (21). This SH–group is conserved in all AAC. Also the arginine triplet is conserved in all twelve AAC sequences known.

The first assay for a change in the AAC function of the mutants was the growth on a non–fermentable source like glycerol which would be indicative of a functionally intact AAC. As shown in Table 1, C73 and P247 mutants were glycerol positive, whereas the R96 and R52 mutants were glycerol negative. The glycerol negative mutants therefore had to be grown on a fermentable source, such as galactose. Glucose was avoided because it suppresses mitochondrial growth. From these cells mitochondria were isolated by the enzymatic method of cell membrane rupture according to (22). In the isolated mitochondria the cytochrome content was spectroscopically determined and the content of AAC molecules by ^{3}H–CAT binding.

Table 1
Growth of Site–Directed Mutants

	mutation	glycerol	generation time (h) lactate	galactose
wild		+	2.5	–
C 73S	(TGT – AGC)	+	2.7	–
R 96D	(CGT – GAC)	–	–	3.2
P247G	(CCA – GGA)	+	2.7	–
R252I	(AGA – ATA)	–	–	2.8
R253I	(AGA – ATA)	–	–	2.8
R254I	(AGA – ATA)	–	–	2.8

From these strains mitochondria were isolated and therefrom the AAC were purified and reconstituted into phospholipid vesicles for measuring the ADP/ATP transport. In these vesicles the transport rate was related to the number of AAC molecules incorporated as determined by ^{3}H–CAT binding, in order to obtain the molecular transport activity of the AAC. The ADP/ATP transport in the isolated mitochondria could not be measured because of the low endogenous adenine nucleotide content and possible damage during isolation.

The results (Table 2) show that in the mitochondria from the R96D mutant there is nearly no cytochrome and no AAC carrier present as measured by ^{3}H–CAT binding. The R252I mutant mitochondria also have a largely reduced content of cytochromes and AAC. On the other hand, in the C73S mutant the cytochrome content is reduced only by about 40%. The P247G mutant had the smallest change in the cytochrome content. The content of AAC as determined by ^{3}H–CAT binding has a different response to the mutations as the cytochromes. It is about the same in mitochondria from the glycerol positive strains, both in wild type and in mutants. The AAC content is strongly decreased in the R96D mutant and partially in the R252I mutant.

Table 2
Effect of Site–Directed AAC2 Mutagenesis on Yeast Mitochondria

Function	Unit	Strain Wild	C73S	R96D	P247G	R252I
Cyt aa_3	nmol/g prot.	220	135	10	120	30
Cyt b		531	210	74	190	102
Cyt c		1250	350	130	300	200
Resp.*	μatom O min x g prot.	610	149	67	157	192
1/Cytaa_3	1/min	2.8	1.1	6.7	1.3	6.4
^{3}H CAT	μmol/g prot.	0.32	0.31	0.03	0.31	0.20
K_D	10^{-6}M	0.16	0.19	–	0.18	–

* Respiration with NADH

The transport activities of the AAC from the various mutants were determined after reconstitution of the purified AAC both with ADP and ATP (Table 3). There are, despite a wide range of activity change, only minor differences in the K_M for ADP and ATP between the AAC of the various mutants. All the changes are manifested in the maximum exchange rate. Whereas there is no measurable activity in the R96D mutant, it is only about 8% in the R252 mutant, but reaches 45% in the C73S and 65% in the P247I mutants.

Table 3
Exchange Activity of Reconstituted Site–Directed Mutants of AAC2

Function		Unit	Wild	C73S	R96D	P247G	R252I
$V_{max}(25^0)$	ADP	μmol/	1150	470	0	780	110
	ATP	min g prot.	840	300		260	–
K_M	ADP	10^{-6}M	28	51	0	24	16
	ATP		6	20		10	–
molecular activity	ADP	1/min	600	220	0	390	40
$1/^3$H CAT	ATP		400	140		130	–

Calculated Exchange Activity in Mitochondria

	Unit	Wild	C73S	R96D	P247G	R252I
V_{max} x ^{3}H CAT–binding in mito.	μmol/min g prot	192	68	0	121	10
./././cyt aa_3	1/min	870	570		890	300

By multiplying the molecular activity measured in the reconstituted system with the amount of carrier molecules as measured by the ^{3}H–CAT binding in mitochondria, one can calculate the exchange activity in mitochondria, which drastically differs between the wild type and these mutants. However, when the exchange activity is correlated to the cytochrome content these large differences mostly disappear. In other words, the exchange activity is quite well correlated to the cytochrome content in mitochondria (see also Fig. 2). Vice versa, the expression of the respiratory chain is not related to the number of AAC molecules, but rather to the ADP/ATP transport capacity. Or stating it differently, the same content of AAC molecules in the mutants, however with a lower molecular activity than in the wild type, is associated with a correspondingly lower cytochrome content. The expression of the respiratory capacity is adjusted to the ATP production capacity of the mitochondria, since the ADP/ATP exchange is a critical step in the oxidative phosphorylation.

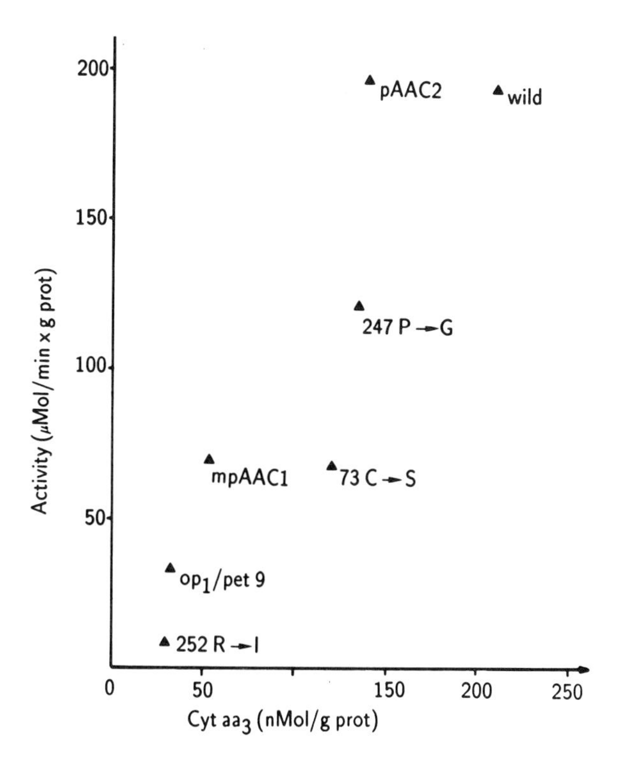

Figure 2. Correlation of exchange activity in various AAC–mutations to the cytochrome content. The curve includes the site–directed mutants of Table 3 and the data from the multiplasmid strain mutation containing only AAC1 and from the mutant op1 (16).

The retention of a considerable exchange activity in the C73S mutant seemed to be surprising. Obviously, C73S is not essential for transport, however its alkylation with the substitution of the ethyl–succinoamido group blocks the transport either by hindrance of the translocation path or, more probable, by breaking a hydrogen bond in which the cysteine is involved. The substitution in the mutant of cysteine by serine still retains the hydrogen bonding capability. The accessibility of C73 to NEM only in the "m–state" of the AAC indicates that the hydrogen bond is loosened by this conformation change. In agreement with other carriers, such as lactose permease (23) or the proline carrier in E. coli (24), also in the AAC, the seemingly "essential" cysteine can be eliminated without much harm to functional activity.

The complete inactivation by the substitution of arginine 96 by aspartic acid may not be surprising in view of the drastic charge change. However, also R96H mutants which are identical with the op1 mutation showed only a poor growth. It is most interesting that in our early search for residues to be changed, R96 was chosen because of its intrahelical localization which subsequently turned out to be also the site of the op1 mutation. In fact, the mutant R96H was constructed (15) before its occurrence in op1 was known (14). Since also the other two arginine substitutions resulted in glycerol negative growth, it appears that all three intrahelical arginines are essential for the transport. It could be suggested that they are part of the binding center for the phosphate groups.

The substitution of intrahelical proline by glycine only slightly decreases the transport activity. Obviously, here the suggestion (25) that prolines are essential for transport by providing extra flexibility along the transmembrane helices is not valid.

The triplet arginine 252 to 254 is obviously essential for AAC function. The substitution of any of the three arginines inhibits largely the activity as judged from the growth in the R253I and R254I mutants. We like to suggest that these arginines which are located at the matrix side of the mitochondrial membrane are not involved in the translocation binding center, but rather in controlling the diffusion of the highly negatively charged solutes into the translocation site.

In conclusion, site–directed mutagenesis of the AAC2 of yeast has provided a first insight into the role of some phenotypically important residues. Most important is the high sensitivity of the substitution of the intrahelical arginines and of the triplet arginines. Strikingly, the only naturally selected mutant of the AAC known, the op1, contains also the substitution R of the intrahelical arginine 96. Whereas the intrahelical arginines may actually participate in the transport, the triplet arginines may rather be located at the mouth of the translocation channel and be involved in steering the substrate into the binding center.

Acknowledgements

This work was supported by a National Institutes of Health Grant GM 36536 (to M.G.D.) and by Grant Kl 134/21 from the Deutsche Forschungsgemeinschaft (to M.K.) and by a NATO Grant RG0744187 (to M.G.D. and M.K.).

References

1 Riccio P, Aquila H, Klingenberg M. FEBS Lett 1975; 56: 133–138.
2 Aquila H, Misra D, Eulitz M, Klingenberg M. Hoppe–Seylers Z Physiol Chem 1982; 363: 345–359.
3 Aquila H, Link TA, Klingenberg M. EMBO J 1985; 4: 2369–2376.

4 Klingenberg M, Aquila H, Link TA. Chemica Scripta 1987; 27B: 41–45.
5 Aquila H, Link TA, Klingenberg M. FEBS Lett 1987; 212: 1–9.
6 Runswick MJ, Powell SJ, Nyren P, Walker JE. EMBO 1987; 6: 1367–1373.
7 Runswick MJ, Walker JE, Bisaccia F, Palmieri F, et al. Biochemistry 1990; 29: 11033–11040.
8 Klingenberg M. Arch Biochem Biophys 1989; 270: 1–14.
9 Klingenberg M. Nature 1981; 290: 449–454.
10 Klingenberg M. In: Martonosi AN, ed. The Enzymes of Biological Membranes: Membrane Transport. New York: London, 1976; 3: 383–438.
11 Klingenberg M, Buchholz M. Eur J Biochem 1973; 38: 346–358.
12 Adrian GS, McCammon MT, Montgomery DL, Douglas MG. Mol Cell Biol 1986; 6: 626–634.
13 Lawson JE, Douglas MG. J Biol Chem 1988; 263: 14812–14818.
14 Kolarov J, Kolarova N, Nelson N. J Biol Chem 1990; 265: 12711–12716.
15 Lawson JE, Gawaz M, Klingenberg M, Douglas MG. J Biol Chem 1990; 265: 14195–14202.
16 Gawaz M, Douglas MG, Klingenberg M. J Biol Chem 1990; 265: 14202–14208.
16a Klingenberg M. In: Kuby SA, ed. A Study of Enzymes, Vol. II: Mechanism of Enzyme Action. Boca Raton, 1991; 367–388.
17 Brandolin G, Boulay F, Dalbon P, Vignais PV. Biochemistry 1989; 28: 1093–1100.
18 Eckerskorn C, Klingenberg M. FEBS Lett 1987; 226: 166–170.
19 Bogner W, Aquila H, Klingenberg M. Eur J Biochem 1986; 161: 611–620.
20 Boulay F, Vignais PV. Biochemistry 1984; 23: 4807–4812.
21 Klingenberg M, Heldt HW. In: S. Sies, ed. Metabolic Compartmentation. London, 1982; 101–122.
22 Knirsch M, Gawaz MP, Klingenberg M. FEBS Lett 1989; 244: 427–432.
23 Viitanen PV, Meninck DR, Sarkar HK, Kaback HR et al. Biochemistry 1985; 24: 7628–7635.
24 Yamato I, Anraku Y. J Biol Chem 1988; 263: 16055–16057.
25 Brandl CJ, Deber, CM. Proc Natl Acad Sci USA 1986; 83: 917–921.

© 1992 Elsevier Science Publishers B.V. All rights reserved.
Molecular mechanisms of transport. E. Quagliariello, F. Palmieri, eds.

A genetic approach to studying the structure of membrane transport proteins. Application to the yeast adenine nucleotide translocator.

D.R. Nelson§, J.E. Lawson¶, M. Klingenberg† and M.G. Douglas§

§Dept. of Biochemistry and Biophysics, University of North Carolina at Chapel Hill Chapel Hill, North Carolina, USA 27599-7260.

¶Dept. of Biochemistry, Clayton Research Foundation, University of Texas, Austin, TX USA 78712

†Institut für Physikalische Biochemie, Universität München, München, Germany

ABSTRACT

The yeast adenine nucleotide translocator(AAC2) is required for respiration but not for fermentation. Therefore, the AAC2 gene can be mutated and its function evaluated by testing for growth on glycerol. In this way, an essential arginine cluster R252-R254 was detected. A representative from this group, R252I, has been characterized biochemically. The isolated mitochondria of the R252I mutant contained only 15% of the wild type cytochrome content, but 60% of AAC molecules as measured by ^{3}H-CAT binding. The isolated reconstituted R252I carrier has only 8% of wild type molecular transport activity. Three other gly- mutants were also found in the membrane portion of the translocator, R294L, R204L and W235F. R96H, the pet9 mutation, grows on glycerol, but very slowly. The gly- mutants R252I, R253I and R254I were subjected to selection for growth on glycerol, and intragenic second site revertants were determined by sequencing the plasmid AAC2 gene. 15 unique second site revertants were found in 14 different codons. R254I gave 11 different revertants, R253I gave four and R252I had none. The mutations were 19-228 amino acids distant from the original arginine mutation. Only one mutant was on the matrix side of the inner membrane, presumably near the R253I mutation. The others were in the membrane(12) or 1-3 residues above the membrane(2) on the cytosolic side. A specific portion of the transmembrane channel appears to be defined by these mutations. Helical wheel analysis shows that charges and second site mutations fall on one side of 5 transmembrane helices. These results have significant implications for the structure of the transmembrane channel of AAC2. Attempts to select revertants of R96H,L,P,D, R204L, W235F and R294L gave no intragenic second site revertants, though R96L may have generated one extragenic mutation.

INTRODUCTION

Membrane transport proteins have been studied by protein chemical, biophysical and molecular biological techniques. These methods have revealed the importance of particular amino acids in these proteins. However, no crystal structure of a membrane transport protein is available to allow interpretation of these data. In this report, we present a genetic approach in yeast that gives a glimpse of how critical mutations affect the structure of a transport protein, in this case the yeast adenine nucleotide translocator AAC2. In this approach non-functional mutants of AAC2, in a suitable yeast host, are selected to regain function by plating the yeast on a non-fermentable carbon source. A functional AAC2 is required for growth, leading to intragenic second site mutations that restore function. Sequencing of these revertant genes allows the sphere of influence of the original mutation to be mapped in the linear sequence of the gene. In the case of the AAC2 protein, the membrane topology is reasonably well established and the mutants can be mapped on a topological model. By integrating revertant mutation data with structural information obtained by other methods, a detailed, testable model of the AAC2 transmembrane channel has been developed. The method should be applicable to other proteins and especially to membrane proteins whose topology is known.

EXPERIMENTAL PROCEDURES

Strains and media. JLY1053 (MATa aac1::LEU2 aac2::HIS3 his 3-11,15 trp1-1 ura 3-1 can 1-100 ade 2-1 leu 2-3,112) was described in (1). Yeast was grown on YPGE media (1% yeast extract, 2% bactopeptone, 3% glycerol, 3% ethanol, 2% agar) during selection for suppressors. Yeast was grown overnight in liquid culture in synthetic minimal media (YNBD) supplemented with 14 amino acids and adenine(2) before plating on YPGE. Yeast was grown overnight on YPD (1% yeast extract, 2% bactopeptone, 2% dextrose) for isolation of plasmids. *E. coli* strain MC1066 was used for rescue and amplification of plasmids from yeast. MC1066 (*E. coli* K-12 F^- Δ(lac IPOZY)X74 galU galK rpsL hsdR trpC9830 [complemented by TRP1] leuB600 [complemented by LEU2] pyrF74::Tn5 (Km^r) [complemented by URA3]) (3,4) nomenclature of Bachmann et al (5). *E. coli* was grown in Luria broth with ampicillin (50μg/ml). Plasmid pSEYc58 was used (4) with AAC2 inserted as described in (1) where it was mislabeled as pSEYc63 (polylinker orientation reversed).
DNA and genetic methods. Site directed mutagenesis was done according to Zoller and Smith (6). Yeast were transformed with the pSEYc58 AAC2 plasmid by the Li acetate procedure of Shiestl and Gietz(7) with a 15 min. heat shock at 42°C. Yeast were washed twice in TE buffer pH 8.0 and resuspended in 400μl YPD before spreading 200μl on YNBD plates supplemented with 14 amino acids and adenine. Plasmids containing URA3 conferred the ability to grow on this ura^- media. Colonies were tested for growth on YPGE plates to determine function of the AAC2 protein. Gly^- colonies imply a non-functional AAC2. Gly^- mutants were grown in selective media overnight and 10^8 - 10^9 cells were plated on YPGE to select for revertants. Plasmids were rescued from 5 ml of overnight culture in YPD by the method of Hoffman and Winston (8). This preparation (0.5 - 1.0μl) was used to transform *E. coli*. It was necessary to use a very small amount of plasmid to avoid effects of an inhibitor in the preparation. Colonies that grew on ampicillin containing plates were screened for the presence of plasmid by restriction digestion of miniprep DNA. 100ml cultures were grown and plasmid isolated on Qiagen columns according to the manufacturers instructions. Sequencing was done by the dideoxy chain termination method. Plasmids were removed from strains by serial culture in YPD for one week without selection for plasmid retention, followed by checking for growth on uracil deficient media.
Protein methods. The ^{3}H-carboxyatractylate binding and transport studies on the R252I mutant were done as described in (9) for the R96H mutant.

RESULTS

The transformation of 16 site directed mutants of AAC2 into a translocator deficient yeast strain JLY1053, showed that six mutants were able to grow on glycerol and ten were not. Nine of the ten mutants that tested gly^- were Arg mutants. Three of these were located in transmembrane segments [R96, R204 and R294] while the remaining three formed an unusual arginine cluster in the mitochondrial matrix [R252, R253 and R254]. R96 was mutated to Leu,

Pro Asp and His, with His being the naturally occurring mitochondrial petite mutant pet9 (1,10). R204 and R294 were changed to Leu and R252-254 were individually changed to Ile. The single gly^{-} mutant that was not initially an Arg residue was W235F, also located in a transmembrane segment.

Each of these mutants except R96D was grown overnight in selective media to prevent plasmid loss and plated on YPGE to isolate suppressors of the mutations. The six membrane sited mutants tested could not be induced to revert by an intragenic second site mutation, though quite a few back mutations were seen. However, the matrix arginine cluster provided suppressor mutations in a gradient of reversion frequency, with R254I giving the most and R252I giving none. Table 1. shows the number of colonies generated and the reversion frequency obtained. With one exception, R96(H,L,P) and R294L revertants were shown to be caused by recombination with a portion of the AAC2 gene still present in the chromosome. Either the mutation on the plasmid was replaced by wild type sequence from the chromosome or the HIS3 disruption of AAC2 was replaced by the plasmid sequence, restoring wild type AAC2 to the chromosome. One R96L revertant appeared to be extragenic, since the HIS3 disruption of AAC2 was still present and there were no other mutations in the plasmid AAC2 gene sequence.

Table 1
Selection of Gly^{+} Revertants

Initial Mutation	Number of Gly^{+} Colonies	Reversion Frequency x 10^{-6}	Colonies Analyzed
R96H	12	1/42	10
R96L	>200	>1/2.5	9
R96P	4	1/125	4
R204L	0	<1/1500	0
W235F	0	<1/1000	0
R252I	0	<1/1500	0
R253I	18	1/83	13
R254I	123	1/12	16
R294L	2	1/250	2
		total	54

As shown in Table 1. plasmids were rescued from 54 suppressor strains. The first 32 plasmids were reintroduced into JLY1053 to see if the gly^{+} phenotype was plasmid linked. All but seven were, indicating an intragenic change. Of the seven potential extragenic mutants, six were missing the HIS3 gene disruption of AAC2 and grew on glycerol in the absence of plasmid. These were presumably restorations of AAC2 to the chromosome. All 54 plasmids were sequenced to see if there was a back mutation to the wild type sequence and if not to find the intragenic second site mutation. This search yielded 29 second site mutants, 15 of which were unique. Figure 1. shows the location of these mutants on a topological map of the AAC2 protein.

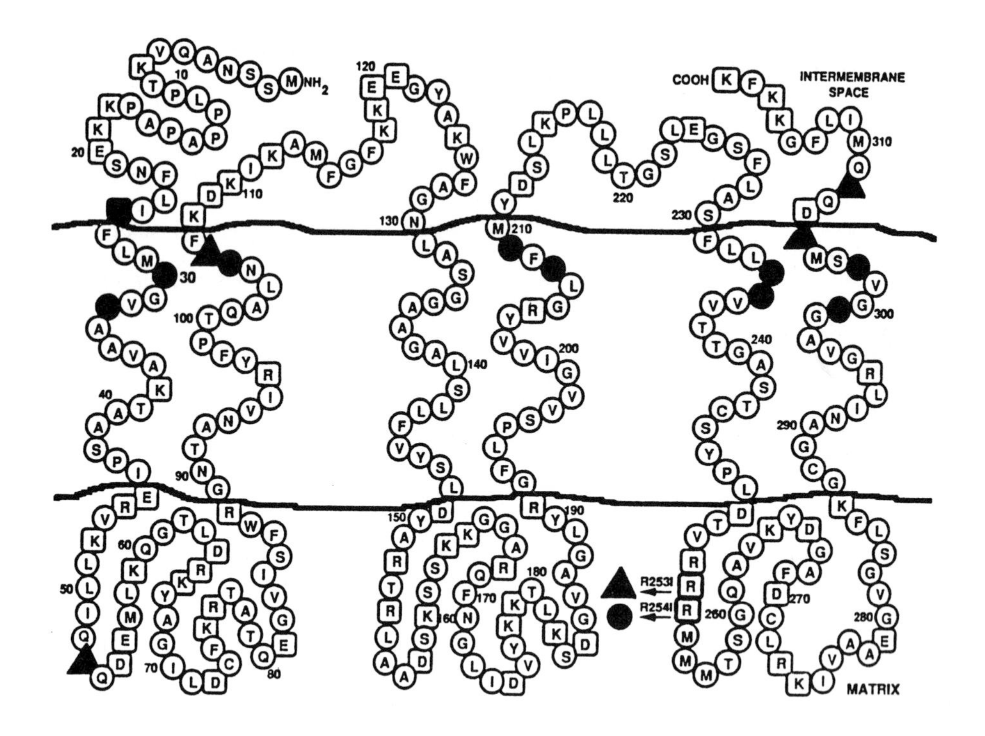

Figure 1. Second site mutants in AAC2 on a three domain topological model of the protein. Charged residues are square. Second site mutants are black. Triangles are revertants of R253I. The others are revertants of R254I. Transmembrane segments are TM1-TM6 from left to right.

Twelve of the revertants were distributed in five of the six transmembrane segments, with two or three in each except for TM3 which had no spontaneous revertants. D26E, at the cytosolic boundary of TM1, was the only charged amino acid affected and the charge was conserved. The only other revertant on the cytosolic side of the membrane was L308P. One other revertant, N53I, was outside the membrane on the matrix side. Since it was on the same side as its parent mutation, these residues may be quite close to one another. The revertants found in the membrane are G30C, S33N, F105L, F105V, A106T, Y207H, G209S, G234S, W235L, A299S, I302T and Y305H. All of these were located 1-7 amino acids down from the intermembrane space in the top third of each helix. The first 100 amino acids are encoded on a Hind III fragment in the AAC2 gene. Isolation of this Hind III fragment allowed ligation of these mutants into wild type AAC2 backgrounds. The phenotypes of D26E, S33N and N53I have been determined and they are gly+. The AAC2 protein with these mutations alone is active.

DISCUSSION

The topological map shown in figure 1 is based on charge distribution, proteolysis experiments on related mitochondrial inner membrane proteins and the internal sequence homology of the translocator. The six transmembrane segments contain only four charges (all positive) while 60 other charged amino acids are found in the external portions of the sequence.

Trypsin digestion of the uncoupler protein in intact mitochondria has shown the C-terminal to be facing the cytoplasm(11). Similar experiments with the bovine phosphate carrier show a cytoplasmic location for the N and C-terminal and a matrix location for residues equivalent to R154 and R168 of AAC2 (12). Since these proteins are believed to share a common ancestor and a common membrane topology with AAC2 (13), the results mentioned for the uncoupler protein and the phosphate carrier should apply to the AAC2 structure. The AAC2 protein belongs to a superfamily of mitochondrial inner membrane transporters with 17 sequenced members. All of these sequences show triplicate internal sequence homology that suggests the ancestral protein arose by two gene duplications (13,14). Comparison of the three domains of AAC2 with each other shows 54/99 amino acid positions are identical in at least one pair of domains. Eight are identical in all three domains. Therefore, each domain is expected to have a similar structure. The topological model in figure 1 is consistent with these facts.

With the exception of N53I, the other 14 revertants appear clustered in a specific region of the transmembrane channel. This area is far (19-228 amino acids) from the starting mutation and suggests a structural alteration has been propagated across the membrane from the R252-R254 matrix Arg cluster. The fact that AAC2 is functional with the single mutations D26E or S33N means that these revertant mutations are not causing major changes. They compensate for R254I so the change induced by this gly- mutation must be fairly small also, yet it is enough to turn off the protein. These results provide a snapshot of what must occur during a transmembrane signaling event. Some effector on one side of the membrane, like a mutation or the binding of a ligand, slightly perturbs the helix arrangement on the other side of the membrane, as evidenced by our second site mutations. The method used here should be equally informative for any membrane protein that has multiple transmembrane segments. Mutations placed in charged residues near the ends of transmembrane segments are the most likely to elicit the desired effect. These residues are probably critical for orientation of the transmembrane segment.

Indeed, this approach has been taken by several groups. Harris *et al.* (15) have used selection of second site revertants to probe the structure of yeast H^+ ATPase. They report finding second site mutations of a cytosolic starting mutant in at least four transmembrane segments, similar to our results with AAC2. di Rago *et al.*(16) did an enormous search for revertants of yeast cytochrome b mutants and found 13 second site mutants. Several were more than 100 amino acids distant from the starting mutation. These studies have also been conducted in proteins with known crystal structures, such as T4 lysozyme and dihydrofolate reductase (DHFR). In lysozyme, crystal structures of two second site revertants suggest that the active site cleft is blocked in the primary mutant, but the active site geometry is restored in the revertants, showing the effect of the second mutation is local (17). Another example of local compensating geometries is seen in tryptophan synthase, where a second site mutation is proposed to restore active site geometry (18). DHFR on the other hand, is an example of a second site mutation being 15Å distant from the active site primary mutation (19). King *et al.* (20) have done revertant selections in lactose permease and found an ion pair that joins two helices in the membrane based on their topological model (see below).

The topological model of Figure 1 provides useful information about AAC2 structure in a side view. Another perspective is gained from a top view. If the transmembrane segments are assumed to be α-helical, constraints on the orientation of each helix should be apparent based on helical wheel projections. The four positive charges must face away from the lipid bilayer and be available to interact with the negative charges on ADP or ATP. Furthermore, second site mutations that restore function should also face away from the lipid. Their role should be in helix:helix contacts or formation of the substrate channel. Figure 2 shows helical wheel diagrams for each transmembrane segment. TM1, TM2, TM5 and TM6 have charges and revertant amino acids on one half of each helix. TM3 has no charges or revertants while TM4 has the three amino acids in a 200° sector, slightly larger than half of this helix.

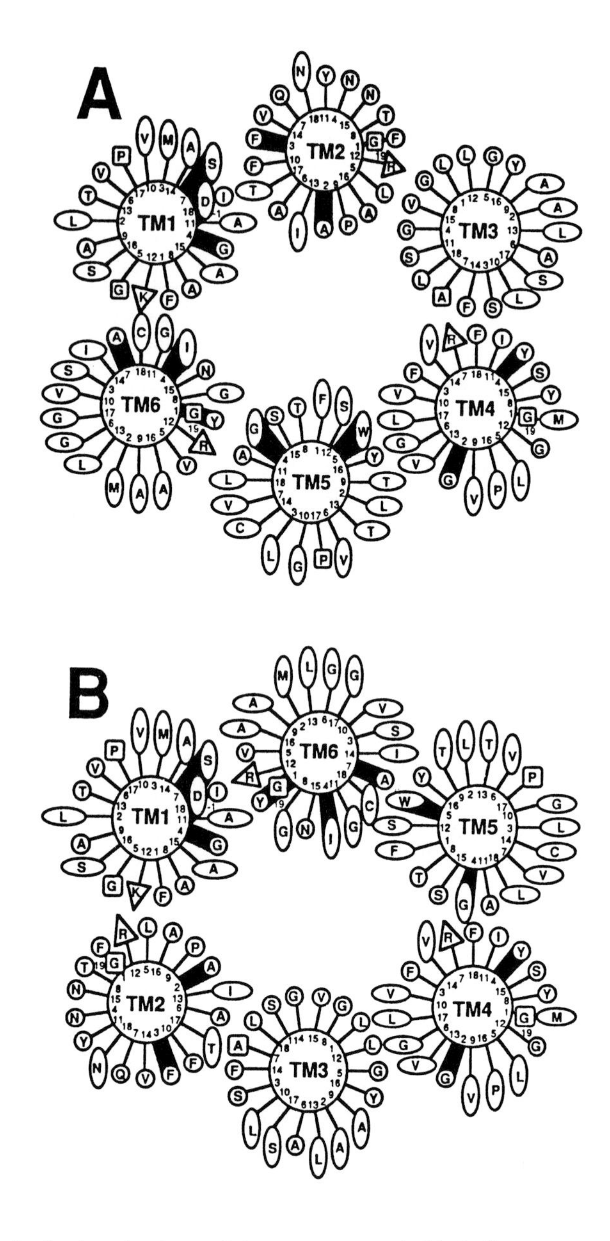

Figure 2. Helical wheel projections of the two most probable helix arrangements. Second site revertants have wide spokes. Positive charges [100% conserved in all translocators] are triangular. Other residues 100% conserved in all 11 translocators are circular. Residues conserved in all 17 carriers are square. Amino acids less than 100% conserved are ellipsoidal. Numbers represent depth in the membrane from the intermembrane space.

Channels A and B depicted in Figure 2 are derived from two assumptions. The first is that transmembrane segments in each of the three domains shown in Figure 1 will be adjacent in the pore structure. This is inherent in the concept of gene duplications giving rise to a three domain ancestor of all 17 of these related carriers. The second assumption is a constraint used by Engelman *et al.*(21) in predicting the bacteriorhodopsin structure. Short connecting peptides between helices were considered unlikely to cross over the channel created by the seven helices in bacteriorhodopsin. This was mainly due to their length being too short to permit this. In the case of AAC2, blockage of the nucleotide channel is also a consideration. Assumption one limits the number of six helix channels to 16 possibilities. Ten of these would have two peptides of 23 and 20 amino acids crossing over the pore, while four others would have one of these two peptides crossing over. Only two arrangements are free of this potential problem. These are shown in Figure 2 as viewed from the intermembrane space. Model A is a clockwise arrangement while model B is counterclockwise. The assumption of TM1 and TM2 being adjacent in a single domain further limits the organization of the pore. Both helices carry a positive charge 12 amino acids down from the surface. In model A, these charges are far apart, but in model B they are in close proximity. The arrangement in model A seems much more likely.

Examination of model A for 100% conserved amino acids shows two categories of conserved residues. One small set of seven amino acids are conserved in all 17 carriers no matter what their substrates are, phosphate, hydroxyl ion, malate or adenine nucleotides. These must be residues essential for the carrier framework. Two are prolines at the ends of TM1 and TM5. Three are glycines at the beginning (19 amino acids down) of TM2, TM4 and TM6. The others are G31 at position 5 of TM1 and A137 at position 7 on TM3. The fact that they are both small suggests a steric constraint present at these positions in all 17 carriers, perhaps related to the conformational changes that must occur during transport. In addition to these seven, there are 53 amino acids 100% conserved in 11 adenine nucleotide translocators. Four of these are positive charges and seven are second site revertants. It is remarkable that half of the second site revertants were in 100% conserved residues. The conserved residues are not evenly distributed among the helices. TM5 and TM6 have only four each, with all non-conserved residues on one side. These helices are easy to orient with the non-conserved amino acids facing the lipid. TM2 and TM3 have 26 conserved amino acids excluding charges and framework residues. These helices cannot be oriented without a large number of conserved residues facing the lipid. One possible rationale for this is dimer formation. AAC2 is purified as a dimer. There must be a site on each monomer that interact to form the dimer. The surface on TM2 and TM3 that faces the lipid may be the site of dimer formation.

The charge location in model A suggests a specific test of the model. Notice that three of the four charges are opposite small residues Gly or Ala. Lys at position 12 on TM1 is opposite Gly at position 11 on TM6. Arg at position 7 on TM4 is opposite Ala at position 7 on TM3. Arg at position 12 on TM6 is opposite Ala at position 11 on TM5. There may be a requirement to pair the bulky Arg and Lys groups with small residues on adjacent helices. Mutation of these charges to Ala and selection for second site revertants may result in a reappearance of the charge on the adjacent helix, effectively switching the pairs. The Arg residues on TM4 and TM6 were already mutated to Leu and these mutants did not revert, but this may be due to the bulky Leu side chain not permitting the switch. If the predicted switches can be found by selection for revertants, it would offer strong support for model A.

The pairing of residues between transmembrane helices by selection for second site revertants has been demonstrated in lactose permease (21). When revertants of K358T were characterized, all 11 contained mutations converting D237 to a neutral amino acid. Twelve revertants of D237N were either back mutants restoring D237 or they were mutations of K358 to a neutral amino acid. The pair had to maintain neutrality or charge complementarity, implying an interaction. In an extensive screening of 210 revertants of yeast cytochrome b mutants, di Rago *et al.* found 16 independent revertants of G137E that were N256K. Again, charge complementarity was always maintained. This indicated a pairing between these two residues. These authors also found other relationships that helped them to orient helices, but they did not find revertants on the opposite side of the membrane based on their topological model.

Our results with AAC2 have shown how useful a genetic approach can be for studying membrane transport proteins. The method of selecting second site revertants yields more information than site directed mutagenesis alone, and it may provide a way to three dimensional models of transmembrane channels without elusive crystal structures.

REFERENCES

1. Lawson JE, Gawaz M, Klingenberg M and Douglas MG. J. Biol. Chem. 1990; 265: 14195-14201.
2. Sherman F. Methods Enzymol. 1991; 194: 3-21.
3. Martinez-Arias AE and Casadaban MJ. Mol. Cell. Biol. 1983; 3: 580-586.
4. Emr SD, Vassarotti A, Barrett J, Takeda M and Douglas MG. J. Cell Biol. 1986; 102: 523-533.
5. Bachmann B, Brooks Low K and Taylor AL. Bacteriol. Rev. 1976; 40: 116-167.
6. Zoller MJ and Smith M. DNA 1984; 3: 479-488.
7. Schiestl RH, and Gietz RD. Current Genetics 1989; 16: 339-346.
8. Hoffman CS and Winston F. Gene 1987; 57: 267-272.
9. Gawaz M, Douglas MG and Klingenberg M. J. Biol. Chem. 1990; 265: 14202-14208.
10. Lawson JE and Douglas MG. J. Biol. Chem. 1988; 263: 14812-14818.
11. Eckerskorn C and Klingenberg M. FEBS Lett.1987; 226: 166-170.
12. Capobianco L, Brandolin G and Palmieri F. Biochemistry 1991; 30: 4963-4969.
13. Runswick MJ, Walker JE, Bisaccia F, Iacobazzi V and Palmieri F. Biochemistry 1990; 29: 11033-11040.
14. Saraste M and Walker JE. FEBS Lett. 1982; 144: 250-254.
15. Harris SL et al. Biophys. J. 1991; 59: 562a .
16. di Rago J-P, Netter P and Slonimski PP, J. Biol. Chem. 1990; 265: 15750-15757.
17. Poteete AR, Dao-Pin S, Nicholson H and Mathews BW. Biochemistry 1991; 30: 1425-1432.
18. Nagata S, Hyde CC, and Miles EW. J. Biol. Chem. 1989; 264: 6288-6296.
19. Howell EE, Booth C, Farnum M, Kraut J and Warren MS. Biochemistry 1990; 29: 8561-8569.
20. King SC, Hansen CL and Wilson TH, Biochim. Biophys. Acta 1991; 1062: 177-186.
21. Engelman DM, Henderson R, McLachlan AD and Wallace BA. Proc. Natl. Acad. Sci. USA 1980; 77: 2023-2027.
22. Supported by grant GM36536 to MGD and National Research Service Award 5 F32 HL08415-02 to DRN. The excellent technical assistance of Yan Li is appreciated.

TRANSPORT OF PROTEIN AND SORTING

© 1992 Elsevier Science Publishers B.V. All rights reserved.
Molecular mechanisms of transport. E. Quagliariello, F. Palmieri, eds.

The structure and function of the *Escherichia coli* Haemolysin A (HlyA) toxin and the membrane localised transport protein, HlyB.

M.A. Blight, B. Kenny, C. Chervaux, A. L. Pimenta, C. Cecchi and I.B. Holland

Institut de Génétique et Microbiologie, Université de Paris-Sud, 91405 Orsay cedex 05, France.

INTRODUCTION

The secretion of the 107KDa haemolysin (HlyA) from *E. coli* depends upon different haemolysin (Hly) determinants [1,2,3], which, together with the *Proteus vulgaris* and *Morganella morganii* determinants [4,5] share an identical genetical organisation: four contiguous genes, *hlyC,A,B,D*. This translocation pathway is an example of a unique and specific export route independent of a classical N-terminal signal sequence and the *secA* gene product.

The 107KDa HlyA polypeptide is synthesized as an inactive precursor [6], proHlyA, activation being mediated by the 20KDa cytoplasmic HlyC protein and endogenous acyl carrier protein (ACP), with the transfer of the acyl group to HlyA, possibly involving an acyl transferase activity inherent to HlyC [7].

The secretion of HlyA is completely independent of its activation [3,6]. Translocation occurs directly across the cell envelope with the absence of a periplasmic intermediate and via a trans-envelope complex consisting of HlyB (66/46 KDa) and HlyD (53KDa) together with endogenous TolC [8,9,10]. Furthermore, secretion requires a novel HlyA C-terminal secretion signal which exhibits no amino acid sequence homology with the classical *sec* dependent N-terminal signal and is uncleaved following translocation [8,11,12,13]. The secretion signal acts independently of the rest of HlyA as has been demonstrated by the secretion of heterologous fusion proteins by the Hly system [14,15].

A NOVEL FAMILY OF TRANSLOCATORS.

It has now become apparent that the haemolysin secretion system, through the participation of the HlyB protein, is a member of a novel family of both pro- and eukaryotic translocation systems involved in the secretion of a wide range of molecules including proteins, peptides, lipophilic drugs, carbohydrate and pigments, secretion being dependent upon a protein with homology to HlyB [see 16 for review].

Cross-complementarity between homologous secretion systems.

The bacterial systems involved in the translocation of even quite diverse cytotoxic polypeptides share considerable genetic organisation typically involving C,A,B and D proteins with the addition of CyaE (a TolC homologue), also required for secretion, in *Bordetella pertussis* and either endogenous TolC, required for HlyA, or its homologue in *Erwinia chrysanthemi* (PrtF). Nevertheless there is little obvious C-terminal amino acid sequence conservation between HlyA,LktA,CyaA,PrtB and protease SM. Surprisingly, therefore, there exists an intriguing cross-complementarity with respect to secretion. The *Pasteurella hamolytica* LktA [17], the *Bordetella pertussis* CyaA [18] and the *Erwinia chrysanthemi* proteases A and B [19] are secreted by *E.coli* when expressed *in trans* with HlyB and D and the *Serratia marcescens* protease SM is secreted from *E.coli* when expressed *in trans* with the *Erwinia chrysanthemi* PrtD,E and F proteins [20]. These cross-complementation experiments indicate that the C-terminal signal sequences must indeed share common characteristics.

Therefore, it is clear that within this diverse and expanding family of novel translocation systems, there are two major features that require investigation. The nature of the substrate specificity and its recognition by the translocator, exemplified by the HlyA C-terminal secretion signal and the membrane localised translocator, HlyB.

The ATP-Dependent membrane localised translocators.

The family of homologous translocation systems may be subdivided into three major categories. Those prokaryotic systems which are involved in the import of small substrate molecules [see 21,22], such as histidine (HisJ,Q,M,P), oligopeptides (OppA,B,C,D,F), maltose (MalG,F,E,K,LamB,MolA) and phosphate (PhoS,PstC,A,B,PhoU), where the ATP-binding polypeptide (HisP, OppD & F, MalK and PstB) and membrane translocator are separate molecules. This class of import system possibly led to the evolution of the HlyB-like export proteins where the ATP-binding and membrane traversing components are together within the same polypeptide. This latter class may be subdivided into the prokaryotic systems which share similar genetic organisation and a C-terminal secretion signal and eukaryotic systems consisting of a protein with homology to HlyB.

Haemolysin B (HlyB).

The *hlyB* gene (2124bp) encodes a 707 amino acid polypeptide (HlyB) with a predicted molecular weight of 79.9KDa [23,24]. HlyB consists of two major domains, an N-terminal, hydrophobic domain, predicted to consist of 6 or 8 membrane spanning regions and a C-terminal ATP-binding consensus motif [16,25] (Figure 1). The latter region is highly conserved between all of the HlyB homologues, whereas the hydrophobic membrane traversing domain shares little amino acid sequence conservation within the family. In fact, the most dramatic example of homology between the membrane domains within the family comes solely from their similar hydropathy profiles [26]. However, at the C-terminal limits of the hydrophobic domain, proximal to the ATP-consensus, the sequences across the family do bear a degree of conservation. This latter observation becomes important with respect to the analysis of the structure and function of this family of polytopic membrane localised transporters (see later).

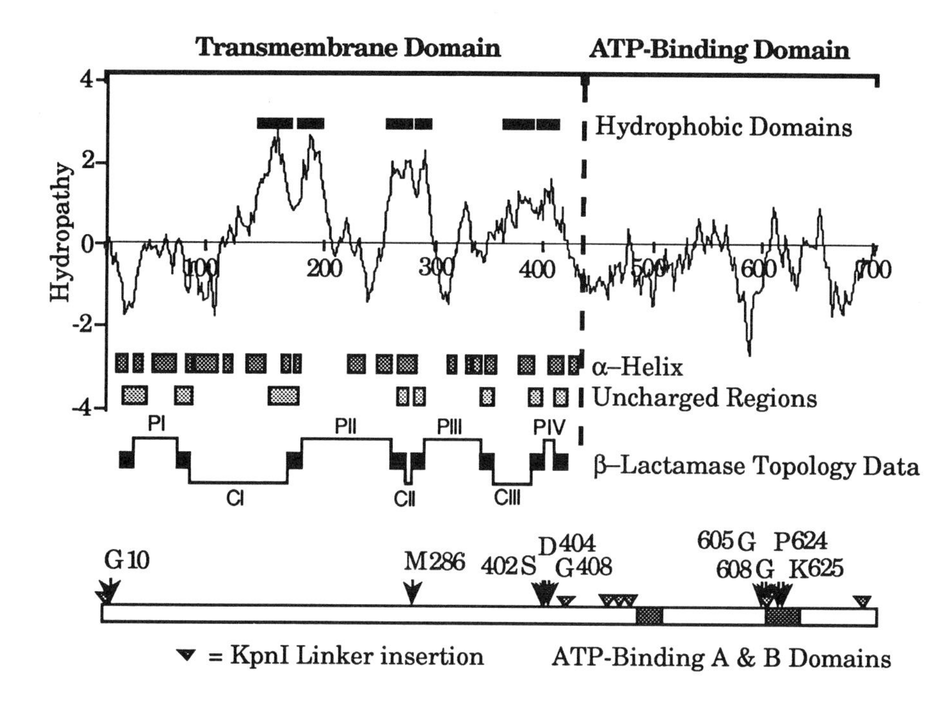

Figure 1. HlyB hydropathy profile, predicted α-helices, transmembrane domains predicted from β-lactamase fusion data, together with the cytoplasmic (CI-CIII) and periplasmic (PI-PIV) loops, uncharged regions and point mutations together with KpnI linker insertions.

Expression of *hlyB in vitro* and *in vivo,* in *E.coli* minicells led to the identification of two polypeptides with molecular weights of 46 and 66KDa [23,24]. Analysis of the open reading frames within *hlyB* reveals a 3' ORF in-frame within the full-length ORF and with the potential to encode a 46KDa polypeptide [23]. This, therefore, suggests that the 46KDa HlyB may correspond to the C-terminus of the full-length protein. However, transposon mutagenesis of *hlyB* [24] and the subsequent analysis of truncated HlyB peptides in radiolabelled *E.coli* minicells and *in vitro* indicated that the 46KDa HlyB originates from the N-terminus of the full-length protein. Furthermore, cloning and expression in *E.coli* minicells of the 46KDa ORF led to the identification of a 42KDa protein (unpublished data). In addition, the expression of HlyB containing a C-terminal epitope from the mammalian multi-drug resistance protein (Mdr) [27], revealed a 66KDa protein on Western blotting of *E.coli* cell membranes. These results again suggest that the observed 46KDa HlyB does not originate from the C-terminus of HlyB, but rather from the N-terminus.

Localisation and membrane topology of HlyB.

Radiolabelling of *E.coli* minicells harbouring plasmids encoding *hlyB* and/or *hlyD* has shown that both polypeptides are localised to the envelope fraction when membranes were analysed by sarkosyl solubilisation or sucrose density gradient centrifugation [24,25]. These fractionation studies further localised the HlyB 66KDa species and HlyD to the inner membrane in minicells and whole cells [27]. Significant amounts of the HlyB 46KDa polypeptide were found in the outer membrane fraction, consistent with the prescence of a trans-envelope complex, which might also involve the outer membrane protein TolC [10]. Protease accessibility studies on *E.coli* minicells and minicell spheroplasts expressing HlyB and HlyD supported the fractionation data, revealing that both HlyB 66KDa and HlyD were protease sensitive in the absence of an outer membrane.

Western blotting of whole cell fractions expressing HlyB with the C-terminal Mdr epitope [27] also identified a soluble, cytoplasmic 65KDa HlyB which suggests that there may be a post-translational modification step prior to membrane insertion.

The membrane topology of HlyB has been investigated using the β-lactamase topology probe system of Broome-Smith & Spratt [28]. A total of 29 fusions were obtained [25], topological organisation of HlyB being indicated by ampicillin sensitivity for fusions where β-lactamase is cytoplasmically localised and ampicillin resistance when periplasmically localised. These studies have led to the formulation of two 6 and 8 membrane domain topology models for HlyB with 11-12 amino acids spanning the membrane. The 8 domain model being constructed with the aid of hydropathy data suggesting the possibility of 2 extra membrane traversing domains between β-lactamase fusions. The 8 domain model has cytoplasmically localised N- and C-terminii and 4 periplasmic (PI-IV) and 3 cytoplasmic (CI-III) extra-membranous domains. Interestingly, a comparison of the experimental topology data with the HlyB hydropathy profile and secondary structural α-helix predictions shows little agreement with what one might expect (Figure 1). The α-helices, hydrophobic domains and experimentally determined transmembrane domains do not coincide consistently.

Similar difficulties have been encountered with the membrane topology analysis of the *E.coli* F_1/F_0 ATPase α subunit. In this case, two different topology models have been proposed from PhoA fusion data. The model of Lewis *et al.*, [29] predicts an 8 transmembrane domain model with 10-15 residues spanning the membrane and with the N- and C-terminii on the periplasmic side of the membrane. Conversely, the model of Bjørbœk *et al.*, [30], predicts 8 transmembrane domains but in different positions within the polypeptide. Moreover, the N- and C-terminii are predicted to be on the cytoplasmic side of the membrane. Furthermore, other polypeptides have been described that are predicted, as HlyB, to contain large extra-membranous domains, such as the *Pseudomonas fluorescens* signal peptidase II [31]and the human TAPA-1 protein [32]. Interestingly, the *P. fluorescens* signal peptidase II is predicted to span the membrane with 4 transmembrane β-sheet structures. These examples are in contrast to polypeptides such as SecY, which does not appear to have large extra-membranous domains and the hydropathy and secondary structure predictions agree well with PhoA fusion topology data [33].

From the foregoing we wish to stress that topological predictions on the basis of hydropathy data and secondary structure predictions should be viewed with caution. This might be particularly so when dealing with polypeptides possessing potentially large extra-membranous domains. Such domains may contain hydrophobic substrate binding pockets, for example. This predictive dilemma is most accutely illustrated when aligning the amino acid sequences of the HlyB homologues across both the pro- and eukaryotic systems. Where there is good hydropathy agreement, suggesting that these proteins may have similar topological organisation within the membrane, there is little overall amino acid sequence homology, little predicted α-helicity and a relatively high distribution of charged residues indicating the prescence of large, extramembranous domains. Therefore, predicting topologies may be quite inefficient within this family of translocators and it will be necessary to obtain experimental evidence of topological organisation, in order to identify and compare domains with potentially similar functions.

HlyB functional domains.

The absence of any strong amino acid sequence conservation within the N-terminii of the HlyB-like family clearly suggests that the specificity for the recognition and transport of a variety of substrates lies within the membrane localised domains. Therefore, cytoplasmic loops may be involved in substrate recognition and periplasmic loops in substrate translocation and interaction with other components such as HlyD and TolC. We have isolated several mutations within HlyB that allow a preliminary functional analysis of this transporter.

Although ATP hydrolysis has been demonstrated as necessary for histidine import [34], for oligopeptide, maltose and glycine-betaine import [35] and for lipophilic drug export by the mammalian HlyB homologue, Mdr [36,37]; it has not been conclusively demonstrated as necessary for HlyA secretion by the HlyB,D,TolC translocator. However, several mutations have been generated within a putative kinase consensus motif (Gly-X-Gly-X_2-Gly-X_{16}-Lys) between residues 603 and 625 of HlyB within the B site of the ATP-binding consensus which dramatically reduce the secretion of HlyA. This motif is conserved in the *Pasteurella haemolytica* LktB but degenerates in the other closely related homologues such as *Bordetella pertussis* CyaB. However, this region is apparently important for HlyB function since substitution of Gly^{605} with Ala, Gly^{608} with Arg and Lys^{625} with Ile give a 60%, 99% and a 98% reduction in HlyA secretion respectively [38].

We have now made several mutations by site-directed and random (hydroxylamine) mutagenesis within HlyB and have located several residues important for secretion activity (Figure 1). Substitution of Gly^{10} with Arg, Gly^{408} with Asp and Pro^{624} with Leu renders HlyB temperature sensitive for secretion of HlyA (unpublished data). Met^{286} to Val leads to a 80% reduction in secretion of HlyA [27], Ser^{402} to Lys and Asp^{404} to Arg together and Ser^{406} to Arg abolish HlyA secretion (unpublished data). In each of the above null mutants, evidence for correct membrane localisation of the defective HlyB polypeptide is still lacking. Therefore, the possibility remains that the Sec^- phenotype is due to proteolytic degradation or aberrant membrane localisation. Insertion of 12mer oligonucleotide KpnI linkers into HlyB HpaII restriction sites (Figure 1) at amino acid residues, 423, 462, 473, 482, 602 and 694 and the introduction of mammalian Mdr epitopes, either by insertion (at residue 1) or site directed mutagenesis (residue 642) also abolished the secretion of HlyA [27]. The only epitope insertion that retained the ability to secrete HlyA was at residue 703. In at least 2 of the above mutants; insertion of and conversion to an Mdr monoclonal epitope of residues 1 and 642 to 650 respectively, the secretion defective phenotype could be attributed to an absence of the HlyB protein within the membrane rather than incorrect membrane assembly. Therefore, it is important to determine the exact nature of the secretion defective phenotype for each mutant and this is currently being undertaken with the HlyB point mutations using antibody raised against over-expressed domains of the polypeptide.

Approaching a tertiary membrane topology for HlyB.

Although direct amino acid sequence alignments between the HlyB-like translocators indicates little sequence conservation, with the exception of the ATP-binding motif, there is apparently greater conservation within the region proximal to the ATP consensus, corresponding to the HlyB periplasmic loop IV (PIV) (Figure 1). In order to determine if this region is important for the structure and function of HlyB we are currently performing mutagenesis experiments followed by both intra-genic and inter-genic (in HlyD and HlyA) suppressor isolation.

It has been demonstrated that first-site mutations within the mitochondrial cytochrome b can be suppressed by intra-genic mutations [39,40]. Moreover, the suppressor mutations occur on the same side of the membrane as the first-site mutation, indicating that suppression is occuring between amino acid residues which are adjacent in the tertiary structure of the molecule. This latter concept is perhaps more easily acceptable with membrane proteins than with soluble proteins, since in the former instance the polypeptide structure is constrained by the membrane traversing domains. First-site mutations are being introduced into HlyB PIV by randomised site-directed mutagenesis [41] which involves "doping" an oligonucleotide at each nucleotide position with the other nucleotides. This approach has already been used upon the secretion signal of HlyA and been shown to generate multiple first site mutations in one *in vitro* mutagenesis reaction (see later).

Using secretion defective first-site HlyB mutants, second site suppressors are being isolated with the subsequent selection of the Sec^+ phenotype. With this approach it is hoped that a map of first site Sec^- and second site suppressor (Sec^+ restoring) mutations within PIV will enable the consolidation of the two-dimensional β-lactamase fusion topology map and its refinement into a three-dimensional map indicating which loops interact with each other.

Such an approach may lead to a more rigourous membrane topology and functional domain map for HlyB. The application of this technique to other members of the family may then enable the determination of topology versus function in this important group of proteins.

THE HLYA C-TERMINAL SECRETION SIGNAL.

Previous deletion analyses of the HlyA C-terminal secretion signal have identified a specific region of the C-terminus required for efficient translocation [8,12,13,14,42]. However, these studies have not yielded a clear definition of the boundaries and critically important domains and/or amino acids of the secretion signal. If cooperatively interacting residues are dispersed throughout this region and are required *in toto* for efficient secretion then deletion analyses alone may be expected to yield contradictory data.

Therefore, we have generated multiple single, double and triple point mutations in the 3' end of *hlyA* corresponding to the last 46 C-terminal amino acids [15] in an attempt to delineate which residues and which secondary structures might be required for targeting to and/or translocation by the HlyB,D and TolC translocator complex. Currently we have 30 mutants within this region, 21 of which give rise to secretion defects.

Secondary structures within the HlyA C-terminal 52 amino acids.

The closely related HlyA homologues [15] bear little amino acid sequence conservation and therefore, direct sequence comparison yields little consistent information regarding the potential secondary structural elements within the C-terminal secretion signal. Having said this, four potential domains have been described (Figure 2).

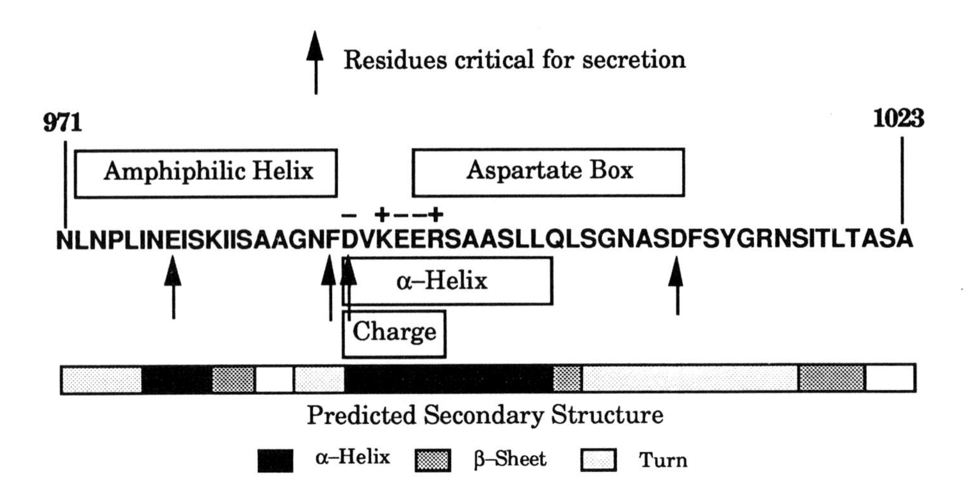

Figure 2. The HlyA C-terminal 52 amino acids showing the predicted secondary structures and the four residues critical for secretion, Glu^{978}, Phe^{989}, Asp^{990} and Asp^{1009}.

An α-helix predicted by 8 secondary structure programs is present in a similar position in HlyA,CyaA, PrtB, NodO and *C. fasciculata* PGK [15] between Asp^{990} and Leu^{1001}. The substitution of Ser^{996} for Pro is predicted to disrupt this helix in HlyA. However, this mutation has no effect upon secretion. Several other single residue changes, predicted to disrupt or perturb the α-helix, again have no effect upon secretion. However, 2 triple residue substitutions, Glu^{994}-Asp, Ala^{997}-Pro, Ser^{999}-Ala and Ala^{998}-Pro, Ser^{999}-Cys, Gln^{1002}-Arg lead to a 3-4 fold reduction in secretion. These data suggest that the putative α-helix is not essential for secretion.

The charge distribution between HlyA Asp^{990} and Arg^{995} follows a pattern of; DVKEER (-0+--+). However, this motif is only conserved in the closely related *Proteus vulgaris* HlyA and is absent from the other homologues. Charge substitutions ranging from -1 to 0 within this region of HlyA led to no significant reduction in secretion, indicating that this domain is not essential for secretion [15].

The region between Leu^{972} and Phe^{989} is predicted to form an amphiphilic α-helix on a Schiffer-Edmunson wheel [13]. However, it is not predicted to be a continuous helical domain by a suite of 8 secondary structure prediction programs. Replacement of Glu^{978} by Lys, although altering the charge, does not alter the amphiphilicity, yet leads to a dramatic 10 fold reduction in secretion. Conversely, substitution of Glu^{978} by Val changes the amphiphilic nature of the helix but only reduces secretion by a factor of 2. This data suggests that it is not the amphiphilicity of this domain, but its charge that is important for secretion [15].

A comparison of the C-terminal secretion signals reveals an apparently conserved feature composed of 12-14 small residues, particularly Ser and Ala, flanked by negatively charged residues, usually Asp. We have called this domain the "Aspartate box" (Glu^{994}-Asp^{1009}). However, despite the promising conservation across the family, point mutations within and deletion of the C-terminal half up to Glu^{1002} do not abolish secretion. Therefore, we have concluded that this domain is not necessary for secretion [15].

Key residues within the secretion signal.

Despite the apparently dispensable nature of the α-helix, amphiphilic helix, charged domain and aspartate box, we have identified a minimal secretion signal domain with a proximal boundary extending to -46 (Glu^{978}) and at least 4 amino acid residues that are critically important for the efficient secretion of HlyA; Glu^{978}, Phe^{989}, Asp^{990} and Asp^{1009}. This data then suggests that, in the light of the poor sequence conservation across the C-terminii of the HlyA-like secreted proteins and the necessity for dispersed amino acid residues for efficient secretion, the secretion signal may consist of a discrete tertiary structure that is only required to position several key residues within three dimensions for efficient recognition and/or translocation by the translocator complex. Providing that these key residues are positioned correctly, many structures might be envisaged that can act as a secretion signal, thereby giving rise to the diverse nature of this domain still consistent with its cross-complementarity between secretion systems.

REFERENCES

1 Goebel W, Hedgpeth J. J. Bacteriol. 1982; 151: 1290-1298
2 Welch RA, Hull R, Falkow, S. Infect.Immun. 1983; 42: 178-184
3 Mackman N, Holland, IB. Mol.Gen.Genet. 1984; 193: 312-315
4 Koronakis V, Cross M, Senior B, Koronakis E, Hughes C. J.Bacteriol. 1987; 169(4): 1509-1515
5 Welch RA, Infect.Immun. 1987; 55: 2183-2190
6 Nicaud J-M, Mackman N, Gray L, Holland IB. FEBS Lett. 1985; 187(2): 339-344
7 Issartel J-P, Koronakis V, Hughes C. Nature(Lond). 1991; 351: 759-761
8 Gray L, Mackman N, Nicaud J-M, Holland IB. Mol.Gen.Genet. 1986; 205: 127-133
9 Gray L, Baker K, Kenny B, Mackman N, Haigh R, Holland IB. J.Cell.Sci. Supplement 11 1989; "The Eigth John Innes Symposium - Protein Targetting." Chater KF, Brewin NJ, Casey R, Roberts K, Wilson TMA, Flavell RB, eds. The Company of Biologists ltd., Cambridge.
10 Wandersman C, Delepelaire P. Proc.Natl.Acad.Sci.USA. 1990; 87: 4776-
11 Felmlee T, Pellet S, Lee EY, Welch RA. J.Bacteriol. 1985a; 163: 88-93
12 Nicaud J-M, Mackman N, Gray L, Holland IB. FEBS Lett. 1986; 204: 331-335
13 Koronakis V, Koronakis E, Hughes C. EMBO J. 1989; 8: 595-605

14 Mackman N, Baker K, Gray L, Haigh R, Nicaud J-M, Holland IB. EMBO J. 1987; 6(9): 2835-2841
15 Kenny B, Taylor S, Holland IB. 1991; Submitted to EMBO J.
16 Blight MA, Holland IB. Mol.Microbiol. 1990; 4(6): 873-880
17 Chang Y-F, Young R, Moulds TL, Struck DK. FEMS Microbiol.Lett. 1989; 60: 169-174
18 Masure HR, Au DC, Gross MK, Donovan MG, Storm DR. Biochemistry. 1990; 29: 140-145
19 Létoffé S, Delepelaire P, Wandersman C. EMBO J. 1990; 9(5): 1375-1382
20 Létoffé S, Delepelaire P, Wandersman C. J.Bact. 1991; In Press
21 Higgins CF, Hiles ID, Salmond GPC, Gill DR, Downie JA, Evans IJ, Holland IB, Gray L, Buckel SD, Bell AW, Hermodson MA. Nature(Lond) 1986; 323: 448-450
22 Ames GFL. Ann.Rev.Biochem. 1986; 55: 397-425
23 Felmlee T, Pellet S, Welch RA. J.Bacteriol. 1985b; 163: 94-105
24 Mackman N, Nicaud J-M, Gray L, Holland IB. Mol.Gen.Genet. 1985; 201: 529-536
25 Wang R, Seror SJ, Blight MA, Pratt JM, Broome-Smith JK, Holland IB. J.Mol.Biol. 1991; 217: 441-454
26 Holland IB, Kenny B, Blight MA. Biochimie. 1990; 72: 131-141
27 Juranka P, Zhang F, Kulpa J, Endicott J, Blight MA, Holland IB, Ling V. 1991; Submitted to J.Bacteriol.
28 Broome-Smith JK, Spratt BG. Gene. 1986; 149: 341-349
29 Lewis MJ, Chang JA, Simoni RD. J.Biol.Chem. 1990; 265(18): 10541-10550
30 Bjørbœk C, Foërsom V, Michelson O. FEBS 1990; 260(1): 31-34
31 Isaki L, Beers R, Wu HC. J.Bacteriol. 1990; 172(11): 6512-6517
32 Levy S, Nguyen VQ, Andria ML, Takahashi S. J.Biol.Chem. 1991; 266(22): 14597-14602
33 Akiyama Y, Ito K. EMBO J. 1987; 6: 3465-3470
34 Bishop L, Agbayani Jr R, Ambudkar SV, Maloney PC, Ames GFL. Proc.Natl.Acad.Sci.USA. 1989; 86: 6953-6957
35 Mimmack ML, Gallagher MP, Pearce SR, Hyde SC, Booth IR, Higgins CF. Proc.Natl.Acad.Sci.USA. 1989; 86: 8257-8261
36 Cornwell MM, Tsuruo T, Gottesman MM, Pastan I. FASEB J. 1987; 1: 51-54
37 Hamada H, Tsuruo T. J.Biol.Chem. 1988; 263: 1454-1458
38 Koronakis V, Koronakis E, Hughes C. Mol.Gen.Genet. 1988; 213: 551-555
39 diRago JP, Netter P, Slonimski PP. J.Biol.Chem. 1990a; 265(6): 3332-3339
40 diRago JP, Netter P, Slonimski PP. J.Biol.Chem. 1990b; 265(26): 15750-15757
41 Hutchison III CA, Nordeen SK, Vogt K, Edgell MH. Proc.Natl.Acad.Sci.USA. 1986; 83: 710-714
42 Hess J, Gentscher I, Goebel W, Jarchau T. Mol.Gen.Genet. 1991; 224: 201-208

© 1992 Elsevier Science Publishers B.V. All rights reserved.
Molecular mechanisms of transport. E. Quagliariello, F. Palmieri, eds.

Membrane translocation during the biogenesis and action of cholera and related enterotoxins

T. R. Hirst, S. J. Streatfield, H. M. Webb, A. Marcello, R. Leece, T. Amin, T. O. Nashar and J. Yu

The Biological Laboratory, University of Kent, Canterbury, CT2 7NJ, United Kingdom

INTRODUCTION

Bacterial species belonging to the genus *Vibrio* exhibit a remarkable capacity to secrete proteins through their double-membraned envelopes into the surrounding medium [1-10]. The most well-characterized of these proteins is cholera toxin (CT) produced by *V. cholerae*, the organism responsible for severe and at times fatal diarrhoeal disease in humans [11-13]. CT is comprised of six non-covalently associated subunits; consisting of a single enzymatically active A-subunit (Mr; 28,000) that possesses ADP-ribosyltransferase and NAD-glycohydrolase activities and five identical B-subunits (Mr; 12,000 each) that bind the toxin to GM1-ganglioside receptors found on the surfaces of eukaryotic cells [11-13]. CT is now recognized as the prototype of a family of structurally and functionally related enterotoxins produced by diarrhoeagenic strains of *E. coli, Salmonella, Campylobacter, and Aeromonas sp.*

The expression of CT or the related *E. coli* heat-labile enterotoxin (LT) in *E. coli* was found to result in both of the toxins being exported to the periplasmic space between the bacterial cytoplasmic and outer membranes [14-16]. In contrast, the expression of LT or CT in *V. cholerae* led to their efficient and selective secretion across the outer membrane into the extracellular medium [17-19]. This finding implicates the existence of a remarkable, but as yet uncharacterized secretory system for translocating oligomeric enterotoxins through the outer membranes of certain Gram negative bacteria [20].

The action of cholera toxin and related enterotoxins on eukaryotic cells involves a further translocation event in which the A-subunit is targeted across the plasma membrane [21]. In this review we describe the three, *very different* membrane translocation events associated with the biogenesis and subsequent action of enterotoxins on eukaryotic cells.

TRANSLOCATION ACROSS THE BACTERIAL CYTOPLASMIC MEMBRANE

The genes that encode the CT and LT subunits are organized as polycistronic operons located either in the bacterial chromosome (for cholera toxin) or on large conjugative *ENT* plasmids (for *E. coli* LT). mRNA translation results in the synthesis of the A- and B-subunits as separate precursor polypeptides that contain at their amino-termini 18 and 21 residue signal peptides, respectively. Translocation of the precursor subunits across the bacterial cytoplasmic membrane is thought to utilize the same mechanism as that involved in the export of all envelope proteins. Much of our knowledge of this mechanism has come from studies in *E. coli* of the export of various periplasmic binding proteins and proteins located in the outer membrane. This has revealed the existence of an efficient export apparatus comprised of SecA, SecD, SecE, SecF, and Sec Y, as well as two peptidases that cleave off the signal sequences of precursor polypeptides [for a review, see Ref. 22]. In addition, a number of other proteins (eg. SecB and GroEL) have been shown to play a role in the targeting of precursors to the cytoplasmic membrane, by maintaining them in a conformation suitable for their entry into the export apparatus [for a review, see Ref. 23]. In other Gram negative bacteria, including *V. cholerae*, the translocation apparatus has as yet to be investigated. Nonetheless its components are widely expected to be very similar, if not identical to those already characterized in *E. coli*.

Biochemical analyses of the kinetics of subunit entry into the periplasm, have revealed that the "mature" processed B-subunits appear there 4 - 6 seconds after polypeptide chain termination [24]. This indicates that B-subunit translocation across the cytoplasmic membrane, with concomitant cleavage of the signal peptide and release of the molecule into the periplasm are rapid events in the overall export process. In contrast, the appearance of mature A-subunits is slower, with processing having a half-time of 30 seconds, and with release of a proportion of A-subunits not being fully completed even after 3 minutes [24]. The reason for this slower processing is unknown, but could be related to ribosomal pausing during mRNA translation or to a slower rate of entry of the precursor A-subunit into the export apparatus.

There are several features of polypeptide transport across bacterial cytoplasmic membranes that remain to be resolved, including the nature of precursor entry into the apparatus, the activity and pathway of interactions with the various Sec-proteins during export and the coupling of energy in the form of ATP-hydrolysis and proton-motive force to the translocation event. In addition, there are several "late" events in protein translocation, and in particular the folding up of the exported molecule as it emerges on the other side of the membrane that are only now beginning to receive attention. The most effective means of evaluating these late events has been through the use of truncated or mutant proteins which fail to complete their export to the periplasm or outer membrane.

Studies on toxin release have focussed on the behaviour of the B-subunit of LT (EtxB) and on various mutant-derivatives of it in which amino acid residues at the carboxyl-terminus have been deleted or modified [25-26]. Deletion of the last two amino acid residues from the B-subunit (designated EtxB215) caused an approximately 50% reduction in the appearance of the molecules in the periplasm, with the remainder being associated with the membrane [26]. A similar observation was made for EtxB191.5 which had had the last four residues replaced by -GLN, although in this case the defect in release was more pronounced. Deletion of the last three residues inhibited B-subunit release altogether, and the molecules could only be detected on the cytoplasmic membrane. Analyses of the susceptibility of the membrane associated B-subunits to proteinase degradation and of their electrophoretic mobility in sodium dodecyl sulphate (SDS) polyacrylamide gels suggested that these molecules were monomers which had failed to be released from the membrane during the export process. There are a number of explanations which could account for this; 1) loss or alteration of carboxyl-terminal amino acids may result in a more hydrophobic molecule which remains associated with the membrane, 2) malfolded B-subunits could accumulate on or in the membrane because of a defect in the folding pathway, or 3) modification might interfere with subunit-subunit assembly which may be a step in either promoting release or in preventing membrane association of monomers after their release from the export apparatus. Analysis of the released periplasmic B-subunits, revealed that both wild-type EtxB and the various mutants had assembled into stable pentamers that were resistant to degradation by proteinase K and migrated as pentamers in SDS polyacrylamide gels [26]. This latter property of the B-subunit has provided a simple means to monitor the formation of assembled B-subunits in *E. coli* and *V. cholerae* and to assess the structural state of the molecule prior to its translocation across the outer membrane of *V. cholerae.*

TOXIN ASSEMBLY

Although the three-dimensional structure of LT has recently been solved [13], the precise pathway of subunit interactions which lead to the formation of this remarkable structure is not known. Studies on the expression of the B-subunits of CT and LT in *E. coli* and *V. cholerae* had revealed that they could form stable pentamers in the absence of concomitant A-subunit synthesis [18]. This led to the hypothesis that enterotoxin assembly proceeded via sequential B- subunit- B-subunit interactions to form a pentamer, followed by association with the A-subunit to give the holotoxin complex [18]. However, a kinetic analysis of the rate of B pentamer formation in the absence or presence of A-subunit synthesis revealed that the A-subunit accelerated the rate of B-subunit pentamerization *in vivo* and increased the yield of pentamers upon refolding and reassociation of denatured B-subunits *in vitro* [27]. This indicated that the A-subunit stabilizes a B-subunit intermediate before the B-subunits attain a stable pentameric structure.

Recently, we have used the crystallographic information on the specific interactions of juxtaposed amino acid residues in adjacent B-subunits to prepare mutants that should be unable form stable B-subunit- B-subunit interactions. In particular, we have concentrated on residues Met-31 and Ala-64. Introduction of an M31D mutation in EtxB had little effect on the amount of assembled B-subunits formed *in vivo*, as determined by a GM1-ELISA assay which detects only B-subunit oligomers. However, the mutation was found to have completely abolished the SDS-stability of any oligomers that had formed. Similarly, an A64D mutation completely abolished the SDS-stability of EtxB and in this instance also caused a 50-fold reduction in the amount of B-subunits detected in the GM1-ELISA. The double mutation, bringing the two aspartate residues into juxtaposition in adjacent B-subunits, resulted in the complete inhibition of B-subunit assembly (H. Webb, T. Sixma, W. Hol & T. R. Hirst - unpublished observations). We anticipate that these assembly-deficient B-subunits will allow us to separate the events of protein folding and release, from the steps associated with toxin assembly.

Assembly of various EtxB mutants in the presence or absence of concomitant A-subunit synthesis has shown that the A-subunit can exert a profound effect on the formation of B-subunit pentamers. This was most clearly exemplified by the assembly behaviour of the mutant B subunit, EtxB191.5.

When the *etxB191.5* gene was recloned into the native *etxAB* operon to yield a plasmid that encoded the wild-type A-subunit (EtxA) and EtxB191.5, it was found that the amount of assembled B-subunits that were produced could be conveniently quantified by a GM1-ELISA. The introduction of an additional mutation early in the *etxA* gene that causes the disruption of A-subunit synthesis, resulted in the complete failure of the encoded EtxB191.5 to assemble into stable oligomers. Quantitatively, there was at least a 100-fold difference in the level of assembled EtxB191.5 when it was expressed in the presence versus the absence of A-subunit synthesis. Thus, under these conditions, the EtxB191.5 subunit is dependent on the A-subunit for its assembly ([28]; M. Sandkvist, S. J. Streatfield, M. Bagdasarian and T. R. Hirst- unpublished results).

This observation has provided a convenient system for further analysing the domains of the A-subunit that are responsible for both interacting with B-subunits and promoting B-subunit assembly *in vivo*. This has led to the remarkable finding that the deletion of the carboxyl-terminal -RDEL sequence from the A-subunit of LT converts the molecule into a molecular chaperone. The loss of the last four amino acid residues from the A-subunit resulted in a molecule that no longer associated with EtxB191.5 to form a hexameric complex, even though this mutant A-subunit still promoted EtxB191.5 assembly ([28];S. J. Streatfield and T. R. Hirst- unpublished results). Indeed the truncated A-subunit was slightly more efficient at promoting B-subunit assembly than the wild-type A-subunit. Thus, removal of the last four residues from EtxA uncovers a "molecular chaperone-like" activity, in which the modified A subunit aids B-subunit assembly, whilst loosing the ability to remain associated with the final complex.

Expression of the truncated A-subunit in the presence of wild-type B-subunits gave a similar result, with the formation of assembled B-subunits which lacked an associated A-subunit . Thus, the loss of the last few residues of the A-subunit prevents stable A-B-subunit interaction. The possible significance of this observation for understanding the translocation of the A-subunit into eukaryotic cells is discussed below.

Assembly of the A and B-subunits is thought to occur in the periplasmic compartment of the bacterial envelope, within a short time after subunit release from the export apparatus. The half-time for the formation of SDS-stable B pentamers in *E. coli* and *V. cholerae* has been analysed by kinetic pulse-labelling techniques and determined to be approximately 1 minute [29]. In the latter case the appearance of assembled toxin in the periplasm is followed by its secretion across the vibrio outer membrane to the surrounding medium.

TRANSLOCATION ACROSS THE BACTERIAL OUTER MEMBRANE

Enterotoxin secretion across the outer membrane of *E. coli* does not occur. Instead the overwhelming majority of toxin remains entrapped within the periplasmic space between the cytoplasmic and outer membranes [15-16], except for a small proportion which has been reported to be released into the medium in association with fragments of outer membrane.

In contrast, enterotoxin production in *V. cholerae* results in the efficient and selective secretion of the toxin across the outer membrane into the surrounding milieu [17-19, 29]. This remarkable difference in the cellular location of enterotoxins in these Gram negative bacteria implies that *V. cholerae* possesses a "secretory machinery" that is absent from *E. coli*. The molecular basis for this phenomenon, and in particular the identity and characterization of any genes that encode the secretory machinery are a major focus of our present studies (see below).

Pulse-labelling experiments on *V. cholerae* strains expressing recombinant LT established that the subunits transiently entered the periplasm prior to their secretion through the outer membrane [19]. Measurements of the rate of radiolabelled-toxin efflux from the periplasm to the medium indicated a half-time of approximately 13 minutes. Thus the apparent rate of toxin translocation through the outer membrane is 100 - 200-fold slower than the half-time for B-subunit translocation across the bacterial cytoplasmic membrane. The reason for this is as yet unknown, but it is likely to be related to several of the aspects of toxin translocation through outer membranes. Firstly, it would appear that subunit assembly precedes translocation since the half-time for assembly is only 1 minute. In this respect, whilst expression of only the B-subunit of CT or LT in *V. cholerae* resulted in their efficient secretion, synthesis of only the LT A-subunit resulted in the molecule remaining associated with the cells. This indicates that in order to achieve entry of the A-subunits into the secretory machinery they must firstly assemble with B-subunits. There is thus compelling evidence that toxin transit through the outer membrane involves the translocation of a quaternary complex of assembled subunits. This makes the mechanism of toxin secretion across bacterial outer membranes fundamentally different from other export processes.

A second feature which distinguishes cytoplasmic and outer membrane translocation is the absence of a clearly defined targeting signal for secretion to the medium. For example, the overwhelming majority of exported proteins, including the A- and B-subunits, carry a typical amino-terminal hydrophobic signal peptide, which targets and facilitates precursor export across the cytoplasmic membrane. These are of course cleaved off during export and therefore cannot serve to target the toxin to the outer membrane. Thus, a further explanation for the slow apparent rate of secretion through the outer membrane may relate to the slower rate of interaction between the toxin and the putative secretory machinery.

Our observation that B-subunits are efficiently secreted from *V. cholerae* in the absence of A-subunit synthesis implies that any molecular recognition between the secretory machinery and the toxin, would be via surface-located residues in the B-subunit (or its pentamer). We are currently exploring this possibility by engineering specific alterations in surface-located residues.

In addition, we have investigated the expression of EtxB in a variety of species belonging to the *Vibrionaceae* family, including a non-pathogenic marine vibrio, designated *Vibrio* sp. 60 [2, 7, 9]. Expression of EtxB in this bacterium resulted in the efficient and selective secretion of the B-subunit into the extracellular growth medium, indicating that although *Vibrio* sp. 60 does not normally produce cholera-like enterotoxins, it nonetheless possesses a secretory machinery able to translocate them across its cytoplasmic and outer membranes [9, 30]. Moreover, expression of EtxB in a secretion-minus mutant of *Vibrio* sp. 60 (MVT1192), which had previously been shown to be defective in the secretion of several extracellular proteins [7], resulted in approximately 95% of the B-subunit remaining entrapped within the periplasm of the bacterial cell-envelope [30]. This implies that the mutation in MVT1192 defines a locus that determines a common step or mechanism for the secretion of extracellular proteins, including oligomeric toxins [7, 10, 30].

If such a pleiotropic mechanism does indeed exist it would have to be able to discriminate between resident periplasmic proteins and those passing through the periplasm *on route* to the outer membrane translocation machinery. This would only be conceivable if all secreted proteins possessed a common structural feature or targeting motif that permitted similar interaction with the outer membrane and/or entry into the secretory machinery. As yet however, no targeting motif for translocation across the outer membrane has been identified; and moreover such a motif would seem unlikely given the almost certain differences in structure between assembled EtxB pentamers and those of the protease, amylase, DNase, soluble hemagglutinin, and aerolysin that are also secreted from *Vibrio* sp. 60 [7, 10].

An alternative explanation, is that there is more than one type of secretory machinery, each determining the secretion of a different protein or sub-set of proteins across the outer membrane. In this case the mutation in MVT1192 must effect the functioning of all of the different translocatory mechanisms in the same way. This could occur if the process of protein translocation across the outer membrane relies on a feature or property of that membrane for efficient translocation to happen. For example, it is conceivable that the constituent lipids, or the integrity of the membrane, or a putative membrane potential [10] could be affected in MVT1192 which causes the pleiotropic inhibition of all outer membrane translocatory mechanisms. Indeed, Ichige et al., [7] reported that all secretion-minus mutants of *Vibrio* sp. 60 were difficult to transduce with phage As3, suggesting a change in cell surface structure. It remains to be established whether the mutations in *Vibro* sp. 60 define genes which encode a pleiotropic secretory apparatus, or the integrity and normal functioning of several membrane processes, including secretion of extracellular proteins.

The use of *Vibrio* sp. 60 as a host for the heterologous expression of EtxB has provided a simple means of secreting and purifying large quantities of B-subunit pentamers, with yields of approximately 25mg of EtxB per litre of culture. In addition, we have found that recombinant fusion proteins consisting of EtxB and various C-terminal extensions, such as B-$(NANP)_3$ are also efficiently secreted into the medium by both *V. cholerae* and *Vibrio sp. 60* (T. R. Hirst, T. O. Nashar, and T. Amin - unpublished results). This finding indicates that the C-terminus of EtxB is not functionally involved in toxin translocation across the outer membrane, and that the translocating mechanism can accommodate a significant expansion in the size of the toxoid that crosses the membrane.

Experiments aimed at identifying the genetic basis of cholera toxin secretion from *V. cholerae*, by either complementation of a sec$^-$ mutant (M14) [19] or by use of TnphoA-mutagenesis have unfortunately, as yet, proved unsuccessful (T. R. Hirst, M. Bagdasarian and J. Yu).

TRANSLOCATION INTO EUKARYOTIC CELLS

The current model for CT entry into eukaryotic cells involves the following steps, 1) binding of CT via its B-subunit moeity to a specific cell-surface receptor, monosialoganglioside GM1; 2) endocytosis, via non-coated membrane invaginations of the toxin receptor complex; 3) acid-triggered membrane insertion of the toxin A-subunit within the endosome; 4) translocation of the amino-terminal 22kDa A_1-peptide across the endosomal membrane; and 5) ADP-ribosylation of the adenylate cyclase stimulatory protein $G_{s\alpha}$ at the plasma/basal membrane surface by A_1 [for a review, see Ref 21]. Binding of CT to GM1 is followed by a lag phase of up to 30 minutes before ADP-ribosylation is observed. The fate of CT during this lag phase is still a matter of investigation, but it is generally accepted that the A-subunit must undergo reduction to yield the A_1-peptide and either that CT or the A-subunit or the A_1-peptide must undergo translocation across the membrane during this lag period. Early observations on several cell lines showed that CT undergoes a time and temperature dependent internalization with morphological evidence of a redistribution of the toxin within the plasma membrane. More recently, electron microscopic studies on the entry of gold- or ^{125}I-labelled CT has led to the proposal that an adsorptive endocytosis through non-coated invaginations of the plasma membrane is responsible for toxin internalization [31]. Our finding that deletion of the C-terminal -RDEL sequence from the A subunit results in its failure to stably associate with B pentamers could provide a possible novel mechanism for releasing the A subunit upon B subunit binding to the cell membrane. A membrane-associated protease might conceivably cleave off the exposed -RDEL sequence and thereby release the A subunit from its interaction with the B pentamer. One of the other major steps in CT action on cells is the generation of the A1-peptide from the A-subunit, which is functionally linked with the activation of adenylate cyclase. This step is believed to happen in endosomes after a drop in the pH of the vesicle. This would be consistent with the results obtained with subcellular fractionation of rat liver cells from animals treated *in vivo* with ^{125}I-labelled CT [32].

Many aspects of the events involved in toxin translocation into eukaryotic cells remain to be confirmed, especially the molecular mechanisms of A-subunit reduction and the translocation of the A1-peptide across the membrane. The process of A-subunit translocation into cells may prove to have striking similarities with the proposed mode of entry of diphtheria toxin [for a review see, Ref 21].

The ability of the A_1-peptide to enter eukaryotic cells heralds the possibility of using non-toxic, but translocation competent A-fragments for the delivery of peptides and drugs to human tissues.

ACKNOWLEDGEMENTS

We thank the Wellcome Trust of Great Britain for their financial support.

REFERENCES

1 Finkelstein RA, LoSpalluto JJ. J Infect Dis 1970; Supp. S63-S72.
2 Oishi K, Yokoshima S, et al. Appl Environ Microbiol 1979; 38: 169-172.
3 Young DB, Broadbent DA. Infect Immun 1982; 37: 875-883.
4 Nishibuchi M, Kaper JB. J Bacteriol 1985; 162: 558-564.
5 Mercurio A, Manning PA. Mol Gen Genet 1985; 200: 472-475.
6 Focareta T, Manning, PA. Gene 1979; 53: 31-40.
7 Ichige A, Oishi K, Mizushima S. J Bacteriol 1988; 170: 3537-3542.
8 Hirst TR, Welch RA. TIBS 1988; 13:265-269.
9 Hirst T R, Leece R. Experientia 1991; 47: 429-431.
10 Wong KR McLean DM, Buckley JT. J Bacteriol 1990; 172: 372-376.
11 Holmgren J. Nature 1981; 292: 413-417.
12 Mekalanos JJ, Swartz DJ, Pearson GDN, et al. Nature 1983; 306: 551-557.
13 Sixma TK, Pronk S, Kalk KH, et al. Nature 1991; 351: 371-377.
14 Pearson GDN, Mekalanos JJ. Proc Natl Acad Sci 1982; 79: 2976-2980.
15 Hirst TR, Randall LL, Hardy SJS. J Bacteriol 1984; 157: 637-642.
16 Hofstra H, Witholt B. J Biol Chem 1984; 259: 15182-15187.
17 Neill RJ, Ivins BE, Holmes RK. Science 1983; 221: 289-291.
18 Hirst TR, Sanchez J, et al. Proc Natl Acad Sci 1984; 81: 7752-7756.
19 Hirst TR, Holmgren J. J Bacteriol 1987; 169: 1037-1045.
20 Hirst TR. In: Alouf JE, Freer JH, eds. A Source Book of Bacterial Protein Toxins. London: Academic Press, 1991; 75-100.
21 Montecucco C, Papini E, et al. In: Alouf JE, Freer JH, eds. A Source Book of Bacterial Protein Toxins. London: Academic Press, 1991; 45-56.
22 Schatz PJ, Beckwith J. Ann Rev Genet 1990; 24: 215-248.
23 Kumamoto CA. Mol Microbiol 1991; 5: 19-22.
24 Hofstra H, Witholt B. J Biol Chem 1985; 260: 16037-16044.
25 Sandkvist M, Hirst TR, Bagdasarian M. J Bacteriol 1987; 169: 4570-4576.
26 Sandkvist M, Hirst TR, et al. J Biol Chem 1991; 265: 15239-15244.
27 Hardy SJS, Holmgren J, et al. Proc Natl Acad Sci 1988; 85: 7109-7113.
28 Hirst TR, Streatfield SJ, et al. In Witholt B et al. eds. Bacterial Protein Toxins, Fifth European Workshop. Stuttgart: Gustav Fischer, (in press).
29 Hirst TR, Holmgren J. Proc Natl Acad Sci 1987; 84: 7418-7422.
30 Leece R, Hirst TR. J Gen Microbiol (submitted)
31 Tran D, Carpentier J, et al. Proc Natl Acad Sci 1987; 84: 7957-7961.
32 Janicot M, Fouque F, Desbuquois B. J Biol Chem 1991; 266: 12858-12865.

© 1992 Elsevier Science Publishers B.V. All rights reserved.
Molecular mechanisms of transport. E. Quagliariello, F. Palmieri, eds.

$F_1\beta$ presequence peptide competes for mitochondrial import after precursor binding.

Douglas M. Cyr and Michael G. Douglas.

Department of Biochemistry and Biophysics, University of North Carolina at Chapel Hill, Chapel Hill, North Carolina, USA 27599-7260.

Abstract

To investigate the mechanism of presequence-dependent protein targeting to mitochondria we synthesized and preformed competition assays with a presequence peptide, $F_1\beta$ 1-32+2, which corresponds to amino acids 1-32 of the F_1-ATPase β subunit precursor ($F_1\beta$). Half-maximal import inhibition occurred at a peptide concentration of 0.12 μM . Treatment of mitochondria with peptide $F_1\beta$ 1-32+2 reduced electrical potential across the inner mitochondrial membrane ($\Delta\Psi$), but this mechanism was ruled out as a cause for import inhibition. Instead, $F_1\beta$ 1-32+2 appeared to inhibit import competitively. However, competition did not occur at the level of precursor binding to the outer mitochondrial membrane as large quantities of bound translocation intermediate accumulated on the mitochondrial surface when import was blocked. Thus, it does not appear that naked presequence is recognized by components of the translocation apparatus which mediate precursor binding events.

Introduction

Nuclear genes encode the majority of mitochondrial proteins. Therefore, assembly and maintenance of mitochondria require a protein import pathway. Proteins destined for the matrix are targeted to this pathway by short, typically transisent, stretches amino acids located at the NH_2-terminus of most mitochondrial precursor proteins known as presequences (for reviews see Douglas et al.,1991 and Pfanner and Neupert, 1990). Despite extensive study, the mechanism for specific intracellular delivery directed by presequences is not clear. This is due in large because presequences contain no consensus targeting signal but instead, based on the abundance and regular spacing of basic amino acids within them, share only the predicted ability to adopt a common amphiphilic structure in hydrophobic environments (von Heijne, 1986) . One model is that proteins are targeted to mitochondria through specific recognition or interaction of the presequence with recently identified outer membrane proteins proposed to be import receptors (Pfanner and Neupert, 1991; Sollner et al., 1989, 1990; Hines, et al.,1990). However, specific interaction of the presequence with these proteins has not been demonstrated and efficient presequence dependent import of some precursors can occur if receptor molecules are removed from mitochondria prior to import experiments (Pfaller et al.,1989; Cumsky and Miller, 1991).

To investigate this problem we have synthesized a series of synthetic presequence peptides corresponding regions within NH_2-terminus of the F_1-ATPase β-subunit precursor ($F_1\beta$) and characterized them as competitors of in vitro mitochondrial import (Cyr and Douglas ,1991). Herein we report results from experiments with peptide $F_1\beta$ 1-32+2 which indicate $F_1\beta$ presequence peptides block import in a competitive manner. However, competition appears to occur beyond initial binding of precursors to mitochondria.

Experimental Procedures

Mitochondria were isolated from yeast strain D273-10B (Mat α) grown in 2% lactate medium as previously described (Daum et al. ,1982). Coupled transcription/translation of plasmid DNA to generate ^{35}S-labeled precursor proteins was as described earlier (Cyr and Douglas, 1991). Measurement of mitochondrial $\Delta\Psi$ was achieved using the fluorescent dye 3',3'-dipropylthiadicarbocyanine (Sims et al., 1974) as previously described (Cyr and Douglas, 1991). Import reactions were carried out and assayed as described previously (Cyr and Douglas, 1991). Peptide F1β 1-32+2 was synthesized and HPLC purified by Immuno Dynamics , La Jolla, Calif. USA. Fluorographed gels were quantitated using a laser densitometer and the results of these measurements are expressed in arbitrary units.

Results

Peptide $F_1\beta$ 1-32+2 blocks import in a competitive manner--- Peptide $F_1\beta$ 1-32 +2 (Fig. 1a), which contains the 19 amino acid presequence of $F_1\beta$ plus basic amino acids within the mature protein shown to increase in vivo import efficiency of full length $F_1\beta$ precursor (Bedwell et al.,1987), was used in this study as an import competitor . Import competition was tested in an in vitro assay which involves incubation of ^{35}S-labeled precursor protein translated in reticulocyte lysate with mitochondria isolated from S.cerevisiae (see "Experimental Procedures") and

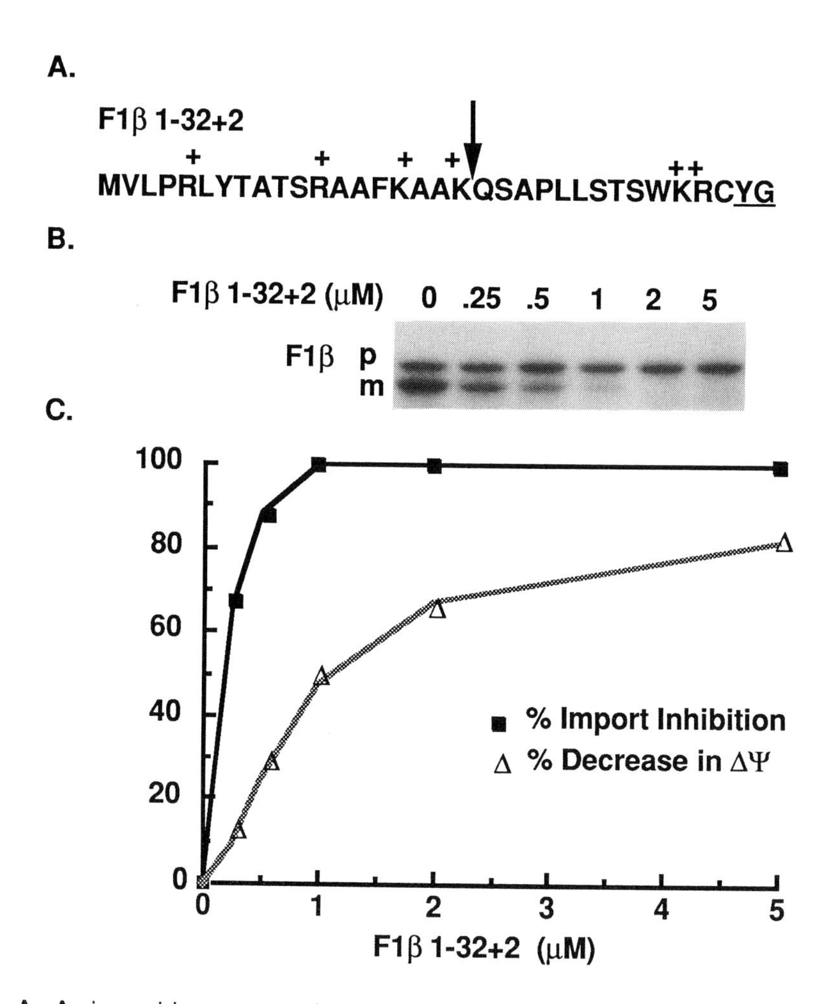

Fig. 1. A. Amino acid sequence of F1β 1-32 . This peptide corresponds to amino acids 1-32 of the F1-ATPase β subunit precursor. The arrow denotes the presequence clevage site. B. Inhibition of F1β precursor import into isolated yeast mitochondria by peptide F1β 1-32+2. Mitochondria (50 μg) were preincubated on ice for 5 min. in 100μl reaction mixtures with or without peptide prior to start of import reactions by addition of 35S-labeled F1β precursor protein in reticulocyte (2 μl, 14,000 cpm) and incubation for 20 min at 23 C. Import was assayed as described in the materials and methods. C. The relationship between dissipation of mitochondrial membrane potential ($\Delta\Psi$) inhibition of import by presequence pepide, Curves represent quantitation of respective decreases $\Delta\Psi$ and inhibition of F1β import observed upon titration of F1β peptide into individual import reactions. See materials and methods for details.

quantitation of precursor processing to a lower molecular weight mature protein to measure import (Gasser et al., 1982). Peptide $F_1\beta$ 1-32+2 inhibited import of $F_1\beta$ precursor in a dose dependent manner and half-maximal inhibition occurred at around 0.12 μM (Fig. 1b). This peptide concentration peptide was similar to the concentration of purified full-length precursors observed to half-maximally block mitochondrial import in other import competition studies (Pfaller et al., 1988 and Sheffield et al., 1990).

Disruption of the inner mitochondrial membrane and dissipation of the electrochemical gradient required to drive protein translocation into mitochondria has been associated with non-specific inhibition of mitochondrial import by synthetic presequence peptides (Gillespie et al., 1885; Roise et al., 1986; Glaser and Cumsky, 1990). To rule out the possibility that $F_1\beta$ 1-32+2 was acting in a similar manner the influence of peptide on both mitochondrial import and $\Delta\Psi$ were compared in companion reactions (Fig 1c). At respective concentrations of $F_1\beta$ 1-32+2 which block import of $F_1\beta$ precursor about 50 and 100% peptide reduced $\Delta\Psi$ by less than 5 and 25% (Fig. 1c). Titration experiments with valinomycin demonstrate that when $\Delta\Psi$ is reduced by 25% import of $F_1\beta$ precursor is inhibited 25% (Cyr and Douglas,1991). Thus at least 75% of import inhibition observed in the presence of peptide $F_1\beta$ 1-32+2 is likely to result from competition for a specific import step.

If $F_1\beta$ 1-32+2 blocks import by saturating a specific import step then addition of extra precursor protein to import reactions overcome import inhibition. To test this idea mitochondria were incubated in reaction mixtures containing 1 and 40ul of reticulocyte containing ^{35}S-labeled $F_1\alpha$ precursor, respectively, and the influence of $F_1\beta$ 1-32+2 on import was tested. $F_1\beta$ 1-32+2 to inhibited import of $F_1\alpha$ 80% in import reactions containing 1ul of precursor (Fig 2 lane 3 vs 4). However, when 40 μl of $F_1\alpha$ precursor was added to import reactions about 15 fold more precursor was imported and import inhibition was reduced to 40% (Fig. 2 lanes 4 vs 6). Thus, import inhibition appears to occur through competition between peptide and full-length precursor for a specific import step.

Import competition by $F_1\beta$ 1-32+2 occurs after precursor binding to mitochondria--- $F_1\beta$ presequence peptides block import without detectably reducing the binding of precursor protein to mitochondria. (Fig 1b, Cyr and Douglas,1991). Additionally, treatment of mitochondria with inhibitory concentrations of peptide $F_1\beta$ 1-32+2 caused a 6-14 fold accumulation of $F_1\alpha$

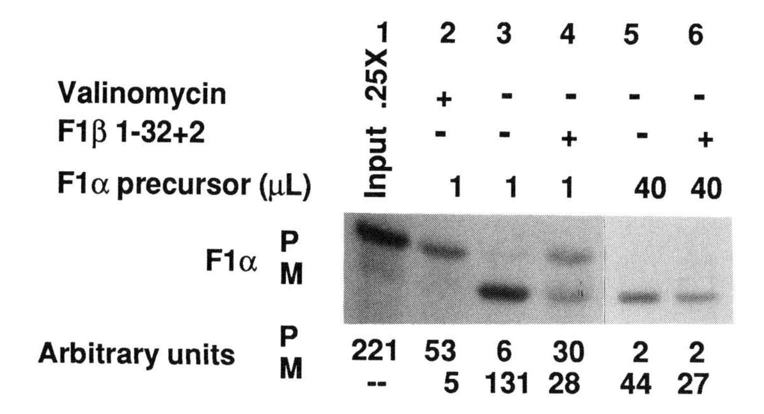

Fig 2. Additional precursor protein in import reactions reduces import inhibition by F1β 1-32+2. F1α precursor was incubated with mitochondria under conditions described in the legend to Figure 1. 1X F1α precursor corresponds to 8,000 cpm of precursor protein in 1μl of reticulocyte lysate (12 μg protein). All reaction mixtures contained a total of 40 μl of reticulocyte lysate. Lane 1, F1α prescursor ,0.25X , was analysed directly. In lanes 2-4 mitochondria form respective reaction mixtures were reisolated, washed and assayed for import. In lanes 5 and 6 reactions mixtures were diluted 40 fold and mitochondria isolated from 1/40 of the reaction mixture were assayed for import. Respective concentrations of F1β 1-32+2 and valinomycin were 1.0 μM and 2.0 μg/ml. Bound precursor was accessable to digestion by protease under all conditions.

precursor on mitochondria (Fig 2 and 3). To obtain evidence that accumulated F1α precursor represented bound translocation intermediates, we asked if bound precursor could be chased into mitochondria from the bound state upon reisolation and reincubation of peptide treated mitochondria in the absence of peptide (Fig 3). In this experiment addition of peptide F1β 1-32+2 to import reactions completely blocked import and 14 times more F1α precursor was found associated with mitochondria (Fig 3. lanes 1 and 2). About 33% of the bound precursor was imported into mitochondria after reisolation and reincubation in the absence of peptide (Fig 3 lanes 2 and 3). That import of precursor in these reactions was occurring from the bound state was indicated by the observation that import was largely insensitive to a 10 fold dilution of mitochondria (Fig. 3 lanes 3 and 4). Thus, in the presence of peptide F1β 1-32+ 2 mitochondria accumulate bound translocation intermediate. Strongly arguing that import inhibition occurs at a site beyond initial binding of precursors to mitochondria.

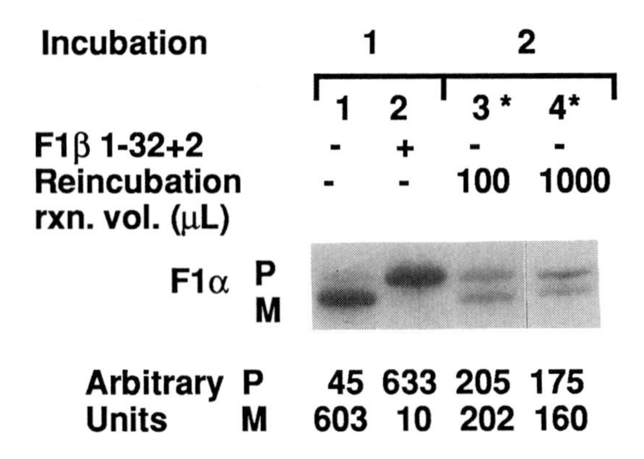

Fig. 3. F1α precursor accumulated on peptide treated mitochondria can be chased into the matrix upon reincubation of mitochondria in the absence of peptide. F1α precursor protein was imported into mitochondria in 300 μl reaction mixtures in the absence and presence of F1β 1-32+2 (1.0 μM) as described in in the legend to figure 1. After import, respective reaction mixtures were divided into 3 aliquots. For lanes 1 and 2, 1/3 of the respective mitochondria were reisolated assayed for import directly. For lanes 3 and 4, 1/3 of the peptide treated mitochondria (*) were reisolated and reincubated without peptide in the indicated import reaction volume. Maturation of the precursor in the second incubation of mitochondria was not observed in the presence of valinomycin indicating precursor processing was $\Delta\Psi$ dependent and resultant from import of the precursor into the matrix and not leakage of the processing protease from damaged mitochondria.

Discussion

The precise role the presequence plays in mediating initial interaction of precursor proteins with mitochondria is not clear. To examine this problem we have studied inhibition of mitochondrial import by synthetic presequence peptides. We find that import inhibition occurs competitively but that competition takes place after precursor binding to mitochondria. Extensive comment on the specificity of import inhibition by F1β presequence peptides has been made elsewhere (Cyr and Douglas,1991) and this discussion will focus on the observation that import competition takes place after precursor binding to mitochondria.

Import inhibition by presequence peptide was accompanied by the accumulation of a significant quantity of bound translocation intermediate on mitochondria (Fig 3). Indicating the import block takes place after precursor binding but before insertion of the presequence across the inner mitochondrial membrane. This result was surprising since binding of precursor proteins to mitochondria is presequence dependent (Reizman et al.,1983; Cyr and Douglas, 1991). One explanation for this result is that naked presequence is not recognized by membrane bound components of the import apparatus during initial binding events.

Instead, a complex between the precursor and a cytosolic translocation factor is probably recognized by import receptors to mediate initial binding events. Consistent with this model, import of $F_1\beta$ and other precursors requires cytosolic ATP and protein factors in reticulocye lysate (Chen and Douglas , 1987 a,b; Pfanner and Neupert, 1990). Additionally, a protein which forms a presequence dependent complex with, and stimulates import of, a purified precursor protein has recently been purified (Murakami et al.,1990).

Not all precursor proteins require cytosolic factors for import (Pfanner and Neupert, 1990 ; Miller and Cumsky, 1991). Failure of synthetic presquence peptides to block binding of precursor proteins to mitochondria indicates they do not present the proper structural features to promote binding of factors necessary for high affinity binding to import receptors. This conclusion is supported by the observation that import of a presequence peptide corresponding to the first 45 amino acids of bovine cytochrome P-450 occurs in a cytosolic and mitochondrial surface factor-independent manner (Furuya et al., 1991). Since naked presequence does not appear to be recognized by import receptors we suggest import receptors serve to recognize and discharge factors associated with precursor proteins which require factors for import and that some precursors which are naturally import competent do not require the assistance of soluble translocation factors or import receptors.

This work was supported by National Institutes of Health Grant GM36537 to M.G.D. and Postdoctoral Fellowship NC-90-F4 from American Heart Association, North Carolina Affiliate to D.C.

References

Bedwell,D.M.,Klinosky D.L. and Emr, S.D. (1987) Mol. Cell Biol. 7, 4038-4047.
Chen, W. J. and Douglas, M.G. (1987a) Cell, 49, 651-658.
Chen, W. J. and Douglas, M.G. (1987b) J. Biol. Chem., 262, 15598-15604.
Cyr, D. M. and Douglas M.G. (1991) J. Biol. Chem. 266, in press.
Douglas, M.G., Smagula, C.S. and Chen, W.J. (1991) Mitochondrial import of proteins. In **Intracellular trafficking of proteins** (ed. C. J. Steer and J.A. Hanover)pp. 658-689. Cambridge, U.K., Cambridge University Press.
Daum, G., Bohni, P.C. and Schatz, G. (1982) J. Biol. Chem. 257,13028-13033.
Furuya, S., Mihara, K., Aimoto, S. and Omura, T. (1991) EMBO J. 10, 1759-1766.
Gasser, S., Daum., G. and Schatz, G. (1982) J. Biol. Chem. 257, 13034 -13041.
Gillespie, L., Argan,C., Taneja, A.T., Hodges, R.S., Freeman, K.B. and Shore, G.C. (1985) J. Biol. Chem. 260., 16045-16048.
Glaser, S.M. and Cumsky M.G. (1990) J. Biol. Chem. 265, 8808-8816.
Hines, V., Brandt, A., Griffiths, G., Hortsmann, H., Brustch, H. and Schatz, G. (1990) EMBO J. 9, 3191-3200.
Miller, B.R. and Cumsky, M. G. (1991) J. Cell. Biol. 112, 833-841.

Murakami, K. and Mori, M. (1990) EMBO J. 9,1129-1135.
Pfaller,R., Steger, H. F., Rassow, J., Pfanner, N. and Neupert, W. (1988) J. Cell. Biol. 107, 2483-2490.
Pfaller, R., Pfanner, N. and Neupert, W. (1989) J. Biol. Chem. 264, 34-39.
Pfanner, N. and Neupert, W. (1990) Ann. Rev. Biochem. 59, 331-353.
Pfanner, N., Sollner, T. and Neupert, W. (1991) Trends Biochem. Sci. 16, 63-67.
Reizman, H., Hay, R., Witte, C., Nelson, N., and Schatz, G. (1983) EMBO J. 2, 1113-1118.
Sheffield, W.P, Shore, G.C. and Randall, S.R. (1990) J. Biol. Chem. 265, 11069-11076.
Roise, D., Horvath, S. J., Tomich, J.H., Richards, J.H. and Schatz, G. (1986) EMBO J. 5, 1327-1334.
Sims, P., Waggoner, A.B. (1974) Biochemistry 13, 3315-3330.
Sollner, T., Griffiths, G., Pfaller, R., Pfanner, N. and Neupert, W. (1989) Cell 59, 1061-1070.
Sollner, T., Pfaller, R., Griffiths, G., Pfanner, N. and Neupert, W. (1990) Cell 62, 107-115.
von Heijne, G. (1986) EMBO J. 5, 1335-13423.

© 1992 Elsevier Science Publishers B.V. All rights reserved.
Molecular mechanisms of transport. E. Quagliariello, F. Palmieri, eds.

TRANSLOCATION OF PREPROTEINS ACROSS THE MITOCHONDRIAL MEMBRANES

J. Rassow, W. Voos and N. Pfanner

Institut für Physiologische Chemie, Universität München
Goethestraße 33, W-8000 München 2, Germany

INTRODUCTION

Mitochondria are assumed to contain about 500 different proteins. More than 90% of the proteins are encoded by the nuclear genome and synthesized in the cytosol. To reach their functional destination the preproteins have to be targeted to the organelle, translocated across the mitochondrial membranes and assembled into their functional form. The basic principles of mitochondrial protein uptake were reviewed in detail (1-3). This report will focus on recent results of our laboratory concerning the mechanisms and structures involved in the translocation of proteins across the membranes.

The proteins which are synthesized in the cytosol of eukaryotic cells are destined for several different locations. How does the cell distinguish between mitochondrial precursor proteins and other polypeptides? Positively charged aminoterminal presequences are a common feature of many mitochondrial precursor proteins. The presequences are sufficient to direct passenger proteins into mitochondria. Recent results suggest that mitochondrial presequences are directly recognized by receptor proteins on the surface of the mitochondrial outer membrane (4, 5). "Cytosolic signal recognition factors" are not required for the binding of precursors to import receptors. However, it has to be emphasized that cytosolic cofactors have chaperone-like functions in mitochondrial protein import by preventing the aggregation and misfolding of preproteins in the cytosol (see below).

After binding to an import receptor, the translocation of presequences into the mitochondrial matrix requires the membrane potential $\Delta\Psi$ across the inner membrane. Due to positively charged amino acids the translocation of the presequences is probably driven by an electrophoretic effect of the membrane potential (6).

In summary, mitochondrial presequences are necessary for two different functions: (i) targeting of precursors to the organelle by direct binding to import receptors, and (ii) initiation of the translocation of the precursors across the inner membrane.

THE MECHANISM OF PREPROTEIN TRANSLOCATION

Unfolding of precursors

The mitochondrial import machinery seems to interact with the mature part of precursor proteins independently of their primary sequence. Hybrid proteins between aminoterminal presequences and non-mitochondrial proteins are recognized and translocated by the import apparatus. While there is no dependence on the primary sequence of the passenger protein, there is an important role of the tertiary structure. Stabilization of the tertiary structure of a precursor protein prevents its import (7). Several lines of evidence suggest that translocation across the mitochondrial membranes is related to an unfolding of the precursor. Two steps of unfolding/conformational changes have to be distinguished: (i) Prior to translocation, precursors are probably kept in a "molten globule state" by attachment to chaperone proteins (8). (ii) The actual unfolding of proteins takes place during membrane translocation.

To which extent are proteins unfolded during translocation? We have addressed this question by the investigation of a series of fusion proteins consisting of aminoterminal parts of the mitochondrial protein cytochrome b_2 and the entire dihydrofolate reductase (DHFR). Fusion proteins of different length were imported into isolated mitochondria *in vitro*. The imported proteins were processed by the processing peptidase inside the matrix and were protected against externally added proteases by the mitochondrial membranes. In the presence of methotrexate (a specific ligand of DHFR), translocation of the DHFR was inhibited. Under these conditions the DHFR moiety stayed outside the mitochondrial membranes and retained its tertiary structure. The translocation of the presequence was not affected by methotrexate and translocation intermediates were accumulated in contact sites between both mitochondrial membranes (9, 10). It turned out that about 50 amino acid residues were sufficient to span both membranes. In α-helical conformation 50 amino acids span a length of 7.5 nm. On the other hand, the thickness of two intact lipid bilayers is 12 to 14 nm (including hydrophobic and hydrophilic parts). Electron micrographs show a space between both membranes also in contact sites and indicate a distance of 18-20 nm to be spanned by translocation intermediates. Considering these values it can be assumed that the membrane-spanning polypeptide segment translocates in a rather extended conformation (10). The length of a polypeptide of 50 amino acids in completely extended conformation is 18.1 nm.

Our results provide evidence for an extensive unfolding of polypetides in transit. We propose that precursor proteins may become unfolded to such a degree that the backbone of the polypeptide chain becomes exposed while being translocated through the translocation sites of mitochondrial membranes.

Role of hsp70s in protein import

A temperature-sensitive mutant of the mitochondrial hsp70 in yeast has been described that accumulates preproteins in mitochondrial contact sites, demonstrating that complete translocation of precursor proteins needs functionally active mt-hsp70 (11). Mt-hsp70 was shown to bind to translocation intermediates spanning mitochondrial contact sites (12, 13). The structure which is recognized by hsp70 may be the covalent backbone of non-native proteins (14). The following model has been proposed: The first step in the translocation of a mitochondrial precursor protein is the electrophoretically driven movement of the presequence into the matrix. The polypeptide in transit becomes completely unfolded. In the matrix, several mt-hsp70 molecules bind to the polypeptide chain and thereby provide the driving force for the complete import of the protein (11). Hydrolysis of ATP allows the release of the bound protein from hsp70 (15).

ATP which is hydrolyzed outside the organelle in the course of protein import is not necessarily related to the actual unfolding of precursor proteins. A completely folded DHFR moiety can be unfolded and translocated (in radiochemical amounts) also in the absence of free ATP (16). In contrast to DHFR, many authentic precursor proteins need ATP prior to translocation, probably because they have to be released from cytosolic cofactors. One of these cofactors has been shown to be cytosolic hsp70 (17, 18). The cytosolic cofactors are assumed to prevent the misfolding and/or aggregation of precursor proteins. In summary, there are two steps in the import of an authentic precursor protein which consume ATP: prior to translocation the release from cytosolic hsp70s and after translocation the release from mt-hsp70.

IMPORT COMPONENTS IN THE MITOCHONDRIAL MEMBRANES

The mitochondrial receptor complex

The mitochondrial outer membrane contains about 25 different polypeptides, termed "mitochondrial outer membrane proteins" (MOMs). The 19 kD protein MOM19 has been shown to be the major protein import receptor of mitochondria (19, 20). Antibodies raised against this protein inhibit the import of nearly all precursors tested so far. In co-immunoprecipitations it turned out that MOM19 is part of a larger protein complex (21). Besides MOM19 the components of this complex are MOM7, MOM8, MOM22, MOM30, MOM38 and MOM72 (named according to their mol. weight in thousands). Are all these proteins involved in import reactions of precursor proteins?

In intact mitochondria only MOM19, MOM22 and MOM72 are accessible to externally added proteases. While the function of MOM22 remains enigmatic, we could show that antibodies raised against MOM72 specifically inhibit the import of the ADP/ATP carrier (22). Interestingly, the ADP/ATP carrier can also be imported via MOM19 albeit with a decreased efficiency, indicating that these two import receptors possess a (partially) overlapping specificity (20). The function of MOM19 and MOM72 has been confirmed by crosslinking to precursor proteins bound to the mitochondrial surface (5).

MOM7, MOM8, MOM30 and MOM38 apparently constitute the general insertion site ("general insertion protein") GIP that mediates the insertion of precursor proteins into the outer membrane (5, 21, 23). This is suggested by an investigation of the products yielded by chemical crosslinking of the ADP/ATP carrier at different steps of its import pathway (5). The ADP/ATP carrier accumulated at the surface of the mitochondrial outer membrane can be crosslinked to MOM19 and MOM72, but not to other components of the receptor complex. After entry into the outer membrane the carrier can be crosslinked to the four components mentioned above, i.e. MOM7, MOM8, MOM30 and MOM38. The efficiency of crosslinking was especially high with MOM7 and MOM8. None of these products was observed after complete import of the ADP/ATP carrier into the inner membrane.

It has been demonstrated that translocation intermediates of the ADP/ATP carrier (and other precursor proteins) accumulate in a hydrophilic environment (24, 25). This environment is probably formed by the components of the receptor complex MOM7, MOM8, MOM30 and MOM38. According to the behaviour of the four components against alkaline extraction, MOM30 and MOM38 are embedded directly in the outer membrane, while MOM7 and MOM8 are bound to the complex by hydrophilic interactions (5). It might be speculated that MOM7 and MOM8 create a hydrophilic channel for translocating precursors within the receptor complex.

A dynamic model of the mitochondrial protein import apparatus

The transfer of precursor proteins from the outer membrane to the inner membrane predominantly occurs at contact sites where both membranes are in close proximity. Several methods have been developed to reversibly accumulate precursor proteins in contact sites *in vitro* and *in vivo* (26, 27). Contact site-intermediates have reached the matrix with their aminoterminus and are processed by the metal-dependent processing peptidase, but remain accessible for externally added proteases (as a carboxyterminal part is still on the cytosolic side). Hence, they span both mitochondrial membranes. Translocation sites for such intermediates are saturable structures (9, 28). These intermediates are apparently not translocated across the outer and inner membranes of mitochondria in two independent steps, but suggest a unique translocation apparatus involving both membranes (Fig.1A).

At least one of the components of the mitochondrial import receptor complex, MOM38 (=ISP42), can be crosslinked to contact site-intermediates (29, 30). We conclude that MOM38 is not only playing a role in the insertion of proteins into the outer membrane, but also in the insertion into the inner membrane. The receptor complex seems to be the outer membrane part of the translocation apparatus in mitochondrial contact sites. By electron microscopy, however, it was demonstrated that MOM38 as well as MOM19 are not only located in contact sites. Both proteins are distributed all over the outer membrane (19, 21 29).

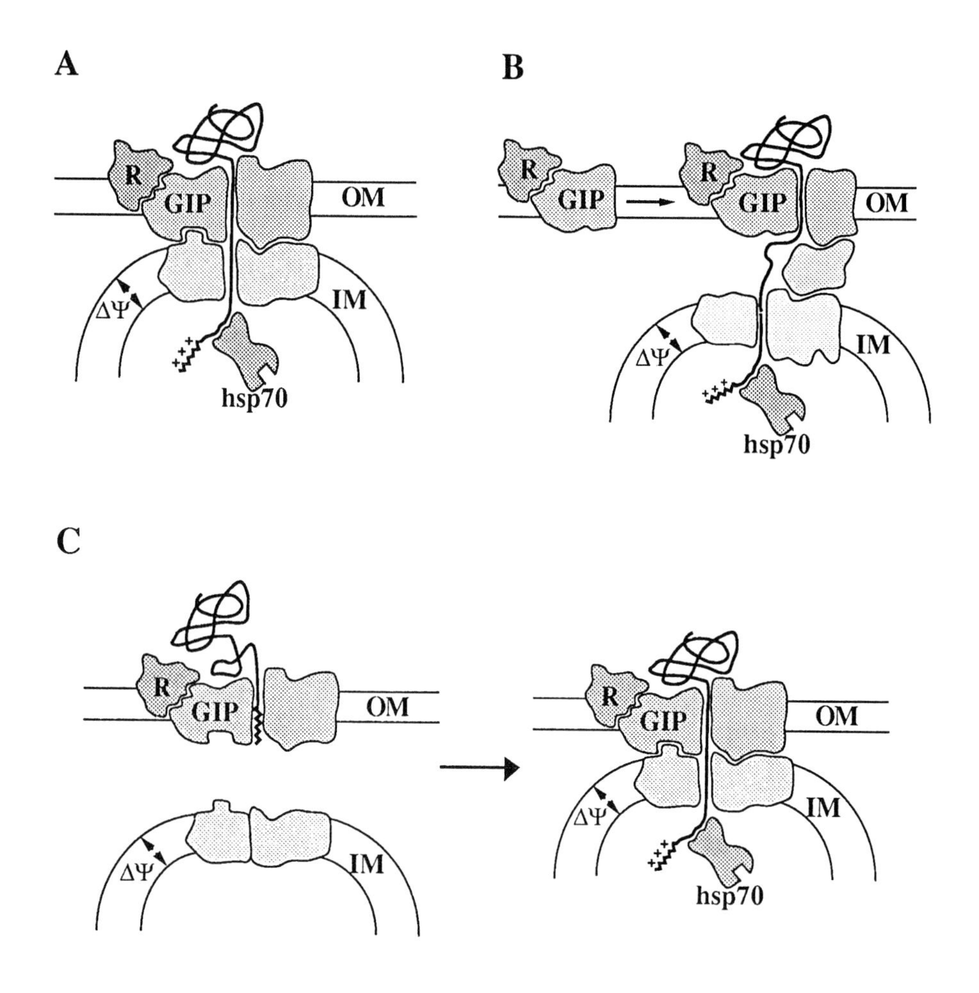

Figure 1. Hypothetical models of the mitochondrial protein import apparatus.
(A) Static model with stable and sealed contact sites.
(B) and (C) Dynamic models. (B) The contact sites are stable, but the membranes are not in direct contact; components of the import apparatus transiently assemble into contact sites. (C) Contact sites are dynamic and can be formed and disassembled.
$\Delta\Psi$, membrane potential; GIP, general insertion site; IM, inner membrane; OM, outer membrane; R, receptor.

As shown for the ADP/ATP carrier, the receptor complex does not need an energized inner membrane to promote the initial steps of protein import, indicating that the import machinery in the outer membrane can act independently of that in the inner membrane.

Up to now, little is known about the protein import machinery of the inner membrane. After inhibition of import by degradation of receptors with proteases, efficient import of some preproteins can be reestablished by opening of the intermembrane space. Translocation of proteins directly across the inner membrane seems to depend on proteinaceous components (30, 31, 32). This could be a hint that the mitochondrial inner membrane contains proteins which play a role in contact site-mediated import but can also promote import reactions directly across the inner membrane. These results favour a model of more or less independently acting "translocases" of both mitochondrial membranes.

We propose that mitochondrial contact sites are not sealed structures, but that polypeptides passing through can get access to the aqueous space between both membranes (Fig.1B and C). In fact, electron micrographs show a distance of about 6 nm between inner and outer membrane in contact sites (9). We thus investigated the accessibility of translocation intermediates to proteases after opening of the intermembrane space. We found that with uncoupled mitochondria the ADP/ATP carrier was accumulated in a site that was protected against externally added proteases, but became accessible to proteases after opening of the intermembrane space. A similar result was obtained for the β-subunit of the mitochondrial ATP-synthase. After pretreatment of mitochondria with apyrase (to lower the levels of ATP) the subunit accumulated as an unprocessed precursor that was partly exposed to the intermembrane space (30).

Figures 1B and 1C show two distinct models that are compatible with the various results. (i) While translocated between both membranes, the polypeptides get access to the intermembrane space since the membranes are not in direct contact (Fig. 1B). Components in the mitochondrial membranes and the intermembrane space constitutively stabilize the structure of the morphologically visible contact sites (9, 33). The components that directly participate in membrane translocation, in particular the outer membrane receptor complex, possess a lateral mobility in the membranes and assemble into and disassemble from stable contact sites. (ii) Translocation contact sites are not necessarily stable structures, but can be formed, e.g. induced by the presence of precursor proteins (Fig. 1C). The protein transport machineries are distributed over outer and inner membranes and the machineries in both membranes can get in close contact to facilitate the rapid transfer of precursors. It is likely that only those parts of the inner membrane which are already relatively close to the outer membrane, the "inner boundary membrane", participate in this dynamic process. Subsequently, the import machineries are separated again. It has to be emphasized that these two models are not mutually exclusive. A combination of both models would be the existence of a fraction of contact sites in a permanent form, while the rest of translocation contact sites is of transient nature.

CONCLUSIONS AND PERSPECTIVES

The initial stages of mitochondrial protein import are triggered by the aminoterminal presequences. Presequences function as signal sequences in that they are recognized by receptors on the mitochondrial surface, and they initiate the translocation of the polypeptide chain across the inner membrane (in that the presequences are electrophoretically driven across the inner membrane). The preproteins become extended and an important portion of the driving force for the rest of the polypeptide appears to be derived from binding to the heat shock protein hsp70 in the matrix. We propose that the translocases of the mitochondrial membranes show a dynamic behaviour: they are laterally mobil and may assemble into pre-existing stable contact sites. Moreover, there is at least the possibility that some translocation contact sites are transient and their formation is induced by the presence of precursor proteins.

Future studies are directed towards elucidating the structure and dynamic of the mitochondrial import machinery. It has to be kept in mind, however, that not only preproteins are translocated, but that there is considerable transport of lipids between both membranes (34) and apparently also of ribonucleic acids.

REFERENCES

1 Hartl FU, Neupert W. Science 1990; 247: 930-938.
2 Baker K, Schatz G. Nature 1991; 349: 205-208.
3 Pfanner N, Söllner T, Neupert W. Trends Biochem Sci 1991; 16: 63-67.
4 Becker K, Guiard B, Rassow J, Söllner T, Pfanner N. (subm).
5 Söllner T, Rassow J, Wiedmann M, Schloßmann J, Keil P,Neupert W, Pfanner N. (subm.).
6 Martin J, Mahlke K, Pfanner N. J Biol Chem 1991 (in press).
7 Eilers M, Schatz G. Nature 1986; 322: 228-232
8 Martin J, Langer T, Boteva R, Schramel A, Horwich AL, Hartl FU. Nature 1991; 352: 36-42.
9 Rassow J, Guiard B, Wienhues U, Herzog V, Hartl FU, Neupert W. J Cell Biol 1989; 109: 1421-1428.
10 Rassow J, Hartl FU, Guiard B, Pfanner N, Neupert W. FEBS Lett 1990; 275: 190-194.
11 Kang PJ, Ostermann J, Shilling J, Neupert W, Craig EA, Pfanner N. Nature 1990; 348: 137-143.
12 Ostermann J, Voos W, Kang PJ, Craig EA, Neupert W, Pfanner N. FEBS Lett 1990; 277: 281-248.
13 Scherer PE, Krieg UC, Hwang ST, Vestweber D, Schatz G. EMBO J 1990; 9: 4315-4322.
14 Rippmann F, Taylor WR, Rothbard JB, Green NM. EMBO J 1991; 10: 1053-1059.

15 Neupert W, Hartl FU, Craig EA, Pfanner N. Cell 1990; 63: 447-450.
16 Pfanner N, Rassow J, Guiard B, Söllner T, Hartl FU, Neupert W. J Biol Chem 1990; 265: 16324-16329.
17 Deshaies RJ, Koch BD, Werner-Washburne M, Craig EA, Schekman R. Nature 1988; 332: 800-805.
18 Murakami H, Pain D, Blobel G. J Cell Biol 1988; 107: 2051-2057.
19 Söllner T, Griffiths G, Pfaller R, Pfanner N, Neupert W. Cell 1989; 59: 1061-1070.
20 Steger HF, Söllner T, Kiebler M, Dietmeier KA, Pfaller R, Trülzsch KS, Tropschug M, Neupert W, Pfanner N. J Cell Biol 1990; 111: 2353-2363.
21 Kiebler M, Pfaller R, Söllner T, Griffiths G, Horstmann H, Pfanner N, Neupert W. Nature 1990; 348: 610-616.
22 Söllner T, Pfaller R, Griffiths G, Pfanner N, Neupert W. Cell 1990; 62: 107-115.
23 Pfaller R, Steger HF, Rassow J, Pfanner N, Neupert W. J Cell Biol 1988; 107: 2488-2490.
24 Pfanner N, Neupert W. J Biol Chem 1987; 262: 7528-7536.
25 Pfanner N, Hartl FU, Guiard B, Neupert W. Eur J Biochem 1987; 169: 289-293.
26 Söllner T, Rassow J, Pfanner N. Meth Cell Biol 1991; 34: 345-358.
27 Wienhues U, Becker K, Schleyer M, Guiard B, Tropschug M, Horwich AL, Pfanner N, Neupert W. J Cell Biol 1991 (in press).
28 Vestweber D, Schatz G. J Cell Biol 1988; 107: 2037-2043.
29 Vestweber D, Brunner J, Baker K, Schatz G. Nature 1989; 341: 205-209
30 Rassow J, Pfanner N. (unpublished).
31 Ohba M, Schatz G. EMBO J 1987; 6: 2117-2122.
32 Hwang S, Jascur T, Vestweber D, Pon L, Schatz G. J Cell Biol 1989; 109: 487-493.
33 Hackenbrock CR. Proc Natl Acad Sci USA 1968; 61: 598-605.
34 Simbeni R, Pon L, Zinser E, Paltauf F, Daum G. J Biol Chem 1991; 266: 10047-10049.

© 1992 Elsevier Science Publishers B.V. All rights reserved.
Molecular mechanisms of transport. E. Quagliariello, F. Palmieri, eds.

MOLECULAR INTERACTIONS BETWEEN MITOCHONDRIAL ASPARTATE AMINOTRANSFERASE AND MITOCHONDRIAL MEMBRANE IN THE IMPORT PROCESS

E. Marra[a], S. Giannattasio[a], M.F. Abruzzese[a], R.A. Vacca[a], M. Greco[a] and E. Quagliariello[b]

[a]Centro di Studio sui Mitocondri e Metabolismo Energetico, Consiglio Nazionale delle Ricerche, via Amendola 165/A, 70126 Bari, Italy.

[b]Dipartimento di Biochimica e Biologia Molecolare, Università di Bari, via Amendola 165/A, 70126 Bari, Italy.

GENERAL ASPECTS OF PROTEIN UPTAKE BY ISOLATED MITOCHONDRIA

The mechanism by which cytoplasmically synthesized proteins enter mitochondria and localize in their specific compartments has been the subject of intensive work in the last 15 years (1-3).

The first experimental evidence that purified protein can enter isolated mitochondria was provided in 1976 in our laboratory where the import of the mitochondrial aspartate aminotransferase (mAspAT) into rat liver mitochondria was shown.

A model system was developed in which rat liver mAspAT uptake by mitochondria was monitored by measuring the increase of the intramitochondrial enzyme activity due to adding mAspAT outside mitochondria (4).

Similarly the capability of purified mitochondrial malate dehydrogenase (mMDH) to enter mitochondria was also shown (5).

In the following years such a model system was used to gain some insight into the mechanism by which mitochondrial protein localization occurs.

The uptake by mitochondria of both purified mAspAT and mMDH shows a hyperbolic dependence on externally added enzyme and reaches equilibrium in less than 2 min. Both imported enzymes are resistant to externally added protease, thus showing that enzyme internalization in the mitochondrial matrix does occur. The protein import proves to be energy dependent; the chemical proton gradient (pH) across the mitochondrial membrane proved to be the energy source used in the process (for ref. see 1).

A few years after our demonstration that mature protein uptake into mitochondria does occur, experimental evidence was provided showing that most mitochondrial proteins are synthesized as precursors with a molecular weight higher than

those of the mature forms. This family includes mAspAT and mMDH which are synthesized as precursors (pmAspAT and pmMDH) with an amino-terminal extrasequence which is cleaved by a matrix protease upon their translocation into mitochondria (for ref. see 1).
Thus these results raised the question as to the role of the presequence of pmAspAT and pmMDH in the mitochondrial import process. In our laboratory we focused our attention on mAspAT.

BOTH mAspAT and pmAspAT SHARE THE SAME IMPORT PATHWAY IN ISOLATED MITOCHONDRIA

In order to perform a comparative study of the import into mitochondria of pmAspAT and mAspAT, we took advantage of recombinant DNA technology.
The full-length cDNA coding for pmAspAT has been cloned into the expression vector pOTS, thus obtaining pOTS-pmAspAT. This plasmid was subsequently modified deleting the nucleotide sequence coding for the presequence of pmAspAT; thus, pOTS-mAspAT was obtained, which contains the nucleotide sequence for mature mAspAT (6). pOTS-pmAspAT and pOTS-mAspAT have been used to synthesize pmAspAT and mAspAT in a cell-free coupled transcription/translation system.
Both in vitro-synthesized forms of mAspAT were incubated with isolated mitochondria, and imported proteins were analyzed through SDS-PAGE and fluorography of the mitochondrial lysate. In both cases the internalization of the protein into mitochondria was demonstrated and a comparative study of the import process of both forms of mAspAT was carried out. Both imported proteins are resistant to externally added pronase. In both cases the import process was found to be energy dependent and the imported proteins showed the same intramitochondrial localization. By studying the time dependence of the uptake, we found that pmAspAT and mAspAT import processes reach equilibrium in about 15 min. Finally pmAspAT import was found to be competitively inhibited by externally added purified mAspAT.
All these findings show that both pmAspAT and mAspAT share the same import pathway in isolated mitochondria (7).
Similar results have been found by Thompson and McAllister-Henn who demonstrated that the presequence for mMDH is not essential for efficient mitochondrial localization and function in yeast. In fact the mature form of mMDH expressed in a suitably engineered yeast strain is able to complement the defect of a MDH1 disruption mutant. The protein product was localized in the mitochondrial matrix and was enzymatically active at a level comparable to that of the

enzyme in a normal yeast strain, with no mature enzyme found in post-ribosomal supernatant fractions (8).
The results reported above, that the presequence of pmAspAT is not directly involved in protein translocation across the mitochondrial membrane, raise the question as to the role of the presequence in the import process.
Thus, a specific role for the presequence of pmAspAT remains to be determined. It may be postulated that it interferes with much of the folding and basic structural properties of the mature protein whose correct dimeric assembly is apparently dependent also upon the N-terminal region (9). However, experimental evidence has recently been provided showing that rat liver pmAspAT, as expressed in Escherichia coli, is soluble and catalytically active (10).
Though properties of selectivity may not be excluded for the presequence in vivo, mature mAspAT does contain sufficient mitochondrial targeting information.

mAspAT DOMAINS INTERACT WITH MITOCHONDRIAL RECEPTOR GROUPS IN THE TRANSPORT PROCESS

In this section a survey will be given of all our results which strongly suggest the existance of certain domains in mAspAT involved in the uptake process as well as the existance of mitochondrial receptor/s for the imported protein.
The first evidence of a chemical reaction between incoming protein and mitochondrial membrane derived from our observation that blockage of a single SH group per monomer of enzyme by mersalyl prevents its uptake into mitochondria (11).
This chemically reactive sulphydryl group is supposed to be the cysteine at position 166 which is apparently exposed in the quaternary structure of the protein (12).
The existence of thiol group/s involved in the uptake of mMDH was also shown (13).
We have also reported that the removal of 31 N-terminal aminoacid residues from mAspAT abolishes its interaction with mitochondria in vitro (14). Consistently certain N-terminal peptides, namely 1-9 and 10-31 fragments, proved to inhibit uptake of mAspAT by isolated mitochondria (15). This clearly shows that the N-terminal portion of mAspAT has a main role in the interaction of the protein with the mitochondrial membrane. Consistently the amino-terminal region has a high degree of homology among mAspATs from different sources and it is more conserved than the same region of the cytoplasmic isoenzyme (16).
In order to study the specific role of the different regions of mAspAT involved in the translocation process, we have

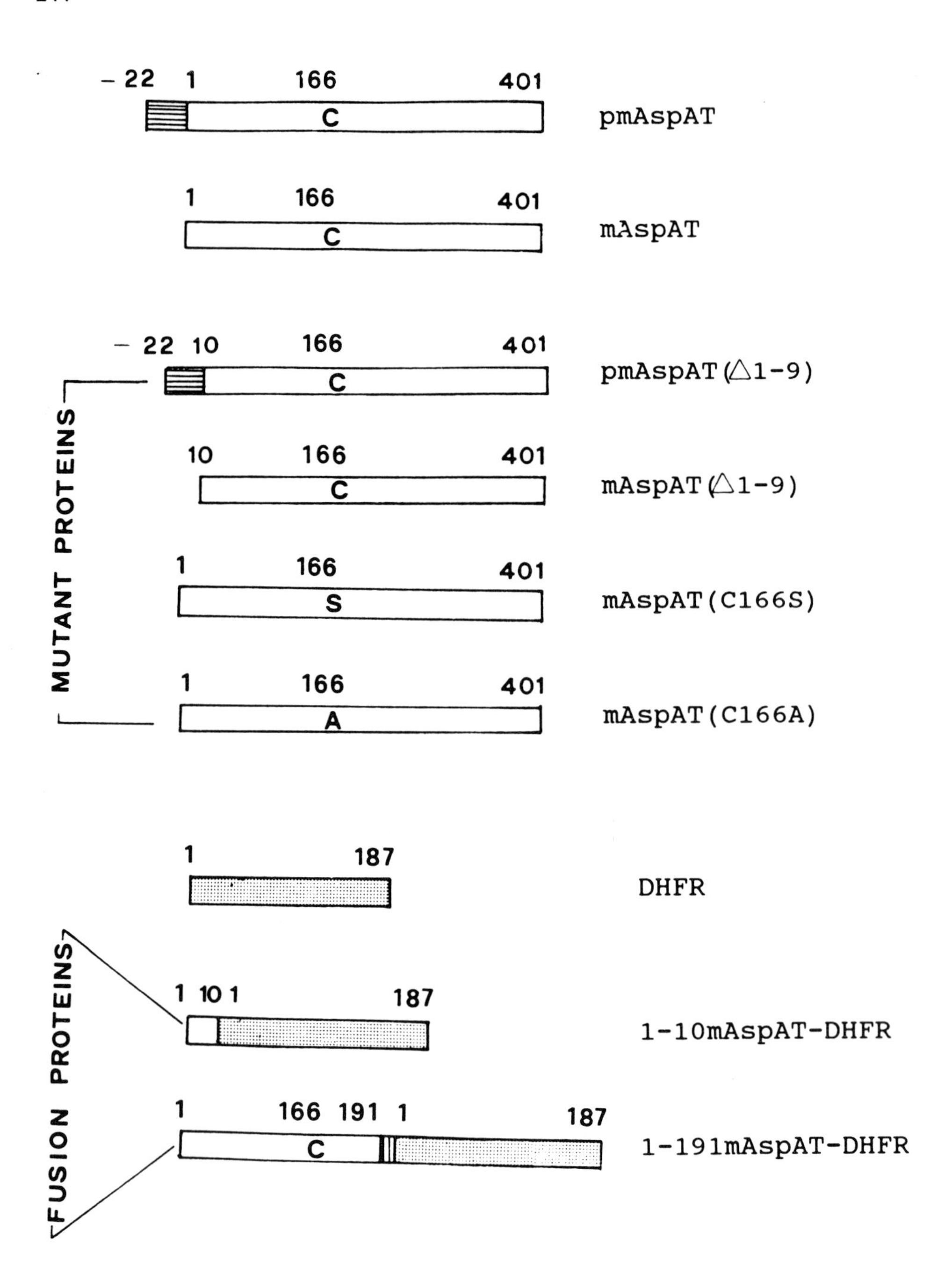

Figure 1. Schematic representation of mutant and fusion proteins.

modified pmAspAT and mAspAT at the level of the amino-terminus of the enzyme and of the cysteine residue at position 166 making use of in vitro mutagenesis techniques.
We have induced mutations of pOTS-pmAspAT and pOTS-mAspAT by Kunkel's method of oligonucleotide-directed mutagenesis (17). The nucleotide sequence coding for the nine N-terminal aminoacids of mature mAspAT has been deleted from both recombinant plasmids, thus obtaining pOTS-pmAspAT(1-9) and pOTS-mAspAT(1-9). Similarly the codon for cysteine 166 of pOTS-mAspAT has been changed into the codon for serine and alanine, thus obtaining two new plasmids, pOTS-mAspAT(C166S) and pOTS-mAspAT(C166A). The four mutant proteins coded for by these new plasmids and schematically shown in Figure 1, have been expressed in a cell-free coupled transcription/ translation system in the presence of the radioactive precursor L-[^{35}S]-methionine.
To further investigate the capability of the N-terminal part of mature mAspAT to determine the mitochondrial localization of a cytoplasmic protein, two fusion genes have also been constructed, by fusing the nucleotide sequence coding for 10 or 191 N-terminal aminoacids of mAspAT to the DNA sequence coding for the entire polypeptide chain of dihydrofolate reductase (DHFR). Both fusion genes have been cloned in a suitable expression vector which may be transcribed in vitro and subsequently translated in a reticulocyte lysate. The fusion proteins 1-10mAspAT-DHFR and 1-191mAspAT-DHFR are schematically shown in Figure 1.
The elucidation of the characteristics of the import into isolated mitochondria of the in vitro-synthesized mutant and fusion proteins will clarify the molecular interactions between mAspAT and mitochondrial membrane.
However, the results reported above are in favour of the existence of a receptor area in the mitochondrial membrane containing certain groups able to recognize the transport competent domains of the incoming proteins.
Direct evidence of such a receptor area was provided by experiments in which two compounds namely 2-mercaptoethanol and the metal complexing agent tiron proved to prevent both mAspAT and mMDH uptake by isolated mitochondria. These reagents were shown to bind to the mitochondrial membrane without any effect on the enzyme molecule (11,13,18). Both the hyperbolic dependence of the uptake and saturation kinetics were consistent with this conclusion. The mitochondrial receptor contains two separate but interacting binding sites for mAspAT and mMDH, in agreement with the different nature of inhibition shown by 2-mercaptoethanol on the rate of uptake mAspAT and mMDH and with mutual non-competitive inhibition (13).

Consistent with the conclusion that the presequence is not strictly required for the transport, the dependence of the rate of pmAspAT uptake on increasing pmAspAT amounts in the absence or presence of externally added mature protein was investigated. Both mature and precursor aspartate aminotransferase bind to the same membrane site as shown by the competitive inhibition of the import of pmAspAT due to externally added purified enzyme (7).
Interestingly, we have recently reported evidence that uptake

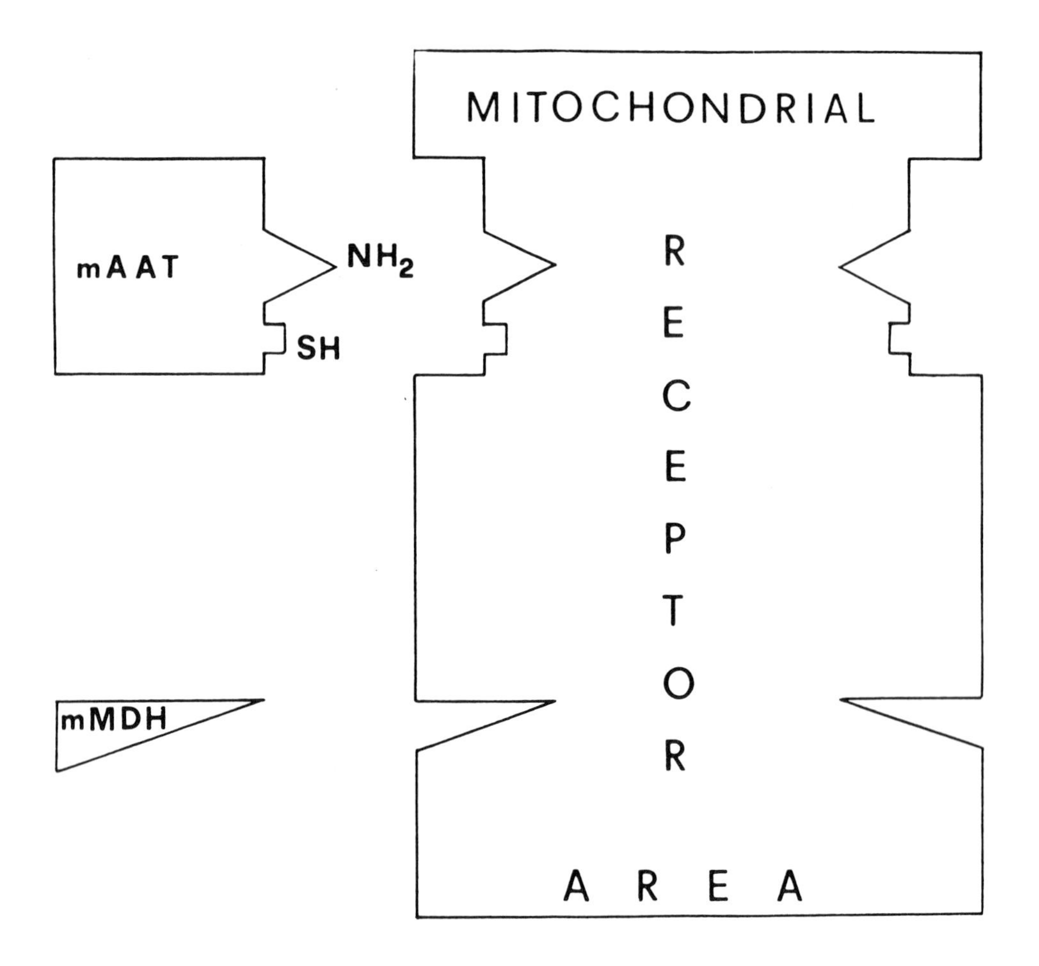

Figure 2. Schematic representation of the mitochondrial receptor area of mAspAT and mMDH.

of mAspAT into mitochondria *in vitro* causes efflux of malate dehydrogenase and *vice versa* (19). Similarly, efflux of malate dehydrogenase also occurs as a result of pmAspAT added outside mitochondria.

Figure 2 shows a schematic representation of the receptor area on the mitochondrial membrane which works in both import and efflux of mAspAT, as deduced from the different experimental findings so far gathered.

In the light of the experimental findings obtained in our laboratory we propose that mAspAT enters mitochondria with a mechanism which involves interaction of certain specific protein domains (probably also including the presequence) with the mitochondrial receptor area; during protein uptake efflux of a modified mMDH/mAspAT occurs as a possible pathway to mitochondrial enzyme degradation.

ACKNOWLEDGEMENTS

We thank Miss. A. Armenise and Mr. G. Sgaramella for skillful technical assistence. This work was supported by C.N.R. Target Project on Biotechnology and Bioinstrumentation.

REFERENCES

1 Doonan S, Marra E, Passarella S, Saccone C, Quagliariello E. Int Rev Cyt 1984; 91:141-186.
2 Douglas MG, Mc Cammon MT, Vassarotti A. Microbiol Rev 1986; 50:166-178.
3 Hartl FU, Pfanner N, Nicholson DW, Neupert W. Biochim Biophys Acta 1989; 988:1-45.
4 Marra E, Doonan S, Saccone C, Quagliariello E. Biochem J 1977; 174:685-691.
5 Passarella S, Marra E, Doonan S, Quagliariello E. Biochem J 1980; 192:649-658.
6 Jaussi R, Behra R, Giannattasio S, Flura T, Christen P. J Biol Chem 1987; 262:12434-12437.
7 Giannattasio S, Marra E, Abruzzese MF, Greco M, Quagliariello E. Arch Biochem Biophys 1991; (in press).
8 Thompson LM, McAlister-Henn L. J Biol Chem 1989; 264:12091-12096.
9 Sandmeier E, Christen P. J Biol Chem 1980; 255:10284-10289.
10 Altieri F, Mattingley JR, Rodriguez-Berrocal FJ, Youssef J, Irarte A, Wu T, Martinez-Carrion M. J Biol Chem 1989; 264:4782-4786.
11 Marra E, Passarella S, Doonan S, Saccone C, Quagliariello E. Arch Biochem Biophys 1979; 195:269-279.

12 Gehring H, Cristen P. J Biol Chem 1978; 253:3158-3163.
13 Marra E, Passarella S, Casamassima E, Perlino E, Doonan S, Quagliariello E. Biochem J 1985; 228:493-503.
14 O'Donovan KMC, Doonan S, Marra E, Passarella S, Quagliariello E. Biochem J 1985; 228:609-614.
15 Barile M, Giannattasio S, Marra E, Passarella S, Pucci P, Sannia G, Quagliariello E. Biochem Biophys Res Commun 1990; 170:609-615.
16 Graf Hausner U, Wilson KJ, Christen P. J Biol Chem 1983; 258:8813-8826.
17 Kunkel TA. Proc Natl Acad Sci USA 1985; 82:488-492.
18 Passarella S, Marra E, Atlante A, Doonan S, Quagliariello E. FEBS lett 1985:235-240.
19 Passarella S, Marra E, Atlante A, Barile M, Doonan S, Quagliariello E. Biochim Biophys Acta 1990; 1022:273-282.

© 1992 Elsevier Science Publishers B.V. All rights reserved.
Molecular mechanisms of transport. E. Quagliariello, F. Palmieri, eds.

TRANSPORT OF THE MAMMALIAN PHOSPHATE CARRIER INTO MITOCHONDRIA

Vincenzo Zara[a], Nikolaus Pfanner[b] and Ferdinando Palmieri[a]

[a]Department of Pharmaco-Biology, Laboratory of Biochemistry and Molecular Biology, University of Bari, Bari (Italy)

[b]Institut für Physiologische Chemie, Universität München, Goethestr. 33, W-8000 München 2, Germany

INTRODUCTION

The phosphate carrier (PiC), an intrinsic protein of the inner mitochondrial membrane, catalyzes an electroneutral symport of phosphate together with a stoichiometric amount of H^+ (or an electroneutral exchange of phosphate against OH^-) (1-3). This transport protein belongs to the carrier family of the inner mitochondrial membrane. The primary structures of four of these carrier proteins, the ADP/ATP carrier, the uncoupling protein, the phosphate carrier and the 2-oxoglutarate carrier, have been determined by amino acid sequence analysis and by cDNA sequencing (4-7). These proteins are about 300 amino acid residues in length and are related in amino acid sequence and structure. All four carriers are encoded by the nuclear DNA and are then synthesized on cytosolic polysomes. Afterwards they are imported into mitochondria to their final destination in the inner membrane. Many mitochondrial precursor proteins that are translocated into or across the inner mitochondrial membrane are synthesized with amino-terminal targeting sequences that are cleaved off by a peptidase in the mitochondrial matrix (8,9). The precursors of the ADP/ATP carrier, of the uncoupling protein and of the 2-oxoglutarate carrier, however, are synthesized without presequence. These proteins therefore contain all the targeting information in the mature protein part. The import of the ADP/ATP carrier and the uncoupling protein into isolated mitochondria has been extensively investigated. These two carriers seem to contain more than one internal targeting signal probably not located at the amino-terminal ends of the proteins (10-12). Surprisingly, the precursors of bovine, rat and human PiC were found to be synthesized with amino-terminal extrasequences of 49 (bovine and human) and 44 (rat) amino acid residues (6,13,14).

In this study we report a summary of the data concerning the import of PiC into isolated mitochondria. We expressed the bovine heart PiC precursor *in vitro* and studied the translocation of the newly synthesized protein into mammalian mitochondria (15). Furthermore we constructed two new proteins, a mature PiC (mPiC) and a fusion protein between the presequence of PiC and mouse dihydrofolate reductase (DHFR), in order to investigate the role of the cleavable presequence in the import of mammalian PiC (16).

IMPORT OF BOVINE HEART PiC INTO MITOCHONDRIA

After construction of an expression vector (pGEM4/PiC) containing a cDNA encoding the bovine heart PiC precursor we performed the *in vitro* expression of the protein by coupled transcription and translation in rabbit reticulocyte lysate in the presence of [^{35}S]methionine (15). We then started to investigate the ability of the PiC precursor to enter mitochondria. Mitochondria isolated from rat liver or bovine heart were incubated with rabbit reticulocyte lysate containing the radiolabeled precursor protein for 15 min at 25 °C. The isolated mitochondria were energized by the addition of 2 mM NADH. After the incubation, the mitochondria were treated with 25 µg/ml proteinase K and finally reisolated by centrifugation. The mitochondrial proteins were then analyzed by SDS-polyacrylamide gel electrophoresis and fluorography. After incubation with energized mitochondria, the bovine heart PiC precursor (apparent molecular mass of 38 kDa) was processed to a protein with an apparent molecular mass of 33 kDa. This apparent molecular mass is in agreement with that of the mature protein isolated from bovine heart mitochondria by hydroxyapatite chromatography. The processed form of PiC was fully imported into mitochondria as it was protected against externally added proteinase K. In contrast, the PiC precursor bound to mitochondria in the absence of protease was completely digested by the addition of 25 µg/ml proteinase K. This means that it was not fully imported into mitochondria. The entry of PiC into mitochondria required an energized inner membrane because the dissipation of the membrane potential by 1 µM valinomycin blocked the import of the protein. In these samples we also included 20 μM oligomycin (in addition to valinomycin) to prevent a decrease of ATP levels due to a generation of a membrane potential via the F_oF_1-ATPase. The dependence of import on an energized inner membrane was also found with all other precursor proteins translocated into the inner membrane or matrix compartment of mitochondria (8,9). We then prepared a matrix fraction from rat liver mitochondria that specifically removed the 49-amino acid residue presequence of bovine heart PiC precursor. This indicates that the amino-terminal extension of this carrier protein reaches the matrix compartment of mitochondria during its import pathway. On the contrary, a purified preparation of MPP and PEP (matrix processing peptidase and processing enhancing protein from Neurospora crassa) was not able to process the PiC precursor. This would indicate that the peptidases from Neurospora crassa and mammalian mitochondria show a different specificity.

Radiosequencing experiments were then performed in order to demonstrate that the correct mature form had been generated after import and processing of PiC precursor by isolated mitochondria. For this purpose, we synthesized the PiC precursor in rabbit reticulocyte lysate in the presence of [^{3}H]glutamic acid or in the presence of [^{35}S]cysteine instead of [^{35}S]methionine, as usual. These newly synthesized proteins were then imported into mitochondria and the mature labeled forms were prepared by electroelution from SDS-polyacrylamide gels. The electroeluted proteins were then subjected to automated Edman degradation and, in the case of [^{3}H]glutamic acid-labeled PiC, we obtained two peaks at position 3 and 4 while in the case of [^{35}S]cysteine-labeled PiC only a peak at position 8, in agreement with the positions of the amino acids in the purified mature protein. We therefore concluded that the amino-terminus of the mature PiC, generated after import of the PiC precursor into isolated mitochondria, corresponds to the amino-terminus of the mature PiC purified from bovine heart mitochondria by hydroxyapatite chromatography (V. Zara, unpublished results).

In our laboratory the PiC can be easily purified from bovine and porcine hearts by one-step chromatography, simply passing a mitochondrial extract over dry hydroxyapatite in the presence of an optimal amount of cardiolipin (17). In contrast, in the absence of cardiolipin at a mitochondrial protein concentration of 1-3 mg/ml, the PiC is absent in the column eluate. We then thought to use this peculiar property of the PiC as a demonstration of the correct assembly of the carrier into its functional location. A similar assay for assembly of low amounts of proteins imported *in vitro* was only reported for the ADP/ATP translocase (18). The PiC imported into energized mitochondria *in vitro* bound to hydroxyapatite in the absence of cardiolipin in the solubilization buffer while it was eluted from the column in the presence of 5 mg/ml of the phospholipid. The precursor of PiC, that was not imported into mitochondria, was not eluted from the column even in the presence of 5 mg/ml cardiolipin.

We then investigated whether the import of PiC showed the characteristics of mitochondrial protein uptake, such as dependence on ATP and involvement of contact sites between mitochondrial outer and inner membranes. A reticulocyte lysate containing the labeled PiC precursor was treated with 0.5 units/ml of apyrase (an ATPase and ADPase from potato). In these conditions, i.e. in the absence of ATP, PiC was not able to enter the mitochondria. The requirement for ATP seemed to be specific because only the addition of ATP but not of GTP was able to restore the import. By performing the import reaction for 10 min at 8 °C, we were able to generate an intermediate of PiC, i.e. an accumulated precursor protein that showed an accessibility to proteases on both sides of the two mitochondrial membranes. On one hand, the PiC precursor with its amino-terminal extension reached the matrix compartment of mitochondria, and, on the other hand, with its carboxy-terminal end was sensitive to externally added proteinase K. The chase of the PiC precursor to the processed and imported form confirmed that a true translocation intermediate had been generated. On the basis of these results we concluded that the import of PiC occurred at translocation contact sites, as already demonstrated with many other precursor proteins imported into mitochondria.

IMPORT OF A PRESEQUENCE-DEFICIENT PiC AND OF A PASSENGER PROTEIN INTO MITOCHONDRIA

The import of PiC from rat liver mitochondria was also studied by Pratt et al. (19). The results obtained with rat liver PiC are in agreement with those obtained with the bovine heart protein in our laboratory. Pratt et al. also investigated the role of the cleavable presequence during the import of mammalian PiC. For this reason they constructed a truncated form of rat liver

PiC that lacked 35 out of 44 amino acid residues of the presequence, but retained the original positive charges of the authentic precursor. This mutant PiC showed only very little ability to enter isolated mitochondria. On the basis of these results Pratt et al. concluded that the cleavable presequence of PiC plays an essential role during the translocation of mammalian PiC into mitochondria. The possibility remained, however, that the accumulation of positively charged amino acids at the amino-terminus of the PiC protein could interfere with the import of a translocation-competent mature PiC. To test this we constructed two new expression vectors encoding two proteins, a mature PiC (mPiC) that completely lacked the 49-amino acid residue presequence and a hybrid protein between the presequence of bovine PiC and mouse DHFR (16). The two plasmids were prepared by modifying the first expression vector (pGEM4/PiC) with PCR experiments and site-directed mutagenesis. The *in vitro* expressed mPiC, having an apparent molecular mass of 33 kDa, exactly corresponded in size to the processed form of the PiC generated after import of PiC precursor into isolated mitochondria. We tested the ability of this presequence-deficient PiC to enter isolated mitochondria. We incubated the reticulocyte lysate containing the labeled protein with rat liver mitochondria for 15 min at 25 °C in the presence or in the absence of a membrane potential. mPiC was indeed imported into mitochondria. As in the case of the authentic PiC (i.e. bovine heart PiC precursor), the import of mPiC required an energized inner membrane. The imported mPiC was protected against proteinase K unless the mitochondria were solubilized by the addition of detergent. The hybrid protein, carrying the presequence of PiC, however, was imported into isolated mitochondria only with a very low efficiency. We compared the import of these two new proteins with that of the authentic PiC at the same conditions, i.e. in the same experiment using the best conditions for import previously investigated for each protein. If we consider the import efficiency of PiC precursor to be 100, that of the mature PiC was about 50-60%, while that of the hybrid protein accounted for less than 5% of that of the authentic precursor. We therefore concluded that the PiC possesses internal targeting signal, as already found in the case of ADP/ATP carrier and uncoupling protein. This is demonstrated by the fact that mPiC is still able to enter mitochondria with a good efficiency. The presequence of bovine PiC is able to direct a passenger protein into isolated mitochondria with a very little efficiency. The more important targeting signals are therefore contained in the

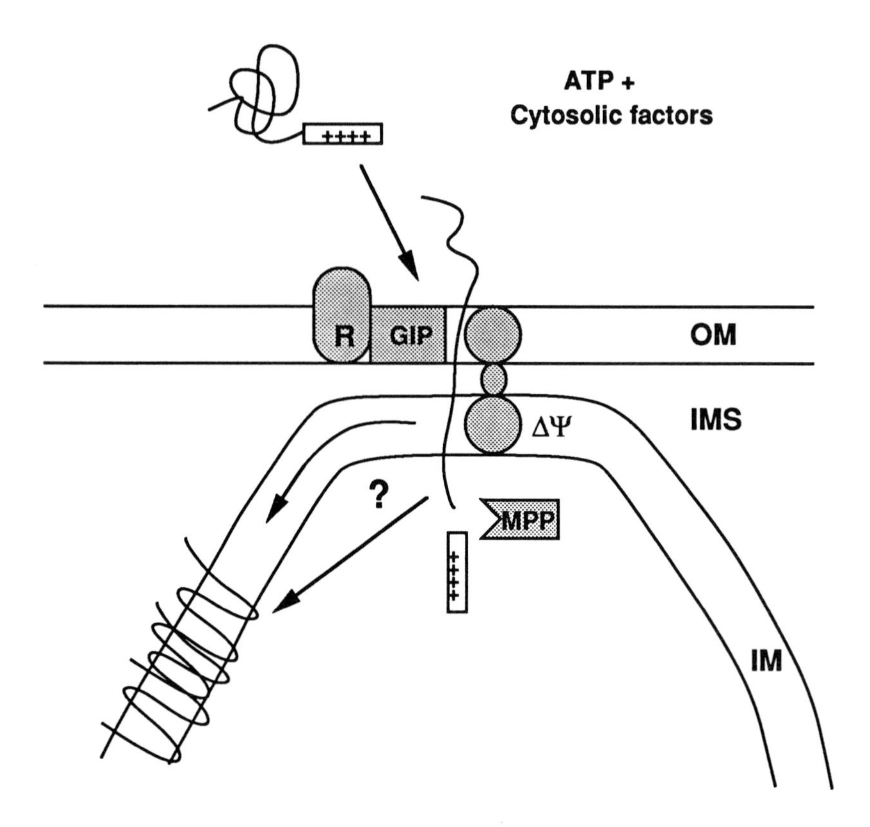

Figure 1. Model for the import and sorting of the mammalian phosphate carrier to the mitochondrial inner membrane

The PiC precursor is synthesized on cytosolic polysomes and interacts with a putative import receptor on the outer mitochondrial membrane. The protein is then translocated through translocation contact sites reaching with its amino-terminus the matrix where a peptidase cleaves off the presequence. The sorting of the protein may occur via the inner membrane or the matrix. Finally, the protein is assembled in the inner membrane.

$\Delta\Psi$, membrane potential; GIP, general insertion site; IM, inner membrane; IMS, intermembrane space; OM, outer membrane; MPP, matrix processing peptidase; R, receptor.

mature protein part of PiC. We think that the 49-amino acid residue presequence of bovine heart PiC has mainly a stimulating effect on the import of mammalian PiC. The exact role of the presequence during the import of PiC is currently under investigation.

CONCLUSIONS AND PERSPECTIVES

On the basis of these results we propose a general model of the pathway followed by PiC during its translocation into the inner membrane (Fig. 1). This model was constructed by assuming that the general import pathway is conserved between higher and lower eukaryotes, i.e., between mammalian and fungal mitochondria. The PiC, after being synthesized on cytosolic polysomes, acquires or is kept in a translocation-competent conformation by ATP-dependent cytosolic cofactors. The targeting signals in the PiC precursor are then recognized by a putative receptor localized on the mitochondrial outer membrane. The import requires the membrane potential $\Delta\psi$ across the inner mitochondrial membrane. The transport of PiC precursor into or across the inner membrane occurs at translocation contact sites between both mitochondrial membranes. The presequence of the PiC precursor is then cleaved off by a peptidase localized in the matrix compartment of mammalian mitochondria. Finally the protein is refolded and sorted to its functional location in the inner membrane. At the moment we do not know if the PiC is completely translocated in the matrix compartment and retranslocated to the inner membrane or if the precursor diffuses laterally from translocation contact sites to its functional location in the inner membrane. The latter pathway has been proposed for the ADP/ATP carrier (20,21). It should be emphasized that future studies will focus on the structural and functional identification of the components involved in the import pathway of such proteins, partly characterized in lower eukaryotes but almost completely unknown in mammalian mitochondria.

ACKNOWLEDGEMENTS

This work was supported by the C.N.R. Target Project "Ingegneria genetica".

REFERENCES

1. LaNoue, K.F. and Schoolwerth, A.C. (1979) Annu. Rev. Biochem. 48, 871-922.
2. Meijer, A.J. and Van Dam, K. (1981) in Membrane transport (Bonting, S. and De Pont, J., eds) pp. 235-256, Elsevier Press, Amsterdam.
3. Kramer, R. and Palmieri, F. (1989) Biochim. Biophys. Acta 974, 1-23.
4. Aquila, H., Misra, D., Eulitz, M. and Klingenberg, M. (1982) Hoppe Seyler's Z. Physiol. Chem. 363, 345-349.
5. Aquila, H., Link, T.A. and Klingenberg, M. (1985) EMBO J. 4, 2369-2376.
6. Runswick, M.J., Powell, S.J., Nyren, P. and Walker, J.E. (1987) EMBO J. 6, 1367-1373.
7. Runswick, M.J., Walker, J.E., Bisaccia, F., Iacobazzi, V. and Palmieri, F. (1990) Biochemistry 29, 11033-11040.
8. Hartl, F.U., Pfanner, N., Nicholson, D.W. and Neupert, W. (1989) Biochim. Biophys. Acta 988, 1-45.
9. Pfanner, N. and Neupert, W. (1990) Annu. Rev. Biochem. 59, 331-353.
10. Pfanner, N., Hoeben, P., Tropschug, M. and Neupert, W. (1987) J. Biol. Chem. 262, 14851-14854.
11. Smagula, C.S. and Douglas, M.G. (1988) J. Cell Biochem. 36, 323-327.
12. Liu, X., Bell, A.W., Freeman, K.B. and Shore, G.C. (1988) J. Cell Biol. 107, 503-509.
13. Dolce, V., Fiermonte, G., Messina, A. and Palmieri, F. (1991) DNA Sequence - J. DNA Sequencing and Mapping, 2 : 131-134
14. Ferreira, G.C., Pratt, R.D. and Pedersen, P.L. (1989) J. Biol. Chem. 264, 15628-15633.
15. Zara, V., Rassow, J., Wachter, E., Tropschug, M., Palmieri, F., Neupert, W. and Pfanner, N. (1991) Eur. J. Biochem. 198, 405-410.
16. Zara, V., Palmieri, F., Mahlke, K. and Pfanner, N., submitted.
17. Bisaccia, F. and Palmieri, F. (1984) Biochim. Biophys. Acta 766, 386-394.
18. Schleyer, M. and Neupert, W. (1984) J. Biol. Chem. 259, 3487-3491.
19. Pratt, R.D., Ferreira, G.C. and Pedersen, P.L. (1991) J. Biol. Chem. 266, 1276-1280.
20. Hartl, F.U. and Neupert, W. (1990) Science 247, 930-938.
21. Mahlke, K., Pfanner, N., Martin, J., Horwich, A.L., Hartl, F.U. and Neupert, W. (1990) Eur. J. Biochem. 192, 551-555.

TRANSPORT AND CELLULAR METABOLISM

© 1992 Elsevier Science Publishers B.V. All rights reserved.
Molecular mechanisms of transport. E. Quagliariello, F. Palmieri, eds.

Mechanisms involved in the regulation of mitochondrial inner membrane permeability and their physiological and pathological significance.

Andrew P. Halestrap, Cathal P.Connern and Elinor J. Griffiths

Department of Biochemistry, School of Medical Sciences, University of Bristol, Bristol BS8 1TD, U.K.

INTRODUCTION

Mitochondrial matrix volume is regulated by the operation of a K^+ uniport and a K^+/H^+ antiport. A net uptake of mitochondrial K^+ and a compensating anion leads to an increase in matrix volume [1,2]. In this article we will describe how, in the liver, these pathways can be regulated by glucogenic hormones and so cause an increase in matrix volume. Such increases in volume are able to cause a substantial stimulation of mitochondrial metabolism which is important in the metabolic response of the tissue to the hormone. In addition we will describe how, in some pathological conditions, mitochondria become overloaded with Ca^{2+} which leads to the opening of a non-specific pore and subsequent massive and damaging swelling of the mitochondria. More detailed reviews of some aspects of this work may be found elsewhere [1,3]

HORMONAL REGULATION OF LIVER MITOCHONDRIAL MATRIX VOLUME

In liver cells stimulated with a Ca^{2+}-mobilising hormone such as adrenaline or vasopressin the matrix volume increases by some 20-30% [1,3,4]. Work in this and other laboratories has established that such relatively small increases in matrix volume are capable of causing a large stimulation of several mitochondrial metabolic pathways that are important for the observed physiological response to the hormones [see 1,3,5]. These effects include stimulation of glutaminase, glycine cleavage enzyme, oxidation of substrates entering the respiratory chain prior to ubiquinone (especially fatty-acyl carnitine), pyruvate carboxylase and citrulline synthesis. In the case of the latter two the effect appears to be secondary to the increase in ATP supply that occurs as a consequence of the stimulation of the respiratory chain [1,3,6]. All the processes listed are known to be increased in the liver in response to glucogenic hormones and show an appropriate sensitivity to the matrix volume over the range that is found within control and hormone treated hepatocytes (see Fig. 1).

Until recently the concept of regulation of mitochondrial function through changes in the mitochondrial volume has been treated sceptically by some, perhaps because of the difficulties involved in measuring mitochondrial volumes *in situ* [1,4,7,8]. A 25% increase in matrix volume is associated with a change in mitochondrial diameter of only 7.5%, and this would be hard to detect by electron microscopy. Furthermore the well established regulation of Ca^{2+}-activated dehydrogenases of the citric acid cycle [3] has tended to eclipse other possible means of regulating mitochondrial function. However three factors have led to a wider acceptance for a role of mitochondrial volume. Firstly, the establishment of a mechanism for the volume changes as outlined below. Secondly, the recognition that stimulation of dehydrogenases alone cannot account for the prolonged hormonal stimulation of respiration in liver that occurs without an increase in $NADH/NAD^+$ or ADP/ATP [1,3,9]. Similar results are also emerging from studies on the perfused heart [3,10; Dr. I. Hassinen personal communication]. Thirdly, it is now known that following perfusion with physiological concentrations of amino acids the liver swells by 10-15% as a result of Na^+-linked amino acid uptake [11]. These small increases in cell volume (less in percentage terms than those seen in mitochondria) are accompanied by profound changes in liver metabolism [11-14]. These include activation of glycine oxidation, glutaminase, urea synthesis, gluconeogenesis and respiration, and a decrease in mitochondrial $NADH/NAD^+$ which could all be explained by an increase in mitochondrial volume, although suitable measurements have yet to be made.

Ca^{2+} and pyrophosphate mediate the hormonal regulation of mitochondrial volume

The hormonally induced increase in mitochondrial matrix volume is dependent on a sustained increase in cytosolic $[Ca^{2+}]$ and can be mimicked in isolated mitochondria by the addition of sub-micromolar $[Ca^{2+}]$. This swelling is caused by a stimulation of the electrogenic uptake of K^+ into the mitochondria, along with a compensating anion such as phosphate [1,15]. A rise in the matrix concentration of pyrophosphate (PPi) is associated with both the hormonally-induced rise in matrix volume *in situ* and the Ca^{2+}-induced increase in volume seen in isolated mitochondria [16,17]. This occurs as a result of inhibition of the matrix pyrophosphatase by micromolar concentrations of

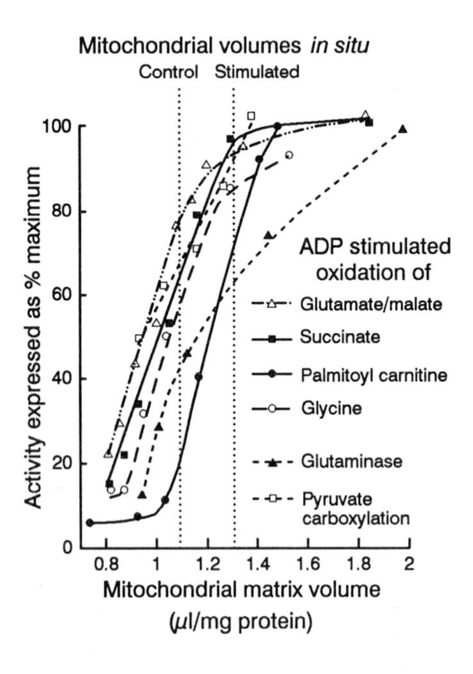

Figure 1. The effects of mitochondrial matrix volume on mitochondrial function. Data are from the sources cited in [1,3,5].

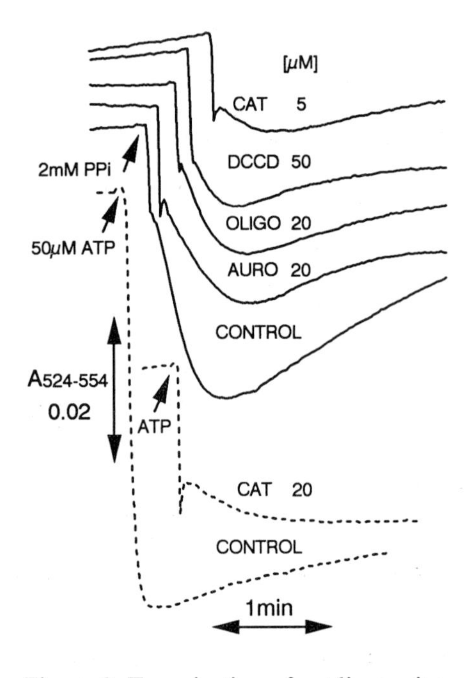

Figure 2. Energisation of rat liver mitochondria by PPi and ATP measured using safranin as described in [53].

Ca^{2+} through the formation of CaPPi which is a very potent competitive inhibitor of the enzyme [18]. It appears to be the increase in [PPi] that stimulates K^+ entry into the mitochondria since other agents which increase matrix [PPi] independently of Ca^{2+}, such as butyrate, also cause swelling [16,17]. We have proposed that this occurs through an interaction with the adenine nucleotide translocase for which PPi is a substrate [1,19,20] but this has yet to be demonstrated directly. However it is of interest that patched clamped mitoplasts possess a K^+-channel that is blocked by ATP [21]. The rise in matrix [PPi] also offers an explanation for the observed increase in mitochondrial adenine nucleotides following hormone treatment [22]. PPi is able to leave the mitochondria in exchange for incoming adenine nucleotides on the translocase [16,19].

Although it is clear that the increase in mitochondrial [PPi] is secondary to an inhibition of matrix pyrophosphatase, it is not known what the source of PPi is. We have used reverse electron flow and safranin measurements of mitochondrial membrane potential to investigate whether PPi might be synthesised by a proton translocating pyrophosphatase such as the one that is present in bacteria, plants and yeast [23,24]. We have shown that PPi is unable to cause energisation of liver sub-mitochondrial particles (SMP), and SMP energised with either ATP or succinate do not synthesise PPi (unpublished data). These data argue against the presence of a proton translocating pyrophosphatase. However we show in Fig. 2 that addition of PPi to mitochondria did cause some energisation that was blocked by inhibitors of the adenine nucleotide translocase, the carrier by which PPi crosses the mitochondrial inner membrane [19]. This might be thought of as evidence for the presence of such a proton translocating pyrophosphatase in intact mitochondria. However the energisation was also blocked by oligomycin and aurovertin, both inhibitors of the ATPase. These data imply that PPi does not energise the mitochondria directly but rather that it must form ATP within the mitochondria which then energises the mitochondria through the proton translocating ATPase. The pathway by which ATP is synthesised from ATP is under investigation.

Mechanisms by which the matrix volume can regulate mitochondrial metabolism

All those mitochondrial processes that are stimulated by an increase in matrix volume are either directly associated with the inner membrane (respiratory chain enzymes, glycine cleavage enzyme and glutaminase) or are stimulated as a result of the increase in mitochondrial ATP/ADP ratio that accompanies stimulation of the respiratory chain (pyruvate carboxylation, citrulline synthesis). The mechanism by which mitochondrial volume can influence membrane associated enzymes remains to be elucidated. The most promising candidate is a change in membrane fluidity which can occur in other systems subject to osmotic stress [25; Dr. T. Dierks, personal communication]. We have shown that benzyl alcohol has inhibitory effects on those processes that are stimulated by an increase in matrix volume, and this is accompanied by an increase in membrane fluidity [26,27]. An additional mechanism would be a change in membrane protein interactions as cristae unfold but recent studies using fluorescence recovery after photobleaching suggest that this is unlikely [28].

MITOCHONDRIAL SWELLING IN PATHOLOGICAL SITUATIONS

When cells are insulted with hypoxia or chemicals that deplete the cell of ATP, cellular [Ca^{2+}] may increase as the ATP-dependent pumps that normally expel excess

Ca^{2+} from the cell are unable to operate. At these higher [Ca^{2+}] concentrations mitochondria become overloaded with Ca^{2+}, especially if they are energised as may occur under conditions of reperfusion following a period of ischaemia. This leads to massive swelling and irreversible damage to mitochondria which precedes cell death [29-31].

Mitochondrial calcium overload leads to the opening of a non-specific pore

The massive swelling of mitochondria that accompanies their uptake of large amounts of Ca^{2+} is a consequence of the opening of a non-specific pore for small molecular weight solutes which become freely permeant [30]. The nature of the pore remains to be established, but several features of it are widely accepted [20,30,32,33]. It accepts molecules of upto about 1500 molecular weight, requires the presence of Ca^{2+} to open, is blocked by the presence of the immunosuppressant cyclosporin A and is associated in some way with both matrix peptidyl-prolyl *cis-trans* isomerase (PPIase) and the adenine nucleotide translocase. The latter conclusion is based on several observations. Pore opening is greatly enhanced under conditions of adenine nucleotide depletion and by the presence of phosphate which can catalyse adenine nucleotide loss from mitochondria. Carboxyatractyloside and fatty acyl-CoA increase pore opening and these agents stabilise the carrier in the "c" conformation, whilst ADP and bongkrekic acid, which stabilise the carrier in the "m" conformation inhibit pore opening. Furthermore Ca^{2+} can interact with the carrier in the "c" conformation to produce a conformational change [20]. The involvement of mitochondrial PPIase has been demonstrated by showing that there are the same number of sites of the PPIase per mg of mitochondrial protein as

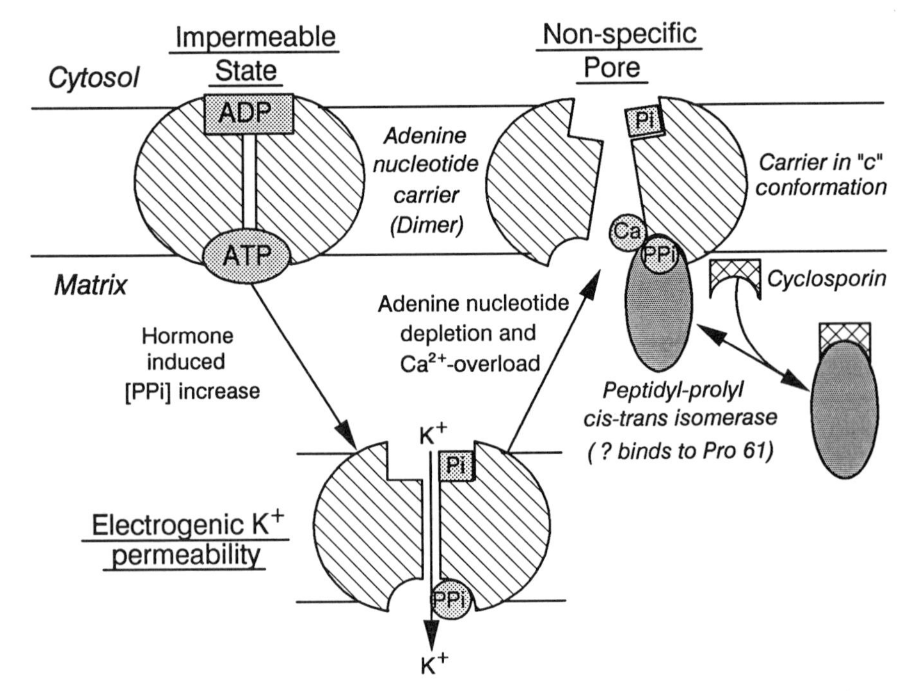

Figure 3. A scheme for the involvement of the adenine nucleotide translocase in physiological and pathological swelling of mitochondria

there are sites for the inhibition of swelling by cyclosporin [20,33]. Furthermore the Ki values for inhibition by a variety of cyclosporin analogues for the mitochondrial PPIase correlate well with their ability to block pore opening [33]. These observations have led us to propose that in the presence of Ca^{2+} and under conditions of adenine nucleotide depletion the adenine nucleotide translocase takes up the "c" conformation and binds the isomerase to a proline residue exposed on the matrix face of the carrier. The resulting conformational change causes the opening of the pore [20,33]. This is illustrated in Fig. 3.

Isolated heart mitochondria are much more stable towards high [Ca^{2+}] than liver mitochondria and we have recently demonstrated that their PPi content increases very little unless high [Ca^{2+}] ($>1\mu M$) is used. As shown in Table 1 this is in marked contrast to the situation in liver mitochondria. The difference was even more apparent when acetate was added to generate additional intramitochondrial PPi through the activation of acetate to acetyl-CoA. The same effects were seen in perfused hearts where neither adrenaline, high extracellular [Ca^{2+}] nor acetate, alone or in combination, gave a significant elevation of tissue PPi (Dr. E.J. Griffiths, unpublished data). This is in marked contrast to the situation in isolated hepatocytes [17] or perfused livers [34] where [PPi] increases several fold under such conditions. Since both liver and heart mitochondria have the same total activity of pyrophosphatase [18], these data imply that some additional regulation of mitochondrial pyrophosphatase may be occurring within the matrix. Our model of the mechanism by which Ca^{2+} causes opening of the non-specific pore requires PPi to bind to the adenine nucleotide translocase, and thus this difference in PPi metabolism may have some relationship to the ability of mitochondria to withstand Ca^{2+} overload.

Table 1
The effects of Ca^{2+} and acetate on liver and heart mitochondrial matrix [PPi]

	PPi (pmol/mg.protein)			
	LIVER		HEART	
[Ca^{2+}] (μM)	Control	5mM Acetate	Control	5mM Acetate
0	62 ± 8(4)	230 ± 39(4)*	106 ± 13(4)	96 ± 11(3)
0.5	121 ± 9(3)*	422 ± 45(3)*	137 ± 21(3)	151 ± 11(3)*
1.0	162 ± 19(3)*	3243 ± 268(3)***	140 ± 30(4)	200 ± 4(3)***
10.0	274 ± 48(3)**	>10 000	266 ± 32(3)*	>10 000

Mitochondria were incubated for 5min in buffer saturated with O_2 at 37°C with [Ca^{2+}] controlled using CaEGTA buffers, before being sedimented through silicone oil into perchloric acid as described elsewhere [16]. Results are expressed as means ± S.E.M. with statistical significance between experimental and control samples being calculated by a paired Student's t-test: *$p<0.05$, **$p<0.02$, ***$p<0.01$

Mitochondrial peptidyl-prolyl cis-trans isomerase

In order to test the hypothesis shown in Fig. 3 we have recently purified the rat liver mitochondrial PPIase (cyclophilin) using a modification of the procedure of Takahashi *et al.* for the purification of cyclophilin from pig kidney [35]. The major form of the active enzyme was found to have a molecular mass of 18.6kDa with a minor active component of 17.5kDa [36]. The total mitochondrial PPIase activity is about 8% of that found in the cytosol. From the specific activity of the pure mitochondrial enzyme it was calculated that isolated liver mitochondria contain approx. 45pmol enzyme per mg total mitochondrial protein. This value is similar to that estimated by others using [^{3}H]-cyclosporin-A binding to liver mitochondria [37]. Higher values estimated in earlier studies from this laboratory [20,33] can be accounted for by the time dependence of PPIase inhibition of the enzyme at low concentrations of cyclosporin A [36]. N-terminal sequencing of the 18.6 and 17.5kDa forms of PPIase show the presence of mitochondrial presequences of 13 and 3 amino acids respectively with the remaining sequence having a strong homology with other cyclophilins including rat cytosolic PPIase [36]. This is shown in Table 2. It is clear that the mitochondrial and cytosolic forms are quite distinct implying that they are products of separate genes unlike the situation in *Neurospora Crassa* [38].

The role of the mitochondrial non-specific pore in health and disease

If mitochondrial Ca^{2+}-overload and consequent pore opening is an important factor in the tissue damage resulting from ischaemia followed by reperfusion it might be

Table 2
N-Terminal sequence homologies of several peptidyl-prolyl cis-trans isomerases

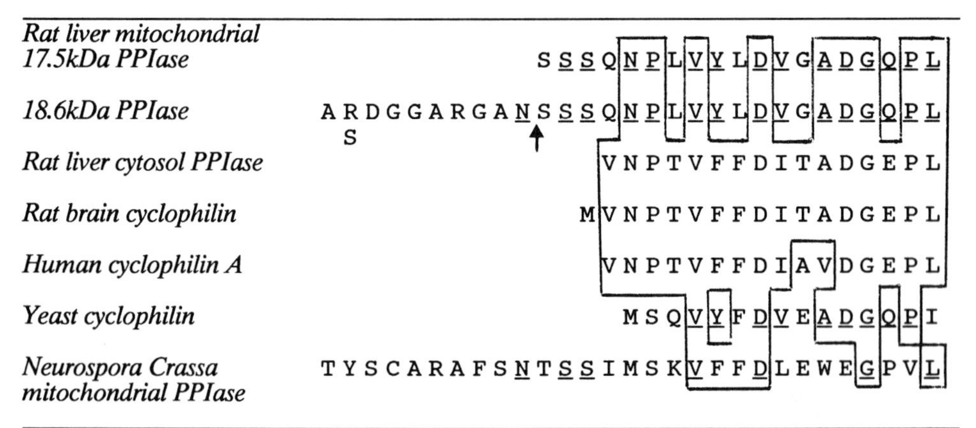

Rat liver mitochondrial 17.5kDa PPIase	S S S Q N P L V Y L D V G A D G Q P L
18.6kDa PPIase	A R D G G A R G A N S S S Q N P L V Y L D V G A D G Q P L
	S ↑
Rat liver cytosol PPIase	V N P T V F F D I T A D G E P L
Rat brain cyclophilin	M V N P T V F F D I T A D G E P L
Human cyclophilin A	V N P T V F F D I A V D G E P L
Yeast cyclophilin	M S Q V Y F D V E A D G Q P I
Neurospora Crassa mitochondrial PPIase	T Y S C A R A F S N T S S I M S K V F F D L E W E G P V L

Data are taken from [36]. The boxed area indicates sequence homology with rat brain cyclophilin whilst underlined sequences show additional homologies between the rat mitochondrial PPIase and the yeast and *Neurospora Crassa* cyclophilins. The two mitochondrial sequences represent data from two identical analyses of the 18.6kDa protein and one analysis of the 17.5kDa protein. Ambiguities in the signal were consistently observed at the point shown. The putative second cleavage site of the 18.6kDa protein is indicated with an arrow.

expected that cyclosporin A should provide protection. There is some evidence to suggest that this occurs [39,40]. However a much better protective regime is lowering of the perfusion pH which also enhances recovery during re-oxygenation [31]. This has now been demonstrated in heart, liver, kidney and tumour cells. Low pH tends to accompany hypoxia because of the accumulation of lactic acid within the cell. However upon reperfusion this lactic acid will be transported out of the cell on the monocarboxylic acid transporter with a resultant alkalinisation of the cytosol [41]. We have demonstrated that the opening of the Ca-sensitive pore within the inner membrane of both rat heart and liver mitochondria is inhibited greatly as the pH drops below 7 [40]. The mechanism of this effect remains unclear but it would seem likely that this may play an important role in the protective effect of low pH against chemical and hypoxic tissue damage.

Peptidyl-prolyl *cis-trans* isomerases are ubiquitous and may represent a family of proteins [42,43]. It is highly probable that their presence in mitochondria has some beneficial effect. One suggested role is in the folding of proteins during synthesis or entry into the mitochondria, but this is unlikely in view of the quite rapid spontaneous rate of isomerisation and the existence of *Neurospora Crassa* mutants lacking cyclophilin [44]. Their function might be to allow occasional opening of the non-specific pore, even under normal conditions as opposed to the catastrophic opening seen in Ca^{2+}-overload. We have provided evidence that the non-specific permeability pathway can open in normal healthy liver cells stimulated with hormones [20,45]. Others have shown that sucrose, which is normally unable to enter mitochondria *in vitro*, can enter the mitochondrial matrix *in situ* [46]. Fluorescent microscopic studies of cells stained with Rhodamine G to report mitochondrial membrane potential show that mitochondria may undergo occasional depolarisation, consistent with pore opening [47]. Thus we may speculate that occasional pore opening is essential for the well being of mitochondria, perhaps to release by-products of metabolism for which no specific transport pathways exist. Permanent closure of the pore with cyclosporin treatment would lead to the build up of such metabolic waste products to the detriment of the mitochondria. It is also possible that the inhibition of mitochondrial peptidyl-prolyl *cis-trans* isomerase is responsible for the toxic effects of cyclosporin A on liver and kidney [42,48]. In support of this hypothesis, toxicity is associated with changes in mitochondrial morphology which are not as severe with cyclosporin G, a less potent inhibitor of PPIase, as with cyclosporin A [33,48,49]. FK-506 appears to lack these side effects [50] and does not inhibit mitochondrial PPIase [33,51].

Acknowledgements

This work was supported by the Medical Research Council.

REFERENCES

1 Halestrap AP. Biochim Biophys Acta 1989; 973: 355-382
2 Garlid KD. In : Lemasters JJ, Hackenbrock CR, Thurman G, Westerhoff HV, eds. Integration of mitochondrial function. New York: Plenum, 1988; 257-276
3 McCormack JG, Halestrap AP, Denton RM. Physiol Rev 1990; 70: 391-425
4 Quinlan PT, Thomas AP, Armston AE, Halestrap AP. Biochem J 1983; 214: 395-404
5 Halestrap AP, Davidson AM, Potter WD. Biochim Biophys Acta 1990; 1018: 278-281
6 Halestrap AP, Armston AE. Biochem J 1984; 223: 677-685
7 Lund P, Wiggins D. Bioscience Reports 1987; 7: 59-66

8 Brown GC, Lakinthomas PL, Brand MD. Eur J Biochem 1990; 192: 355-362
9 Quinlan PQ, Halestrap, AP Biochem J 1986; 236: 789-900
10 Moreno-Sanchez R, Hogue BA, Hansford RG. Biochem J 1990; 268: 421-428
11 Hue L, Baquet A, Lavoinne A, Meijer AJ. In: Grunnet N, Quistorff B, eds. Regulation of Hepatic Function - Metabolic and structural interactions. Copenhagen: Munksgaard, 1991; 447-457
12 Häussinger D, Hallbrucker C, Vomdahl S, et al. Biochem J 1990; 272: 239-242
13 Häussinger D, Lang F, Bauers K, Gerok W. Eur J Biochem 1990; 193: 891-898
14 Häussinger D, Lang F, Bauers K, Gerok W. Eur J Biochem 1990; 188: 689-695
15 Halestrap AP, Quinlan PT, Whipps DE, Armston AE. Biochem J 1986; 236: 779-787
16 Davidson AM, Halestrap AP. Biochem J 1987; 246: 715-723
17 Davidson AM, Halestrap AP. Biochem J 1988; 254: 379-384
18 Davidson AM, Halestrap AP. Biochem J 1989: 258: 817-821
19 Halestrap AP, Davidson AM. In: Azzi A et al, eds. Anion Carriers of Mitochondrial Membranes. Berlin: Springer Verlag, 1989; 337-348
20 Halestrap AP, Davidson AM. Biochem J 1990; 268: 153-160
21 Inoue I, Nagase H, Kishi K, et al. Nature 1991; 352: 244-247
22 Aprille JR. FASEB J 1988; 2: 2547-2556
23 Baltscheffsky M, Nyren P. N Comprehen Biochem 1984; 9: 187-197
24 Mansurova SE. Biochim Biophys Acta 1989; 977: 237-247
25 Curtain CC, Looney FD, Regan DL, Ivancic NV. Biochem J 1983; 213: 131-136
26 Armston AE, Halestrap AP, Scott RD. Biochim Biophys Acta 1982; 681: 429-439
27 Halestrap AP. Biochem J 1982; 204: 37-47
28 Chazotte B, Hackenbrock CR. J Biol Chem 1991; 266: 5973-5979
29 Orrenius S, McConkey DJ, Bellomo G, Nicotera P. Trends Pharmacol Sci 1989; 10: 281-285
30 Crompton, M. In: Langer GA, ed. Calcium and the Heart. New York: Raven, 1990; 67-198
31 Herman B, Gores GJ, Nieminen AL, et al. Critical Rev Toxicol 1990; 48: 759-764
32 Gunter TE, Pfeiffer DR. Am J Physiol 1990; 258: C755-C786
33 Griffiths EJ, Halestrap AP. Biochem J 1991; 274: 611-614
34 Inoue T, Yamada T, Furuya E, Tagawa K. Biochem J 1989; 262: 965-970
35 Takahashi N, Hayano T, Susuki M. Nature 1989; 337: 473-475
36 Connern CP, Halestrap AP. Submitted to Biochem J
37 McGuinness O, Yafei N, Costi A, Crompton M. Eur J Biochem 1990; 194: 671-679
38 Tropschug M, Nicholson DW, Hartl F-U, Köhler, H, et al. J Biol Chem 1988; 263: 14433-14440
39 Goto S, Kim YI, Kamada N, Kawano K, Kobayashi M. Transplantation 1990; 49: 1003-1005
40 Halestrap AP. Biochem J 1991; In Press
41 Halestrap AP, Poole RC, Cranmer SL. Biochem Soc Trans 1990; 18: 1132-1135
42 Ryffel B. Pharmacol Rev 1989; 41: 407-422
43 Harding MW, Galat A, Uehling DE, Schreiber SL. Nature 1989; 341: 758-760
44 Schönbrunner ER, Mayer S, Tropschug M, Fischer, G, et al. J Biol Chem 1991; 266: 3630-3635
45 Tropschug M, Barthelmess IB, Neupert W. Nature 1989; 342: 953-955
46 Davidson AM, Halestrap AP. Biochem J 1990; 268: 147-152
47 Tolleshaug H, Seglen PO. Eur J Biochem 1985; 153: 223-229
48 Bereiter-Hahn J, Seipel K-H, Voth M. Cell Biochem Function 1983; 1: 147-155
49 Foxwell BMJ, Ryffel B. Immunology Allery Clinics N America 1989; 9: 79-93
50 Hiestand PC, Gunn HC, Gale JM, Ryffel B, Borel JF. Immunology 1985; 55: 249-255
51 Shapiro JJ, Fung AB, Jain AB, Parks P, et al. Transplant Procs 1990; 22: 35s-36s
52 Kay JE, Moore AL, Doe SEA, Benzie CR, et al. Transplant Procs 1990; 22: 96-99
53 Åkerman KEO, Wikström MKF FEBS Lett 1976; 68: 191-197

© 1992 Elsevier Science Publishers B.V. All rights reserved.
Molecular mechanisms of transport. E. Quagliariello, F. Palmieri, eds.

ANION TRANSLOCATION AND LIPID COMPOSITION IN RAT HEART MITOCHONDRIA: THE EFFECT OF AGING

G. Paradies, F.M. Ruggiero, P. Dinoi and E. Quagliariello

Department of Biochemistry and Molecular Biology and C.N.R. Unit for the Study of Mitochondria and Bioenergetics, Bari, Italy

INTRODUCTION

Mitochondrial anion transport proteins are responsible for the flux of negatively charged metabolites that occurs across the inner mitochondrial membrane (1). These fluxes are required steps for the normal functioning of several cytosolic and mitochondrial metabolic pathways and are therefore essential for the bioenergetics of eucariotic cell (2).

Aging is a biological phenomenon characterized by impairment of various aspects of cell functions. At mitochondrial level aging causes changes in biochemical pathways involved in the energy metabolism (3). These changes have been related to molecular and functional changes in the properties of the mitochondrial membranes. Aging has a profound effect on cardiac performance. The well kown age-dependent decrements in heart performance may be related, in addition to other factors, to changes in the activity of mitochondrial anion carrier systems (4-6).

Pyruvate, together with fatty acids, is the major energy source in the heart. Phosphate is essential for the synthesis of ATP during the oxidative phosphorylation and for the uptake of several important substrates (1). The transport of these two substrates across the mitochondrial inner membrane is mediated by specific transporting systems. The kinetic properties and the substrate specificity of these two carrier systems have been studied in detail (7-10). More recently, the pyruvate and the phosphate carrier proteins have been isolated and purified and their activity reconstituted in artificial membranes such as liposomes (11-14). These reconstitution experiments have shown that lipids, in particular cardiolipin, may have an important role in the regulation of the activity of these two carrier systems (11,14).

In this work, the effect of aging on the activity of the pyruvate and the phosphate carriers and on the lipid composition in rat heart mitochondria was investigated. The results obtained demonstrate that the transport activity of these two carrier systems is significantly depressed with aging. The age-dependent decrements in the activity of the pyruvate and the phosphate carriers in heart mitochondria, appear to be due to a modification in the lipid microenvironment (decrease in the cardiolipin content) surrounding these two carrier molecules in the membrane.

MATERIALS AND METHODS

Animals

Male Fisher 344 rats of 4-5 months or 26-28 months were used throughout these studies.

Rat heart mitochondria were prepared by differential centrifugation of heart homogenates essentially as described previously (15).

The standard medium used in the measurements of pyruvate and phosphate transport, binding experiments, ΔpH measurements and respiratory activities usually contained: 100 mM sucrose, 50 mM KCl, 20 mM Tris-HCl and 0.5 mM EDTA.

Mitochondrial metabolite transport

The transport of pyruvate and phosphate by rat heart mitochondria was measured by the stop inhibitor method essentially as described in refs. 16 and 17, respectively.

Measurements of binding

The binding of cyanocinnamate to rat heart mitochondria was assayed essentially as described in ref. 15

Transmembrane ΔpH measurements

The transmembrane ΔpH values in mitochondria from young and aged rats were determined as described in ref. 18.

Mitochondrial swelling

Mitochondrial osmotic volume changes were measured by the apparent absorbance changes at 540 nm with a spectrophotometer linked to a suitable recorder.

Mitochondrial respiration

Rates of oxygen uptake were measured polarographically with a Clark oxygen electrode connected to a suitable recorder, in a water-jacketed closed chamber at 30 °C.

Analysis of lipids

Phospholipids and cholesterol were analyzed by HPLC, using a Beckman 344 gradient liquid chromatograph. Extraction and analysis of phospholipids and cholesterol were carried out essentially as described in ref. 19.

Determination of phosphate and protein

The endogenous level of phosphate was determined chemically (20) in perchloric acid extracts.

Protein concentration was measured by the usual biuret method.

Statistical analysis

Results are expressed as mean values ± S.E. Statistical significances were determined by the Student t test.

RESULTS AND DISCUSSION

The kinetic parameters of the pyruvate carrier in heart mitochondria from young and aged rats are reported in Table I. While the affinity of the carrier for pyruvate was practically the same in the two types of mitochondria, the maximal velocity of the pyruvate translocation was markedly depressed in mitochondria from aged rats.

TABLE I

KINETIC PARAMETERS OF THE PYRUVATE CARRIER AND THE NUMBER OF CYANOCINNAMATE BINDING SITES IN HEART MITOCHONDRIA FROM YOUNG AND AGED RATS

The K_m and the V_{max} values were calculated from double reciprocal plots of the rates of pyruvate uptake versus pyruvate concentrations (ranging from 0.05 to 1 mM). The number of cyanocinnamate binding sites were determined as described in (15). Each value represents the mean ± S.E. obtained for five experiments.

Animals	Pyruvate carrier		Cyanocinnamate binding sites
	K_m (mM)	V_{max} (nmol/min/mg prot	(pmol/mg protein)
Young	0.210 ± 0.021	6.8 ± 0.7	48.5 ± 4.7
Aged	0.221 ± 0.023	4.2 ± 0.5[a]	47.9 ± 4.9

[a] $P < 0.005$

The age-related decrement in the activity of the pyruvate carrier in rat heart mitochondria cannot be ascribed to a decrease in the mitochondrial transmembrane pH gradient, toward which the pyruvate carrier is very sensitive (7-8). In fact, almost similar transmembrane ΔpH values were measured in both mitochondria from young and aged rats.

The depressed activity of the pyruvate carrier with aging could also be due to a decrease in the total number of the pyruvate carrier molecules. However, as reported in Table I, the total number of the pyruvate carrier molecules, titrated as cyanocinnamate binding sites (15,21), was the same in mitochondria from young and aged rats. This excludes the possibility that the depressed activity of the pyruvate carrier with aging is due to a lower content of pyruvate carrier molecules.

The kinetic parameters of the phosphate carrier in mitochondria from young and aged rats are reported in Table II. Both mitochondria from young and aged rats exhibited the same K_m value for phosphate translocation. How-

TABLE II

KINETIC PARAMETERS OF THE PHOSPHATE CARRIER AND THE CONTENT OF ENDOGENOUS PHOSPHATE IN HEART MITOCHONDRIA FROM YOUNG AND AGED RATS

The K_m and the V_{max} values were calculated from double reciprocal plots of the rate of phosphate uptake versus phosphate concentrations (ranging from 0.5 to 5 mM). The endogenous level of phosphate was determined chemically. Each value represents the mean ± S.E. obtained for five different experiments.

Animals	Phosphate carrier		Endogenous phosphate (nmol/mg protein)
	K_m (mM)	V_{max} (nmol/min/mg protein)	
Young	1.72 ± 0.21	284 ± 29	28.5 ± 3.2
Aged	1.78 ± 0.24	202 ± 24[a]	27.3 ± 3.1

[a] $P < 0.05$

ever, the maximal velocity of this process was significantly reduced in mitochondria from aged rats when compared to that obtained with mitochondria from young rats.

The lower activity of the phosphate carrier in heart mitochondria from aged rats is further documented by swelling experiments. In fact, it was found (results not shown) that both the rate and the final extent of the swelling in ammonium-phosphate were significantly decreased in mitochondria from aged rats as compared to the same values obtained with mitochondria from young control animals. These results were compared with the rates of swelling of the same preparations of mitochondria when suspended in ammonium-acetate. No difference in both the rate and the final extent of the swelling in ammonium-acetate could be observed between these two types of mitochondria. It should be recalled that the uptake of phosphate and acetate by mitochondria depends on the transmembrane ΔpH (1). Thus, the lower activity of the phosphate carrier in mitochondria from aged rats is not due to a decrease in the mitochondrial transmembrane ΔpH. This conclusion is also supported by direct measurements of the transmembrane ΔpH values, as reported above.

The reduced activity of the phosphate carrier in mitochondria from aged rats could also be due to a lower content of endogenous exchangeable phosphate. As shown in Table II, no difference in the amount of endogenous phosphate could be observed in these two preparations of mitochondria.

A lower content of functional phosphate translocase could also be responsible for the age-dependent decrease in the activity of the mitochondrial phosphate carrier. However, inhibitor titrations of the rate of phosphate transport with mersalyl (see also ref. 17) have shown that mitochondria from aged rats require the same concentrations of this inhibitor to produce total inhibition of the rate of phosphate transport as mitochondria from young control rats (results not reported). This suggests that the amount of functional phosphate translocase is unaffected by aging as found for pyruvate carrier (see above) and for the adenine nucleotide translocase (5).

Deterioration of the mitochondrial membrane with aging may represent another possible factor responsible for the reduced activity of the pyruvate and the phosphate carriers. However, as reported in Table III,

TABLE III

RESPIRATORY ACTIVITIES OF CARDIAC MITOCHONDRIA ISOLATED FROM YOUNG AND AGED RATS

Rates of mitochondrial respiration were measured polarographically in the presence of malate and pyruvate as respiratory substrates.

Animals	Respiratory control ratio	ADP/O
Young	8.3 ± 0.8	2.8 ± 0.2
Aged	8.1 ± 0.7	2.7 ± 0.2

neither the respiratory control ratios, nor the ADP/O ratios were changed in both mitochondria from young and aged rats. These results exclude alterations in the intactness of the mitochondrial membrane with aging.

The lower translocating activity of both the pyruvate and the phosphate carriers with aging appears, therefore, to be a consequence of a modification of the lipid surrounding of these two carrier molecules in the mitochondrial membranes. In fact, the analysis of the mitochondrial membrane lipids (see Table IV) reveals significant changes in the lipid composition in aged rats. Particularly, the total cholesterol and the cholesterol/ phospholipid molar ratio increased and the phosphatidylethanolamine/phosphatidylcholine ratio decreased in the mitochondrial membrane from aged rats. These changes are known to be associated with an increase in the membrane viscosity which, in turn, may reduce the mobility of the carrier molecules in the membrane and hence their functional activity.

The change in the mitochondrial membrane lipid composition with aging may have a more specific and localized effect on the activity of the pyruvate and the phosphate carriers. In fact, as demonstrated by

TABLE IV

EFFECT OF AGING ON CHOLESTEROL AND PHOSPHOLIPID CONTENT IN HEART MITOCHONDRIA FROM YOUNG AND AGED RATS

For cholesterol and phospholipid extraction and analysis see Materials and Methods. Cholesterol is expressed as nmol per min per mg protein and phospholipid as nmol lipid Pi per mg protein. Each value represents the mean ± S.E.M. obtained for five experiments.

Lipids	Young	Aged
Cholesterol	11.0 ± 1.4	15.3 ± 1.0[a]
Phospholipids	275.0 ± 9.8	240.9 ± 9.9[a]
Ratio cholesterol/phospholipid	0.04 ± 0.006	0.063 ± 0.008[a]

[a] $P<0.001$

TABLE V

PHOSPHOLIPID COMPOSITION IN HEART MITOCHONDRIAL MEMBRANE FROM YOUNG AND AGED RATS

For phospholipid extraction and analysis see Materials and Methods. Values are expressed as nmol per mg protein. Each value represents the mean ± S.E.M. obtained for five experiments. PE= phosphatidylethanolamine, PC= phosphatidylcholine.

Phospholipid	Young	Aged
Cardiolipin	38.4 ± 2.1	23.5 ± 2.9[a]
Phosphatidylethanolamine	96.5 ± 5.4	78.1 ± 5.0[a]
Phosphatidylinositol	3.4 ± 0.9	3.2 ± 0.7[a]
Phosphatidylserine	6.7 ± 1.0	7.6 ± 1.1
Phosphatidylcholine	130.1 ± 7.5	128.5 ± 7.0
PE/PC	0.74 ± 0.05	0.61 ± 0.04[a]

[a] $P<0.01$

reconstitution experiments in artificial membranes such as liposomes, the phospholipid cardiolipin is required for full activity of the phosphate (14) and the pyruvate carrier (11). Specific cardiolipin requirement has been demonstrated for other mitochondrial proteins and enzymes (22-24). As shown in Table V, the content of cardiolipin was significantly reduced in mitochondrial membrane from aged rats. It should be recalled that cardiolipin is mostly exclusively concentrated on the matrix side of the inner mitochondrial membrane, where the pyruvate and the phosphate carrier molecules are presumably located. It is, therefore, conceivable that a lower content and/or a different composition of the cardiolipin molecules surrounding the pyruvate and the phosphate transporting systems in the mitochondrial membrane, may be directly responsible for the observed lower activity of these two carrier proteins with aging.

Pyruvate and phosphate are essential substrates for the heart mitochondrial energy metabolism. In fact, pyruvate represents the main energy source for the heart, while phosphate is involved in the regulation of oxidative phosphorylation and in addition it is an obbligatory requirement for net uptake of other important metabolites such as malate, citrate and oxoglutarate (1).

Thus, the age-dependent decrements in the activity of both the pyruvate and the phosphate carriers in heart mitochondria can account, at least in part, for the well known decline in cardiac functional competence with aging.

AKNOWLEDGMENTS

This work was supported by Consiglio Nazionale delle Ricerche (Target Project on Aging) and from MURST (60%). The authors acknowledge the excellent technical assistance of Mr. L. Gargano

REFERENCES

1 Lanoue K.F., Schoolwerth A.C. Annu. Rev. Biochem. 1979; 48: 871-922
2 Meijer A.F., Van Dam K. New Comprehensive Biochem. 1981; 2: 235-256
3 Hansford R.G. Biochim. Biophys. Acta 1983; 726: 41-80
4 Hansford R.G. Biochem. J. 1978; 170: 285-295
5 Nohl H., Kramer R. Mech. Ageing Dev. 1980; 14: 137-144
6 Kim J.H. Shrago E. Elson C.E. Mech. Ageing Dev. 1988; 46: 279-290
7 Paradies G. Papa S. Biochim. Biophys. Acta 1977; 462: 333-346
8 Halestrap A.P. Scott R.D. Thomas A.P. Intern. J. Biochem 1980; 11:97-105
9 Fonyo A. Pharmacol. Ther. 1979; 7: 627-645
10 Ligeti E. Brandolin G. Dupont Y. Vignais P.V. Biochemistry 1984; 24: 4423-4424
11 Nalecz K.A. Bolli R. Wojtczak L. Azzi A. Biochim. Biophys. Acta 1986; 851: 29-37

12 Wohlrab H. Collins A. Costello D. Biochemistry 1984; 23: 1057-1064
13 Kramer R. Palmieri F. Biochim. Biophys. Acta 1989; 974: 1-23
14 Kadenbach B. Mende P. Kolbe H.V.J. Stipani I. Palmieri F. FEBS Lett. 1982; 139: 109-112
15 Paradies G. Biochim. Biophys. Acta 1984; 766: 446-450
16 Paradies G. Ruggiero F.M. Biochim. Biophys. Acta 1988; 932: 1-7
17 Paradies G. Ruggiero F.M. Biochim. Biophys. Acta 1990; 1019: 133-136
18 Paradies G. Ruggiero F.M. Arch. Biochem. Biophys. 1989; 269: 595-602
19 Ruggiero F.M. Landiscina C. Gnoni G.V. Quagliariello E. Lipids 1984; 19: 171-178
20 Wahler B.E. Wollenberg Biochem. Z. 1958; 329: 508-520
21 Paradies G. Ruggiero F.M. Biochim. Biophys. Acta 1986; 850: 249-255
22 Ernster L. Sandri G. Hundal T. Carlsson C. Nordembrand K. In: Structure and Function of Energy-Transducing Membranes (Van Dam K. Van Gelder B. F.,eds.) Elsevier, Amsterdam/New York, 1977; 209-222
23 Robinson N.C. Stry F. Talbert L. Biochemistry 1980; 19: 3656-3661
24 Schlame M. Beyer K. Hayer-Hartl M. Klingenberg M. Eur. J. Biochem. 1991; 199: 459-466

© 1992 Elsevier Science Publishers B.V. All rights reserved.
Molecular mechanisms of transport. E. Quagliariello, F. Palmieri, eds.

REEVALUATION OF THE ROLE OF THE ADENINE NUCLEOTIDE TRANSLOCASE IN CONTROL OF RESPIRATION

Kathryn F. LaNoue[a], Deborah A. Berkich[a], Peter T. Masiakos[a], Gerald D. Williams[b], and Michael B. Smith[b,a]

[a]Departments of Cellular and Molecular Physiology and [b]Radiology, The Milton S. Hershey Medical Center, The Pennsylvania State University, PO Box 850, Hershey, PA 17033

INTRODUCTION

Numerous studies carried out in the last 20 years have been aimed at elucidating the control of phosphorylating respiration. Classical studies by Chance and coworkers (1,2) initially demonstrated that respiration of isolated mitochondria is sensitively controlled by media ADP and phosphate and that synthesis of ATP from these substrates is stoichiometric with oxygen consumption. The discovery of the mitochondrial adenine nucleotide translocase (3-5) stimulated speculation that kinetic control of ATP synthesis may reside in the translocase rather than the synthase. On the other hand, a large school of thought held that neither the translocase nor the synthase exerts kinetic control, but instead the translocase and the ATP synthase catalyze a near-equilibrium between the external phosphorylation potential and the electron transport chain, between NADH and cytochrome c (6,7).

Although it now appears clear that control of ATP synthesis is exerted kinetically (8,9) under physiological conditions, the controversy about the site of kinetic control has not been resolved. To help resolve the issue of whether the ATP synthase or the adenine nucleotide translocase exerts major rate limitation when ADP stimulates respiration we developed techniques to separately and independently monitor fluxes through the two proteins when both were active. The translocase flux has been assessed by ^{31}P nuclear magnetic resonance (NMR) saturation-transfer (10) and the ATP synthase monitored, independent of the translocase, by measuring $P^{18}O_4$ exchange with $H^{16}OH$ (11).

MATERIALS AND METHODS

The methods used have been outlined in detail in previous publications. Use of ^{31}P nuclear magnetic resonance spectroscopy to measure matrix nucleotides, inorganic phosphate and translocase flux is described in detail, by Hutson et al. (12,13) and by Masiakos et al. (10). The use of $P^{18}O_4$ to monitor ATP synthase activity was also described recently (11).

^{31}P NMR spectra were acquired at 162.0 MHz with a Bruker AM-400 wide-bore spectrometer equipped with a 20 mm $^{13}C/^{31}P$ double-tuned probe. In the

NMR studies, 8 ml mitochondrial samples were placed in 15 mm NMR tubes that were inserted into 20 mm NMR tubes. The basic suspension media for the mitochondria in the NMR experiments included 130 mM KCl, 15 mM succinate, 15 mM glutamate, 20 mM MOPS (pH 6.8), 1 mM CDTA (*trans*-1,2,-diaminocyclohexane-*N,N,N',N'*-tetraacetic acid) and in some cases 30% fluorocarbons (Fluoronert Electronic liquid FC433M, Cordova, IL) and 3% Pluronic F108 Prill (BASF Corp., Parsippany, NJ).

RESULTS AND DISCUSSION

The mitochondrial proton pumping ATP synthase is easily reversible. When net ATP synthesis by isolated mitochondria is limited by the supply of external ADP, isotopic exchange between ATP and $^{32}P_i$ is faster than net ATP synthesis.

$$ADP+P_i \underset{V_2}{\overset{V_1}{\rightleftharpoons}} ATP$$

Thus forward (V_1) and reverse reactions (V_2) are catalyzed at the same time by the synthase, but when net synthesis (V_1-V_2) occurs, the forward reaction is faster. Using intact mitochondria, in the absence of added ADP, forward (V_1) and reverse (V_2) fluxes are equal. The value (V_1=V_2) at 37°C is 0.17 µmoles/min•mg using rat liver mitochondria and 1.10 µmoles/min•mg using rat heart mitochondria (8). In both liver and heart mitochondria, the reverse reaction declines to zero as external ADP increases whereas the forward reaction doubles. Nevertheless in intact cardiac (14), kidney (15) and brain tissue (16), ^{31}P NMR saturation-transfer measurements show that the total forward flux (V_1) from P_i to ATP is equal to net ATP synthesis, independent of respiration rates. This suggested that in the cell, the reverse reaction (V_2), catalyzed by the mitochondrial ATP synthase is zero. The difference between the *in vivo* data and the isolated mitochondrial data, quantitating forward and reverse fluxes suggest that control of respiration *in vivo* is not based on cytosolic ADP availability.

In order to rule out the possibility that methodological differences produced the apparent discrepancy between *in vivo* results and *in vitro* results with isolated mitochondria attempts were made to measure V_1 and V_2 in isolated mitochondria using both ^{31}P NMR saturation-transfer, and ^{32}P isotopic methods. Liver mitochondria were used in this study because monitoring intramitochondrial phosphorus is technically easier in liver compared to heart because of slower respiration relative to the amount of intramitochondrial nucleotides.

Figure 1 shows the ^{31}P NMR spectra of mitochondria under control conditions and during saturating irradiation of the γ-phosphate of the intramitochondrial ATP. The internal and external ATP phosphorus signals can be observed at different frequencies because the ATP inside the mitochondria was saturated with Mg^{2+} whereas the external media contained 1 mM CDTA (a potent Mg^{2+} chelator). Mg^{2+}-bound ATP has a ^{31}P NMR spectrum which is distinct from free ATP. Previous studies (12) have shown that there is a 1:1 correlation of total biochemically measured matrix ATP and the NMR spectral peak that corresponds to the unique Mg^{2+}-bound β-phosphorus peak of ATP. This implies that the intramitochondrial ATP is homogeneous and metabolically active and thus potentially useful for saturation-transfer experiments. The experiment of Figure 1 was performed, saturating the γ-phosphate of ATP and observing the effect of that saturation on the internal and external inorganic phosphate signals and the external γ-phosphate of ATP. Decreases in the magnetization (area under the spectral peak) of either entity would permit a flux measurement.

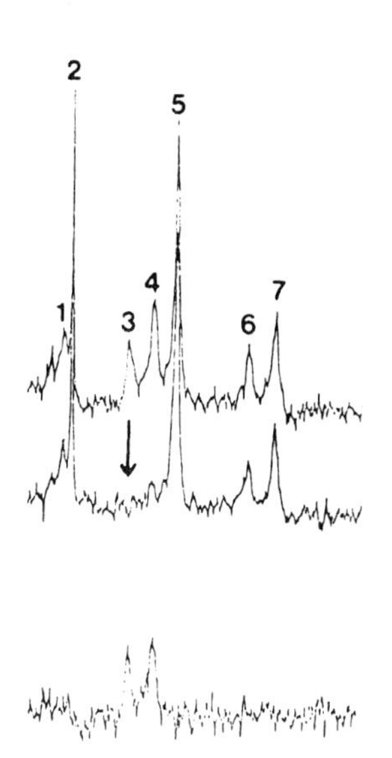

Figure 1: NMR spectra of rat liver mitochondria in the presence and absence of saturating NMR irradiation of the γ -phosphate position of ATP_{in}. Rat liver mitochondria (8 ml, 20 mg/ml) were incubated at 25°C with continuous bubbling with 100% O_2. The basic media contained 1 mM ATP, 1 mM inorganic phosphate and 30% fluorocarbons. A control spectrum was obtained in the uppermost trace, a spectrum with saturating irradiation is shown in the middle trace and the difference spectrum is shown in the bottom trace. Peak assignments are 1) $(P_i)_{in}$, (2) $(P_i)_{out}$, (3) γ-ATP_{in} + β-ADP_{in}, 4) γ-ATP_{out} + β-ADP_{out}, (5) α-ATP, α-ADP and NAD(H), (6) β-ATP_{in} and (7) β-ATP_{out}.

The first-order rate constant of flux may be obtained using the equation derived from Forsen and Hoffman (17).

$$k_1 = (T_{1m}^{-1})(1-M^+/M^\circ) \qquad \text{Eq. [1]}$$

where M° is the equilibrium magnetization of the spectral peak of interest (inorganic phosphate or external ATP γ-phosphate) and is determined from the control spectra. M^+ is the magnetization during saturation of the internal γ-ATP phosphate peak and T_{1m} is the spin-lattice relaxation time of the inorganic phosphate or external γ-ATP. The T_{1m} of external γ-ATP is about 1.1 s (12) whereas the external inorganic phosphate T_{1m} is 3.0 (13). The data of Figure 1 indicate that no flux to inorganic phosphate could be observed under these conditions whereas the flux of external ATP to the internal ATP was so rapid that the external peak disappeared entirely. Inspection of equation 1 reveals that k_1, the first order rate constant for the turnover of external phosphate, would have to be greater than 0.066 sec^{-1} in order to observe a significant decrease in the magnetization of external P_i.

To achieve $k_1 > 0.066$ sec^{-1} with 1 mM inorganic phosphate and 20 mg/ml mitochondria in the buffer, the flux from the γ-phosphate of ATP had to be greater than 0.20 μmoles/min•mg protein. This exchange rate is somewhat higher than that measured isotopically using $^{32}P_i$ at 30°C (8). Therefore it seemed reasonable that no transfer could be detected. On the other hand the complete disappearance of the external γ-phosphate of ATP did appear unreasonable since this led to the conclusion that the transport of adenine nucleotides exceeds a value of 2.00 μmoles/min•mg protein. This value exceeds estimates of flux obtained isotopically by at least 2-3 fold.

Further experiments were undertaken to confirm and expand this observation. Because the separation of the internal from the external ß-ATP peaks is larger than the separation of the γ-ATP's, further measurements were made measuring saturation-transfer between the former. An example of the data obtained is shown in Figure 2, taken from (10). The figure demonstrates that it is possible to measure the transport rate with reasonable accuracy using this method. Arrhenius plots of transport rates measured over a temperature range from 8-25°C were used to calculate activation energies, which were in good agreement with those reported previously by Klingenberg (18). Moreover, the break in the Arrhenius plot indicating a change in activation energy at 14°C was also observed in the present set of data. The absolute values of transport however were much higher than those found using isotopic techniques.

The discrepancy in the two sets of data may be due to different conditions used to make the measurements. The isotopic experiments were routinely carried out in the presence of oligomycin, in order to inhibit the ATP synthase. Under these conditions matrix ATP levels (~2 nmoles/mg protein) are lower than under the conditions used in the NMR experiment (~10 nmoles/mg) where the ATP synthease was active. When matrix ATP levels were deliberately varied using the Aprille technique (19) to alter total matrix nucleotide levels, transport varied with matrix ATP levels in a distinctly sigmoidal fashion with a Hill coefficient of 4.07 ± 0.27. Half maximal transport rates at 18°C were observed at 6.6 nmoles ATP/mg protein. These data all suggested that the translocase has a much higher V_{max} than the ATP synthase and contrary to our expectations, the ATP/$^{32}P_i$ exchange data probably reflects rate limitation by the synthase rather than the translocase.

Recent studies carried out in the collaboration with Dr. Paul Boyer (11) have tended to confirm this conclusion. The studies were performed largely with rat heart mitochondria, but liver mitochondria behave similarly. In these studies ATP/$^{32}P_i$ exchange rates were measured over a wide range of

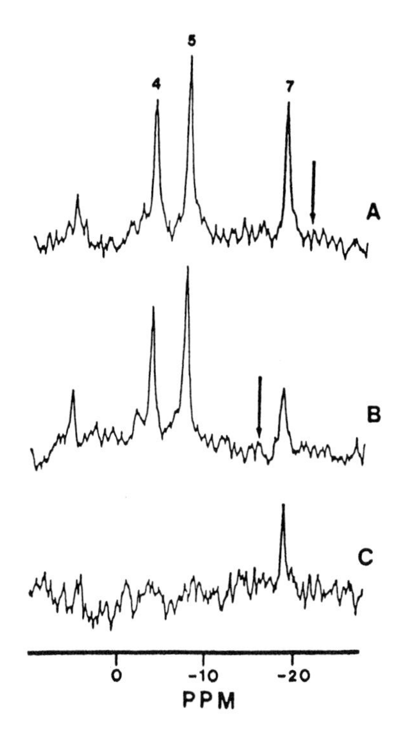

Figure 2: ^{31}P-NMR spectra of liver mitochondria showing saturation transfer. Mitochondria were incubated at 15°C in basic media without fluorocarbons, the external ATP concentration was 1.4 mM. The position of selective irradiation is indicated by the arrows. (A) is the control spectrum; (B) the irradiation was applied at the position of the internal β-phosphate of ATP; and (C) is the difference spectrum (A-B). Peak assignments are as in Figure 1. The internal ATP cannot be seen because of the long delay times (3.34s) used in this experiment to maximize quantitation of the external peaks. Conditions are described more fully in (10).

temperatures, ATP, P_i and ADP concentrations. Likewise mitochondrial membrane electrochemical potential gradients were varied over a wide range of values using dinitrophenol. The relationship of overall ATP/$^{32}P_i$ exchange rates with the partial reactions of the synthase were monitored as well. For example measurement of the disappearance of the $P^{18}O_4$ isotopomer of inorganic phosphate in suspensions of mitochondria provides a measure of the entry and release of inorganic phosphate from the active catalytic site of the ATP synthase. This partial reaction does not involve the translocase but does follow a strict stoichiometry with the overall ATP/$^{32}P_i$ exchange rate. The measured stoichiometry follows the predictions of the rotating catalytic site molecular model of the synthase which predicts a 1:1 relationship between ATP entry into the catalytic site and phosphate release and visa versa (11). The theoretical value would not be observed if the ATP/$^{32}P_i$ exchange were dominated by the translocase. This disappearance of $P_i{}^{18}O_4$ from the media has been termed V_3, and as stated previously the forward reaction of the ATP/P_i exchange is labeled V_1, and the back reaction, V_2. In the absence of added ADP (i.e. state 4), but with widely varied levels of ATP and P_i and different temperatures, V_2/V_3 was consistently equal to 2, since half of the $P^{18}O_4$ entering the catalytic site continued on to become ATP while the other half reversed and was released from the catalytic site as ^{16}O labelled phosphate. However on adding ADP to the media, V_2 rapidly decreases and approaches zero as the ATP/ADP ratio approaches ~20. At ATP/ADP ratios lower than 10, V_2 = O and net ATP synthesis is maximal. Since V_2 appeared to be an important determinant of net ATP synthesis, it seemed important to learn whether its decline was due to inhibition of the back reaction of the synthase or inhibition of ATP entry into the mitochondria by external ADP. Therefore we measured V_3 to monitor flux through the synthase inside the mitochondria. The data are shown in Figure 3. V_3 decreases from 2.00, in state 4 to much lower values in state 3, but its decrease is not as precipitious as that of V_2. In state 3 with 5 mM glutamate and 5 mM malate as substrates, and an initial ATP/ADP ratio of 0.2 the value of V_2 was zero, V_1 was 2.00 μmoles/min•mg and V_3 was 0.72 μmoles/min•mg. Lowering the initial concentration of ATP and providing a hexokinase trap to maintain ATP levels near zero during the experiment lowered V_3 further to 0.50 and using 5 mM pyruvate rather than 20 mM glutamate in the incubation media increased V_1, and further decreased V_3 to 0.30 μmoles/min•mg.

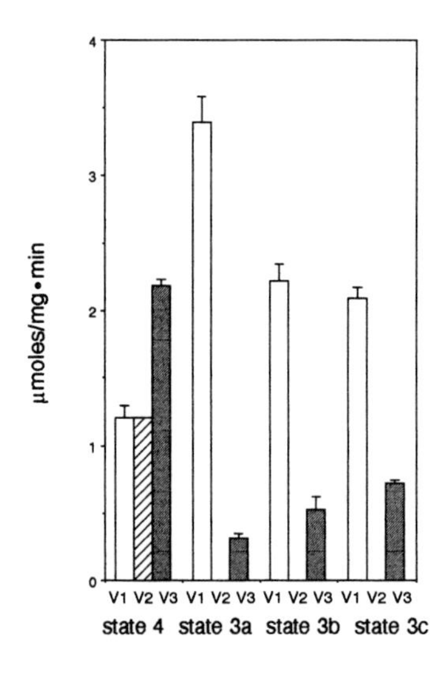

Figure 3: The relationship between external ADP concentration, the rate of the forward reaction of ATP synthase and the backward reaction of the synthase as monitored by release of phosphate from the catalytic rate of the synthase (V_3). State 4 conditions included 20 mM glutamate, 1 mM malate 5 mM ATP, 5 mM P_i and no added ADP. State 3a included 5 mM pyruvate, 1 mM malate, 5 mM P_i, 1 mM ADP and a hexokinase/ADP regenerating system. State 3b conditions were the same as 3a only 20 mM glutamate replaced pyruvate as a substrate. State 3c included 20 mM glutamate, 1 mM malate, 1 mM ATP initially, 5 mM ADP and 5 mM P_i. For more experimental details see (11).

CONCLUSIONS

Inhibition of the back reaction of ATP synthesis by ADP is an important component of the control of oxidative phosphorylation by ADP. As media ADP increases the back reaction decreases to zero. Although the capacity of translocase (V_{max}) is much higher than the synthase, control by ADP of the translocase appears to be somewhat more stringent than control by ADP of the synthase back reaction. It is nonetheless important to note that the back reaction of the synthase in the presence of excess ADP is far slower than the forward reaction. This underscores the conclusion that the ATP synthase itself is kinetically controlled and thus likely to be far from equilibrium under most physiological conditions.

^{31}P NMR saturation-transfer measurements made in intact tissues (9,14-16) suggest that net ATP synthesis is approximately equal to V_1, the forward reaction of the synthase. Thus it would appear that V_2 is frequently equal to zero in these tissues. Therefore control of respiration by substrate supply may be more important than previously supposed and the translocase flux (ADP_{in}/ADP_{out}) may be limited under many physiological situations by the supply of matrix ATP.

REFERENCES

1 Chance B, Williams GR. J Biol Chem 1955; 217: 409.
2 Chance B, Williams GR, Holmes WF, Higgins J. J Biol Chem 1955; 217,439.
3 Klingenberg M. In: Mortimore AH, (ed.) The Enzymes of Biological Membranes: Membrane Transport. New York: Plenum Publishing 1976; 3: 383-438.
4 Klingenberg M. J Membr Biol 1980; 56: 97-105.
5 LaNoue KF, Schoolwerth AC. In: Ernster I, (ed.) New Comprehensive Biochemistry, Bioenergetics. Elsevier Amsterdam; 9: 221-268.
6 Ericinska M, Wilson DF. J Membr Biol 1982; 70: 1-14.
7 Ericinska M, Stubbs MM, Miyata Y, Ditre CM, Wilson DF. Biochem Biophys Acta 1977; 462: 20-35.
8 LaNoue KF, Jeffries FMH, Radda GK. Biochemistry 1986; 25: 7667-7675.
9 Kingsley Heckman P, Sato EY, Andreone PA, St Cyr JA, Michurski S, Foker JE, From AHL, Petein M, Ugurbil K. FEBS Lett 1986; 198: 159-163.
10 Masiakos PT, Williams GD, Berkich DA, Smith MB, LaNoue KF. Biochemistry 1991; 30: 8351-8357.
11 Berkich DA, Williams GD, Masiakos PT, Smith MB, Boyer PD, LaNoue KF. J Biol Chem 1991; 266: 123-129.
12 Hutson SM, Berkich DA, Williams GD, LaNoue KF, Briggs RW. Biochemistry 1989; 28: 4325-4332.
13 Hutson SM, Williams GD, Berkich DA, LaNoue KF, Briggs RW. Biochemistry (in press).
14 Matthews PM, Bland JL, Gadian DG, Radda GK. Biochem Biophys Res Commun 1981; 103: 1052-1059.
15 Freeman D, Bartlett S, Radda G, Ross B. Biochem Biophys Acta 1983; 762: 325-336.
16 Shoubridge EA, Briggs RW, Radda GK. FEBS Lett 1982; 140: 288-292.
17 Forsen S, Hoffman RA. J Chem Phys 1963; 39: 2892-2901.
18 Pfaff E, Heldt HW, Klingenberg M. Eur J Biochem 1969; 10: 484-493.
19 Austin JJ, Aprille JR. J Biol Chem 1984; 259: 154-160.

© 1992 Elsevier Science Publishers B.V. All rights reserved.
Molecular mechanisms of transport. E. Quagliariello, F. Palmieri, eds.

ROLE OF AMINO ACID TRANSPORT ACROSS THE PLASMA MEMBRANE IN THE CONTROL OF METABOLIC PATHWAYS IN HEPATOCYTES

L. Boon and A.J. Meijer

E.C. Slater Institute for Biochemical Research, Academic Medical Centre, Meibergdreef 15, 1105 AZ Amsterdam, The Netherlands

SUMMARY

Under physiological conditions hepatic amino acid metabolism is controlled, at least in part, at the level of amino acid transport across the plasma membrane. Important factors affecting the rate of plasma membrane amino acid transport are the extracellular pH and the position of the hepatocyte in the liver acinus. Interactions between hepatic metabolism of alanine and ammonia are discussed in relation to the transport of alanine across the plasma membrane.

RESULTS AND DISCUSSION

The liver plays a crucial role in whole body nitrogen metabolism. Amino acids, produced either by proteolysis of dietary proteins in the fed state or by proteolysis of body proteins in the starved state, are ultimately oxidized to CO_2 and H_2O. Most of the nitrogen (>90%) is excreted as urea which is exclusively synthesized by the liver. Under all conditions there is a continuous flow of amino acids to the liver which, depending on the conditions, extracts amino acids for the synthesis of proteins, fat, ketone bodies, and glucose or they can be used for oxidation to CO_2 and H_2O to yield ATP.

There is now considerable evidence that plasma membrane amino acid transport systems can exert significant control on the rate at which amino acids are metabolized by the liver. In reviewing the literature on this issue we recently concluded that, at physiological amino acid concentrations (<1 mM), many of the amino acid transport systems are potentially rate controlling for amino acid catabolism [1]. In order to illustrate this we refer to a recent paper by Fafournoux et al [2]. In their study, rats were exposed to low and high-protein diets. In spite of the fact that the supply of amino acids to the liver was strongly enhanced by the high-protein diet, the intrahepatic concentration of most gluconeogenic amino acids decreased, indicating that the induction of

Table 1
Effect of pH on amino acid metabolism in isolated rat hepatocytes. Hepatocytes from 20-24h-starved rats (10-15 mg dry mass/ml) were incubated for 1 h with 10 mM glucose, 2 mM octanoate and a physiological mixture of all amino acids. The final concentration of each of the amino acids was equal to three times their plasma concentration in the starved rat (cf. [3] for composition). In order to avoid overestimation of ornithine cycle flux arginine was replaced by ornithine. The extracellular concentrations of the amino acids in the medium, depicted in the Table, at the end of the incubation period were: serine, 610 µM; glutamate, 336 µM; glutamine, 1090 µM; glycine, 745 µM; alanine, 915 µM; asparagine, 275 µM. Intracellular concentrations of other amino acids were below the limit of detection. Values (means ± S.E.) are from 3 different hepatocyte preparations. *Significantly different from the value at pH 7.4 ($P < 0.05$).

Metabolites and amino acids	pH 7.0	pH 7.4
	Production rate (µmol/ g dry mass/h)	
Urea	54.5 ± 6.2*	100.5 ± 8.2
Ammonia	21.2 ± 2.8*	11.5 ± 1.7
	Intracellular concentration (µM)	
Serine	937 ± 172*	1566 ± 70
Glutamate	2930 ± 530*	5697 ± 554
Glutamine	2874 ± 196*	4792 ± 294
Glycine	3052 ± 460	3223 ± 366
Alanine	271 ± 10*	362 ± 32
Asparagine	181 ± 20*	238 ± 24

From ref. 3.

intracellular amino acid catabolism had exceeded that of the transport systems.

In the past the importance of plasma membrane amino acid transport in the control of amino acid metabolism has been underrated. The reason for this was that many studies on amino acid metabolism in either the perfused liver or isolated hepatocytes had been performed with unphysiologically high concentrations of amino acids. Under such conditions, plasma membrane amino acid transport is unlikely to control amino acid catabolism [1]. When, however, isolated hepatocytes are exposed to mixtures of amino acids in which each amino acid is present at concentrations normally occurring in the portal vein, the importance of plasma membrane amino acid transport becomes readily apparent. In Table 1 the effect of a change in extracellular pH on urea synthesis under such conditions is shown. At low pH urea synthesis decreased: this phenomenon was accompanied by a fall in intracellular amino acid concentrations and also by decreased consumption of the added amino acids (not shown in Table 1, but see [3]). Although the decrease in urea synthesis was accompanied by an increase in ammonia accumulation this was by no means stoichiometric (Table 1). Clearly, at low pH transport of amino acids across the plasma membrane must have been decreased. This phenomenon is likely to be one of the major mechanisms contributing to the fall in urea synthesis in acidosis which is necessary to spare HCO_3^- for neutralisation of the excess H^+ in the blood under these conditions (cf. [1] for review).

In a similar type of experiment we were able to show that plasma membrane amino acid transport is faster in hepatocytes from the periportal area of the liver acinus than in those derived from the pericentral area : synthesis of urea was higher, but intracellular amino acid concentrations were lower in periportal hepatocytes as compared to pericentral hepatocytes (Table 2). This property, in addition to the higher activities of enzymes involved in amino acid catabolism [6], contributes to higher rates of amino acid catabolism in the portal area of the acinus.

Table 2
Amino acid metabolism in periportal and pericentral hepatocytes. Hepatocytes (3 preparations in each group) from 20-24h-starved rats from the periportal and pericentral zone of the acinus, respectively, were isolated by digitonin-collagenase perfusion [4]. The cells were incubated for 1 h at 37°C with 2 mM octanoate and a physiological mixture of amino acids. Values (means ± S.E.) are from 3 different preparations in each group. *Significantly different from the corresponding values in pericentral hepatocytes ($P<0.05$).

	Periportal hepatocytes	Pericentral hepatocytes
	Production rate (µmol/g dry mass/h)	
Urea	153.0 ± 12.9*	51.4 ± 7.7
	Intracellular concentration (µM)	
Serine	3063 ± 420*	757 ± 264
Glutamate	12929 ± 1096*	2660 ± 778
Glutamine	8822 ± 545*	4032 ± 638
Glycine	5486 ± 384*	1666 ± 299
Alanine	549 ± 54*	237 ± 42
Asparagine	513 ± 46*	238 ± 53

From ref.5.

Another interesting example of control of amino acid catabolism by plasma membrane amino acid transport was found when we examined the interaction between ammonia and alanine metabolism. *In vivo*, these two compounds together account for more than 50% of total urea synthesis [1].

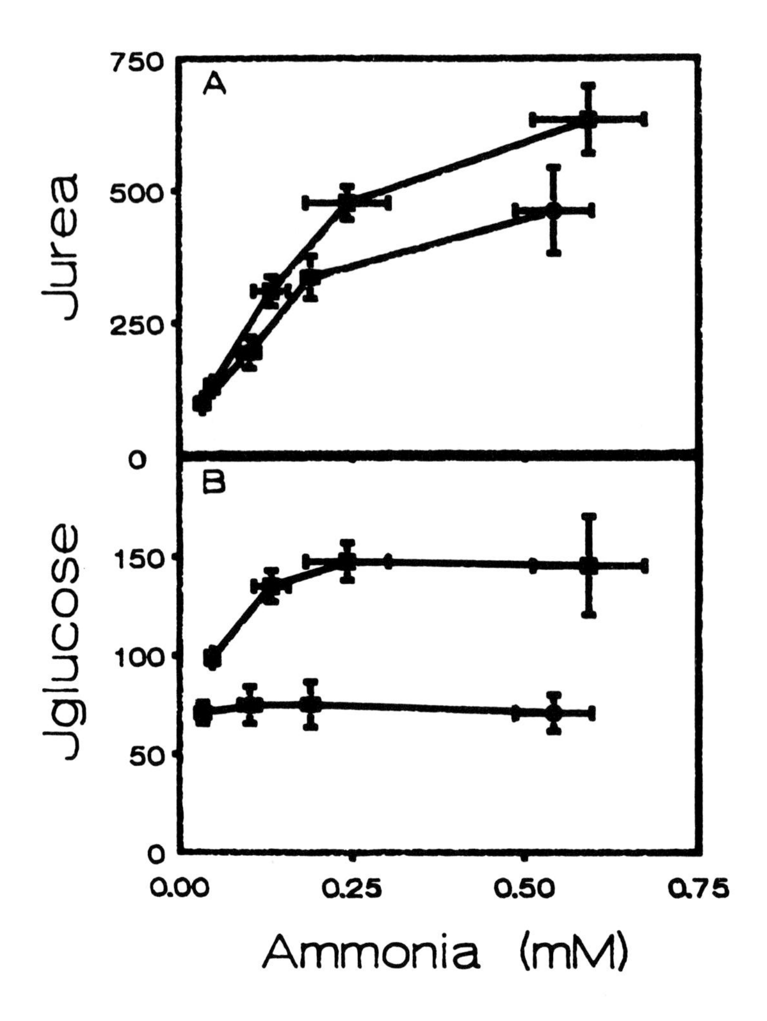

Figure 1. Rate of urea and glucose synthesis from alanine in perifused rat hepatocytes in the absence and presence of ammonia. Hepatocytes from 20-24h-starved rats were perifused [8] at 37°C in the presence of 0.2 mM octanoate, 2 mM ornithine, increasing concentrations of ammonia and either 1.2 (●) or 5 mM (■) alanine. Ammonia, urea and glucose were assayed in the perifusate. Values for urea and glucose are expressed in µmol/g dry mass/h and are the means (± S.E.) of 3 different hepatocyte preparations. Stimulation of urea synthesis by ammonia at 1.2 mM alanine was due to consumption of ammonia; in the presence of 5 mM alanine stimulation of urea synthesis by ammonia was more than could be accounted for by consumption of the ammonia added (not shown).

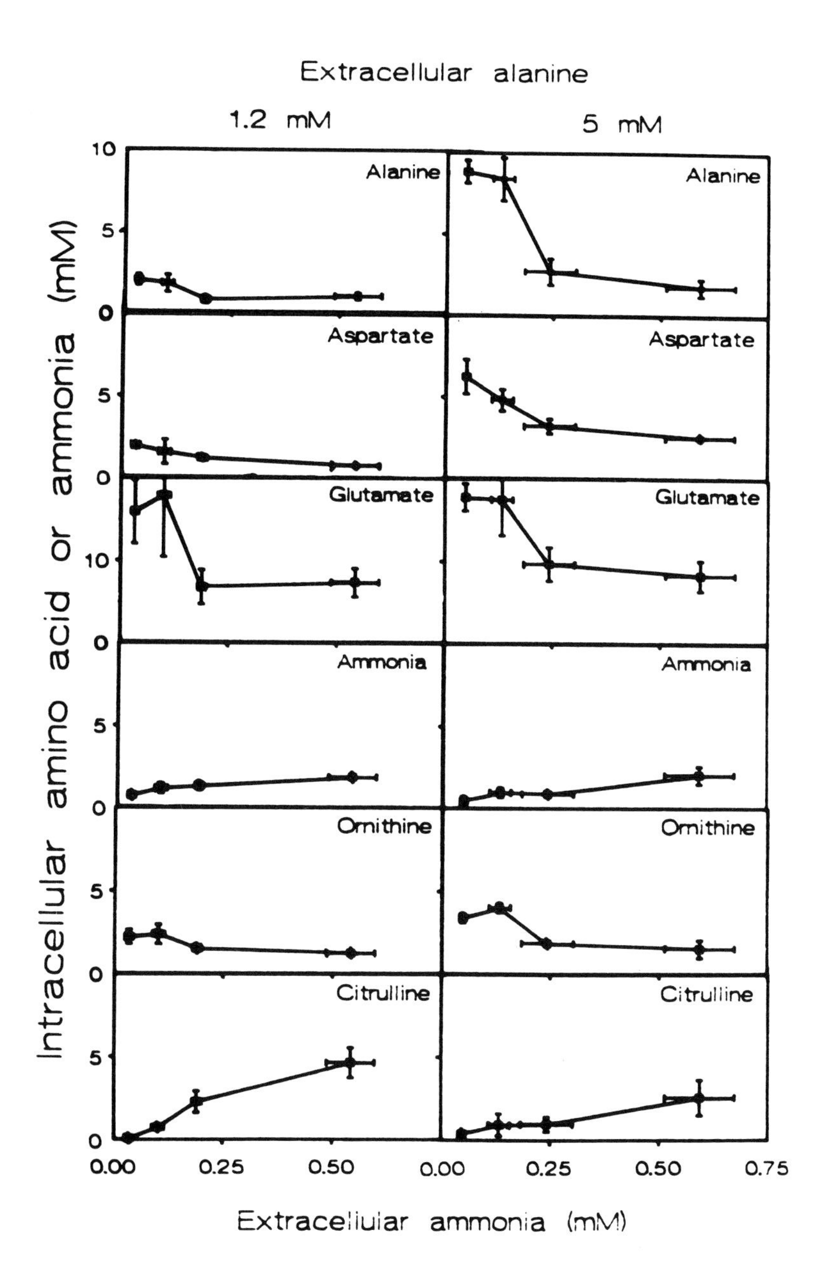

Figure 2. Intracellular concentration of metabolites of alanine degradation in perifused hepatocytes. Experimental conditions as in Fig. 1.

Ammonia is an essential activator of hepatic glutaminase and is known to accelerate hepatic glutamine metabolism ([7] for review). Less well known is the fact that ammonia also accelerates hepatic alanine catabolism as first shown in 1976 by one of us [8] and recently confirmed by Bohnensack and Fritz [9]. In order to analyze the mechanism of this effect, we have studied the interaction between alanine and ammonia in perifused rat hepatocytes under steady state conditions. Infusion of ammonia (0.2, 0.5 and 1 mM) in the presence of a high concentration of alanine (5mM) stimulated the formation of glucose and urea from alanine (Fig. 1), simultaneously decreased the intracellular concentrations of alanine, aspartate, glutamate and ornithine while increasing that of citrulline (Fig. 2). Infusion of ammonia in the presence of a low concentration of alanine (1.2 mM) caused similar changes in intracellular metabolites (Fig. 2) but, in this case, the rate of alanine catabolism remained unaltered (Fig. 1). Apparently, only at high, but not at low, extracellular alanine concentrations plasma membrane alanine transport is sensitive to a decrease in intracellular alanine concentration brought about by an increase of flux through the ornithine cycle. It is concluded that plasma membrane alanine transport controls alanine catabolism at low, but not at high, alanine concentrations. This conclusion is at variance with that of Groen et al. [10] according to whom alanine transport controls alanine catabolism at all alanine concentrations. However, it must be stressed that the experimental conditions used by these authors were different from ours in that in their studies hepatocytes were perifused with alanine in the absence of ornithine.

The analysis of Fig. 2 indicates that the activating effect of ammonia on alanine catabolism can be easily explained by the coupling of the ornithine cycle with the citric acid cycle [11]. Increased production of citrulline from ornithine and ammonia pulls aspartate away from the aspartate aminotransferase equilibrium resulting in increased production of 2-oxoglutarate which then becomes available for transamination with alanine. Indeed, the intracellular concentration of 2-oxoglutarate increased upon infusion of ammonia (not shown).

If the above described mechanism is correct it can be predicted that activation of amino acid degradation by ammonia is not specific for alanine but may also apply to other amino acids. Stimulation of hepatic amino acid catabolism by ammonia also occurs <u>in vivo</u>. Deutz et al. [12] recently demonstrated that infusion of ammonia in the portal vein of pigs resulted in a stimulation of urea production that was more than could be accounted for by the consumption of ammonia alone.

As our data with isolated hepatocytes indicate, at low, but not at high extracellular alanine concentrations, metabolism of alanine is controlled by the transport of alanine across the plasma membrane. The question now arises: what controls alanine metabolism at high extracellular alanine concentration? An answer to this question can be found in the experiment of Fig. 3 in which the relationship between intracellular alanine, glutamate and flux through the ornithine cycle was examined. In order to manipulate the intracellular concentration of alanine, hepatocytes were incubated with increasing concentrations of alanine in the absence or presence of α-methylaminoisobutyrate, an inhibitor of alanine transport. Both urea production and intracellular glutamate became maximal at intracellular alanine concentrations above 5 mM (Fig. 3A and B). By contrast, there was a linear relationship between intracellular glutamate and urea synthesis (Fig. 3C). From these data we conclude that at high extracellular alanine concentrations, alanine catabolism is controlled by the conversion of alanine to glutamate. This conclusion is consistent with the observation, discussed above, that stimulation of intracellular alanine catabolism occurs at the level of alanine aminotransferase.

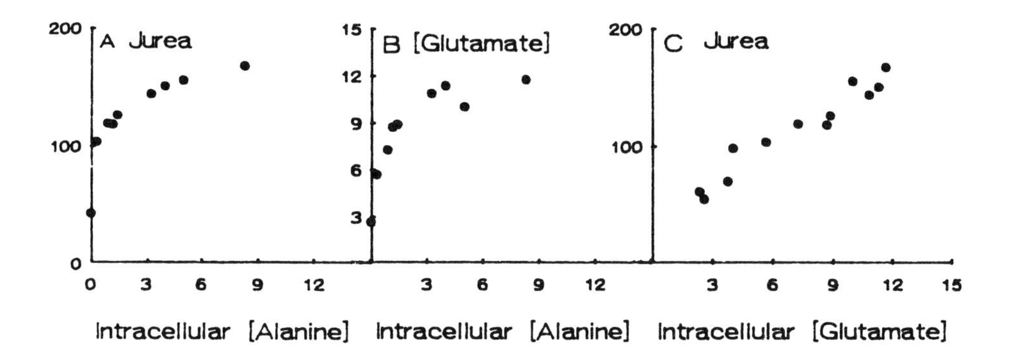

Figure 3. Relationship between urea synthesis and the intracellular concentration of alanine and glutamate. Hepatocytes from 20-24h-starved rats were incubated for 1 h at 37°C in the presence of different concentrations of alanine (0-10 mM), in the absence or presence of 5 or 10 mM α-methylaminoisobutyrate. Rates of urea synthesis are expressed in μmol/g dry mass/h. Intracellular alanine and glutamate are expressed in mM.

REFERENCES

1 Meijer AJ, Lamers WH, Chamuleau RAFM. Physiol Revs 1990; 70: 701-748.
2 Fafournoux P, Rémésy C, Demigné C. Am J Physiol 1990; 259: E614-E625.
3 Boon L, Meijer AJ. Eur J Biochem 1988; 172: 465-469.
4 Lindros KO, Penttilä KE. Biochem J 1985; 228, 757-760.
5 Boon L, Meijer AJ. In: Bengtsson F, Jeppsson B, Almdal T, Vilstrup H, eds. Progress in Hepatic Encephalopathy and Metabolic Nitrogen Exchange. Boca Raton: CRC Press, 1991; 471-478.

6 Jungermann K, Katz N. Physiol Revs 1989; 69: 708-763.
7 Meijer AJ, Verhoeven AJ. Biochem Soc Trans 1986; 14: 1001-1004.
8 Williamson JR, Gimpel JA, Meijer AJ et al. In: Tager JM, Söling HD, Williamson JR, eds. Use of Isolated Liver Cells and Kidney Tubules in Metabolic Studies. Amsterdam: North-Holland Publishing Company, 1976; 339-349.
9 Bohnensack R, Fritz S. Biochim Biophys Acta 1991; 1073: 347-356.
10 Groen AK, Sips HJ, Vervoorn RC, Tager JM. Eur J Biochem 1982; 122: 87-93.
11 Wahren J. In: Soeters PB, Wilson JHP, Meijer AJ, Holm E, eds. Advances in Ammonia Metabolism and Hepatic Encephalopathy. Amsterdam: Excerpta Medica, 1988; 121-129.
12 Deutz NEP, Dejong CHC, Reijven PLM, Soeters PB. In: Bengtsson F, Jeppsson B, Almdal T, Vilstrup H, eds. Progress in Hepatic Encephalopathy and Metabolic Nitrogen Exchange. Boca Raton: CRC Press, 1991; 329-339.

© 1992 Elsevier Science Publishers B.V. All rights reserved.
Molecular mechanisms of transport. E. Quagliariello, F. Palmieri, eds.

Properties and regulation of chloride channels involved in transepithelial chloride secretion.

J. Disser, W. Krick, A. Hazama and E. Frömter

Zentrum der Physiologie, Klinikum der J.W.Goethe-Universität, Theodor-Stern-Kai 7, 6000 Frankfurt am Main 70, FRG

INTRODUCTION

Epithelia can either absorb (in direction towards their vascular surface) or secrete (in opposite direction) chloride ions and some are even able to achieve both, absorption or secretion, depending on the kind and level of nervous or hormonal stimulation. Transepithelial chloride fluxes do either proceed through the tight junctions between epithelial cells (those flxues are mainly absorptive) or across the two membranes of the epithelial cells: the apical and basolateral cell membrane. In contrast to transepithelial sodium or potassium transport, the molecular mechanisms of chloride absorption and secretion are less well known today. We know, however, that these chloride fluxes are passive in nature and can either proceed as symport or antiport or as uncoupled conductive fluxes through specific ion channels. The by now classical model of transepithelial chloride secretion is depicted in Fig 1. It says that chloride is taken up from interstitium (B) into the cell (C) via an electroneutral Na-K-2Cl

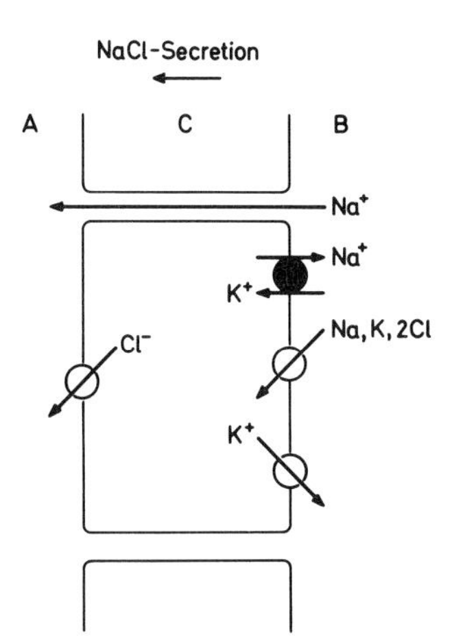

Fig. 1. Model of transepithelial chloride secretion [1]

cotransporter and is expelled across the apical cell membrane via a chloride conductance pathway [1]. The uptake into the cells is energized by the sodium concentration gradient (which in turn is generated by the Na/K pump) and the exit from the cell is energized by the electrochemical chloride gradient across the apical cell membrane which is mainly determined by the amount of chloride entering via the cotransporter from the blood side. At rest, the apical chloride conductance appears to be downregulated to stop secretion, after stimulation, however, it is activated so that transepithelial chloride transport can commence. The present paper tries to summarize our present ideas about the nature and the intracellular regulation of this apical chloride conductance.

SURVEY OF CHLORIDE CHANNEL PROPERTIES

In recent years a great number of patch clamp experiments have been performed on the apical cell membrane of various chloride secreting epithelia (natural or monolayer cell cultures) to analyze the nature of this chloride conductance, however, a great diversity of observations has been made. As shown in Table 1, channels with a wide range of conductances from sub-pico-Siemens (sub-pS) levels to approximately 360 pS and more have been reported with often two or more channels being observed in parallel. Besides different conductances, the channels exhibit different current/voltage curves, when studied in symmetrical plasma-like NaCl-solutions on either side of the patch membrane. Most peculiar in this respect is the outwardly rectifying 30 pS chloride channel. In addition the channels

Table 1
Electrophysiological characterization of epithelial chloride channels

origin	a/b	g (pS)	i./v.	stimulat.	I/Cl	ref
stomach	?	0.35	lin	?	>1	2
lacrimal gl.	?	1.2	?	(Ca)	>1	3
A6 cells	a	3	outw. rect.	cAMP	1	4
pancr. duct	a	4	lin	cAMP	1	5
A6 cells	a	8	lin	?	?	4
airway cells	a	15	lin	?	?	6
HT29 cells	a	15	lin	?	?	7
airway cells	a	20	lin	?	?	8
airway cells, HT29, T84 cells	a	35	outw. rect.	cAMP ?	>1	9
salt gland	a	45	(lin)	cAMP	?	10
urin. bladder	b	64	inw. rect.	?	1	11
T84 cells	a	75	outw. rect.	swelling	>1	12
A6 cells	a	360	lin	?	1	13

a/b indicates origin from apical or basolateral cell membrane
g is the single channel conductance in pico Siemens;
i./v. indicates the shape of the current/voltage relation;
I/Cl indicates channel selectivity for iodide over chloride;
Please note that it is impossible in the present context to list data from all publications, however, the data shown are representative of the spectrum of channel properties described.

differ in their selectivity for iodide over chloride, and appear to respond differently to stimulation by either cAMP or ionophor-induced calcium-influx or by other stimuli. In view of this diversity of channels and their properties a main task today appears to consist of sorting the phenomena and clarifying the relation between channel type and stimulation pathway.

THE CHLORIDE CHANNEL THAT IS DEFECTIVE IN CYSTIC FIBROSIS

Much about chloride channel structure and function has been learned recently from a cruel experiment of mother nature: the lethal inherited, recessive, autosomal disease called cystic fibrosis. This disease which affects one in approximately 2000 newborn Caucasians may be characterized as polyexocrinopathy. Up to a few years ago the children would die before reaching adolescence mainly from recurrent infections of the respiratory tract. The central pathogenetic mechanisms seems to consist of an imbalance between epithelial fluid absorption and secretion which leads to viscous obstructions of the bronchial tree and the pancreatic duct system with subsequent inflammations. Similar symptoms may also develop in the intestinal tract and genital organs. In addition salt absorption is defective in the sweat ducts. The common cause of these alterations seems to be a defective chloride permeability [14,15] which prevents salt secretion in the respiratory epithelium and salt absorption in the sweat duct. From studies on the secretory coil of sweat glands as well as from studies on respiratory epithelial cells it had been found that cAMP-stimulated chloride secretion was defective but that the signal transmission chain was intact up to and including proteinkinase A [16,17]. This pointed to a defect in the chloride channel itself, but for years the researchers were unable to identify which type of chloride channel was functional in control tissue but defective in cystic fibrosis tissue. This situation changed after the gene had been identified. Fig 2 shows the supposed structure of the gene product (named CFTR) derived from hydropathy plots [18].

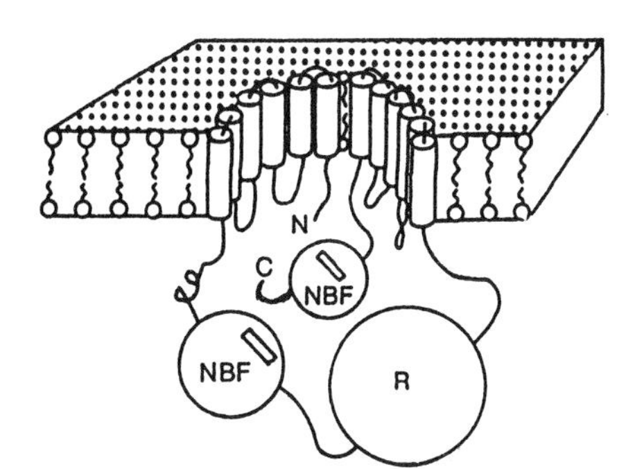

Fig. 2. Proposed structure of CFTR gene product [18]

It consists presumably of a membrane protein with 12 membrane spanning domains, two nucleotide binding folds and a central hydrophylic domain, the so-called R-domain. Because of a considerable structural homology with the multi-drug-resistance protein and because of the presence of nucleotide binding folds similar to those operative in ATP-splitting enzymes, the hypothesis evolved that CFTR might have other functions, e.g. it might transport a chloride channel regulator into or out of the cell. This hypothesis, however, may already be dismissed today, after molecular biologists succeeded in expressing the gene product in different cells. Transfection with CFTR was able to restitute the normal chloride transport function in cystic fibrosis tissue and generated a cAMP-stimulated chloride conductance in different types of cells which had never expressed such a conductance before [19]. Moreover, it was shown that such transfected cells expressed a cAMP-stimulated chloride channel of 7 pS conductance which was not present in the specific control cells nor in cells transfected with other genes [20]. In addition, very recently it was shown that substitution of two arginins in the mouth region of the membrane spanning domains by glutamate changed the selectivity of the newly expressed chloride conductance [21], while deletion of the R-domain resulted in the expression of a stable chloride conductance, suggestive of the expression of permanently open channels [22]. Although these brilliant experiments constitute virtually irrefutable proof for the hypothesis that CFTR is actually a chloride channel, it remains unexplained, why for years so many patch clamp investigators missed this channel in their experiments with the only exception of pancreatic duct, where a cAMP regulated 4 pS chloride channel has been described two years ago [5]. A possible explanation might be that the CFTR-channel assumes other single channel conductances (sub-pS conductance?) in the membrane environment of airway and intestinal epithelial cells. Some recent observations from Dr. Gregers laboratory and from our own laboratory would point into this direction.

THE OUTWARDLY RECTIFYING 30 PS CHLORIDE CHANNEL

This channel was one of the first to be identified in respiratory epithelia, but it has also been found in a great number of other absorptive and secretory epithelial cells as well as in non-epithelial cells. Initially this channel was thought to be responsive for the defect in cystic fibrosis [6,9], this interpretation, however has become more and more unlikely recently. As can be seen in Table 1, this channel, unlike the CFTR channel, is more permeable to iodide than to chloride and again, unlike the CFTR channel, it can be inhibited by a number of more or less specific chloride channel inhibitors. A most peculiar property of this channel is its silence in on-cell patches and its activation after excision of a patch from the cell [6,9,23]. This finding suggested that the cytoplasm contained an inhibitory signal which dissociated and diffused away after excision. In parallel with Dr. Gregers laboratory and in collaboration with Dr. W. Krick and Dr. G. Burckhardt we have recently demonstrated the existence of such a cytosolic inhibitor by exposing excized membrane patches to

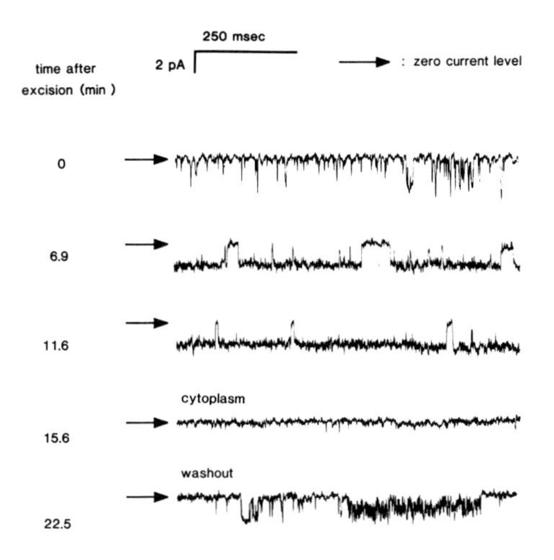

Fig. 3. Inhibition of chloride channel activity by cytosol extract [25]

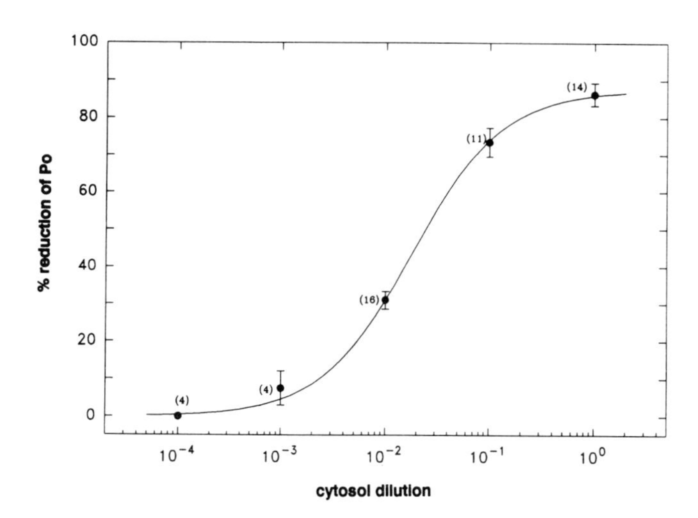

Fig 4. Inhibition of channel open probability (P_o) as function of cytosol dilution [25]

cytosol fractions [24.25], see also Figs 3 and 4. The chemical nature of the inhibitor, however is not yet known. All that we know at present is that we are dealing with a heat stable, amphiphilic substance, of low molecular weight (< 1000). The cytosolic inhibitor is not only found in respiratory cells but also in a colon carcinoma cell line, in placenta and in kidney cortex. It appears therefore, that the inhibitory principle is as wide spread among tissues as the outwardly rectifying chloride channel.

Whether and how the outwardly rectifying chloride channel can be stimulated in an intact cell is presently not known. Many investigators think that it is activated by calcium ion influx into cells through activation of a calcium-calmodulin dependent proteinkinase II. However, while the stimulation of chloride conductances in response to calcium influx and mediation of this effect by the above-mentioned proteinkinase is not subject to dispute, the involvement of the outwardly rectifying chloride channel in this event has not yet been proven.

OTHER CHLORIDE CHANNELS

At present it is rather difficult to decide which other types of chloride channels exist and to what extent the observations of Table 1 can be explained by relatively small alterations in only a few types of basic channel proteins. This answer will have to await further progress in molecular biology approaches including immunocytochemical tagging of specific protein structures. A distinctly different molecule, however, appears to be the giant chloride channel of 300 to 400 pS conductance, which has a peculiar voltage dependent open probability and has also been found in a large number of epithelial and non-epithelial cells. Thus far, however,its mode of stimulation and its physiological functions have remained obscure.

CONCLUSIONS

The molecular characterization of epithelial ion channels has just begun. Thus far a great number of different channels have been described in patch clamp experiments, however, the electrophysiological characterization is still incomplete and the intracellular regulation of channel activity is only poorly understood. The only exception is the cAMP regulated chloride channel which is defective in cystic fibrosis and seems to represent the gene product of the CF gene.

ACKNOWLEDGEMENT

The authors gratefully acknowledge the collaboration and devote technical assistance of U.Schröder, A. Rabe, I. Doering-Hirsch and U. Fink (all Frankfurt/Main), as well as the most helpful discussions with Dr. G. Burckhardt (Frankfurt) and Drs R. Greger and K. Kunzelmann (Freiburg). Funds for the experimental work were obtained from the Deutsche Gesellschaft zur Bekämpfung der Mukoviszidose (Bonn, FRG) and from the Deutsche Forschungsgemeinschaft (Bonn, FRG)

REFERENCES

1 Greger R, Schlatter E. Pflügers Arch 1984; 402:63-75
2 Sakai H, Okada Y, Morii M, Takeguchi, N. In: Murakami M, Seo Y, Kuwahara A, Watari H, eds. Ionic Basis and Energy Metabolism of Epithelial Transport. Okazaki: Kenbun Printing, 1991;177-180
3 Marty A, Tan YP, Trautmann A. J Physiol 1984; 357:293-325
4 Marunaka Y, Eaton DC. Am J Physiol 1990; 258:C352-C368.
5 Gray MA, Greenwell JR, Argent BE. J Membrane Biol 1988; 105:131-142.
6 Frizzell RA, Rechkemmer G, Shoemaker RL. Science 1986; 233:558-560.
7 Hayslett JP, Gögelein H, Kunzelmann K, Greger, R. Pflügers Arch 1987; 410:487-494.
8 Duszyk M, French AS, Man SFP. Biophys J 1990; 57:223-230
9 Welsh MJ. Science 1986; 232:1648-1650.
10 Greger R, Schlatter E, Gögelein H. Pflügers Arch 1987; 409:114-121.
11 Hanrahan JW, Alles WP, Lewis SA. Proc Natl Acad Sci 1985; 82:7791-7795.
12 Worrell RT, Butt AG, Cliff WH, Frizzell RA. Am J Physiol 1989; 256:C1111-C1119.
13 Nelson DJ, Tang JM, Palmer LG. J Membrane Biol 1984; 80:81-89
14 Schulz I, Frömter E. In: Windorfer A, Stephan U, eds. Mukoviscidose. Stuttgart: Thieme, 1968;12-21
15 Quinton PM. Am J Physiol 1986; 251:C649-C652.
16 Sato K, Sato F. J Clin Invest 1984; 73:1763-1771
17 Barthelson R, Widdicombe J. J Clin Invest 1987;80:1799-1802
18 Riordan JR, Rommens JM, Kerem BS, Alon N, R Rozmahel R, Grzelczak Z, Zielenski J, Lok Si, Plavsic N, Chou JL, Drumm ML, Iannuzzi MC, Collins FS, Tsui LC. Science 1989; 245:1066-1073.
19 Rich DP, Anderson MP, Gregory RJ, Cheng SH, Paul S, Jefferson DM, McCann JD, Klinger KW, Smith AE, Welsh MJ. Nature 1990; 347:358-362.
20 Kartner H, Hanrahan JW, Jensen TJ, Naismith AL, Sun S, Ackerley CA, Reyes EF, Tsui LC, Rommens JM, Bear CE, Riordan JR. Cell 1991; 64:681-691.
21 Anderson MP, Gregory RJ, Thompson S, Souza DW, Paul S, Mulligan RC, Smith AE, Welsh MJ. Science 1991; 253:202-205.
22 Rich DP, Gregory RJ, Anderson MP, Manavalan P, Smith AE, Welsh MJ. Science 1991; 253:205-207
23 Kunzelmann K, Pavenstädt H, Greger R. Pflügers Arch 1989; 415,172-182
24 Kunzelmann K, Tilmann M, Hansen CP, Greger R. Pflügers Arch 1991; 418:479-490.
25 Krick W, Disser J, Hazama A, Burckhardt G, Frömter E. Pflügers Arch 1991; 418:491-499.

© 1992 Elsevier Science Publishers B.V. All rights reserved.
Molecular mechanisms of transport. E. Quagliariello, F. Palmieri, eds.

EXPRESSION OF mRNA CODING FOR FROG URINARY BLADDER UREA CARRIER IN *XENOPUS LÆVIS* OOCYTES.

Sonia Martial [a], Cristina Ibarra [b], Luisa Rebelo [a], Véronique Berthonaud [a] and Pierre Ripoche [a].

[a] Dept. de Biologie Cellulaire et Moléculaire, C.E.N. Saclay, France.
[b] Dept. Fisiologia, Faculdad de Medicina, UNBA, Buenos Aires, Argentina.

SUMMARY

Urea movement across the plasma membrane of mammalian kidney collecting ducts plays an important physiological role in maintaining the osmotic pressure in this tissue. Urea transport has also been described in the amphibian urinary bladder, which may therefore be used as a model of the kidney collecting duct. The protein(s) involved in this transport have not yet been isolated and characterized because of the lack of either specific labeled inhibitors of the transport, antibodies raised against the protein, or probes of the gene coding for the carrier. We describe here a new approach for studying this protein, using the strategy employed by Hediger et al [1], that is the expression of amphibian urinary bladder mRNA in Xenopus oocytes.

50 nl of mRNA (1 µg/µl) were injected into experimental oocytes, and water into control oocytes. The uptake of urea was measured by a radiotracer technique up to 7 days after the injection. Between 2 days and at most 4 days following injection, the J_{urea} of oocytes injected with mRNA was about 4 times greater than in oocytes injected with water. The urea permeability of the oocytes injected with mRNA was reduced in the presence of the urea analogue, nitrophenyl-thiourea (NPTU) at a concentration of 10^{-4} M. J_{urea} was comparable with the J_{urea} of not-injected oocytes when they were incubated in the presence of 8-Br-cAMP (10^{-4} M) or when injected with mRNA isolated from rat brain, suggesting that the increase of J_{urea} corresponds to a direct expression of the urea carrier via the injected mRNA.

Size fractionation of the mRNA by electrophoresis under denaturing conditions led to the identification of a fraction able to express the urea carrier. In our conditions, the efficient fraction had a size between 6 and 10 kbases. The injection of this fraction into oocytes increased the J_{urea} four-fold and this increase was inhibited by NPTU (10^{-4} M). This mRNA fraction identification is a useful starting point for isolating and characterizing the cDNA coding for the urea transporter.

INTRODUCTION

The urea permeability of human red blood cells [2] or some ADH-stimulated epithelia (inner medullary collecting duct [3] or toad urinary bladder

[4]) is larger than the permeability predicted from the Overton's rule (see Cohen [5]) which states that the membrane permeability coefficient of solutes can be correlated to the lipid solubility of these solutes.

The higher permeability of the plasma membrane of these cells to urea has been taken as evidence of the presence in these tissues of hydrophilic pathways for this molecule [6]. Several studies have reinforced this hypothesis. In 1970, Macey and Farmer [7] showed that the urea fluxes across erythrocytes were inhibited by parachloromercuribenzene sulfonate and by phloretin both of which alter protein structures. On the other hand, Brahm [8] described a saturation of the urea transport and a competitive inhibition by urea analogues of this transport in red blood cells. A similar saturation as well as an inhibition by phloretin was also observed in toad urinary bladder [9,10].

We describe here a new approach to the study of the urea carrier by the use of the strategy employed by Noma et al. [11] for interleukin expression and by Hediger et al. [1] for intestinal Na^+/glucose cotransport. We have attempted to express the urea carrier in Xenopus oocytes. The first step in this approach was to establish assay conditions for specific detection of the transport protein concerned. In this, we confirmed the work of Zhang and Verkman [12] and our previous report [13] that no specific urea transport exists in native *Xenopus lævis* oocytes. The second step was to demonstrate that mRNA isolated from frog urinary bladder epithelial cells, a tissue known to express the transporter, is able to express a functional protein for urea transport when microinjected into Xenopus oocytes.

MATERIALS AND METHODS

RNA preparation.

Epithelial cells were isolated as described by Pisam and Ripoche [14]. Female frogs, *Rana esculenta*, originating from Central Europe were pithed and their bladders immediately filled with a Ca^{2+} free Ringer solution (100 mM NaCl, 5 mM KCl, 1 mM $MgCl_2$, 2.5 mM $NaHCO_3$, 1 mM EGTA, pH 8.1) supplemented by either 1000 U/ml RNAsin (Promega Biotec, Madison, WI, USA) or 10 mM vanadylribonucleoside (Gibco BRL, Uxbridge, U.K.). 15 to 20 bladders were isolated as bags and incubated for 20 minutes in this medium at room temperature. They were then gently rubbed between thumb and forefinger for 2 min to detach epithelial cells from the basal membrane, after which they were excised and the cell suspension, chilled to 4°C, was centrifuged at 2000 x g for 5 minutes. The cell pellet was homogenized at 4°C in 10ml of a solution containing 100mM NaCl, 10 mM EDTA, 50 mM Tris-HCl, pH 9.0 and 5% SDS. An equal volume of saturated phenol/ chloroform/ isoamylic alcohol (50:48:2 v/v) solution [15] was added. The mixture was ground using a polytron (Kinematica, Luzern, Switzerland) for 4 x 20 seconds at dial "5" and the suspension centrifuged at 5000 x g for 20 minutes. The aqueous phase was extracted. The operation was repeated three times [15]. Total RNA was obtained by precipitation from the aqueous phase by adding 0.1 volume of 2.5 M sodium acetate (pH 5.0) and 3 volumes of cold ethanol. The preparation was maintained at -70 °C for at least 30 min.

Poly(A^+)-RNA was purified by oligo(dT)-cellulose chromatography. The precipitated nucleic acids were centrifuged at 10000 x g at 4°C and the pellet dissolved in 10 mM Tris-HCl, pH 7.5 and 1 mM EDTA. The nucleic acid solution was passed twice over the column [16]. Around 20 µg mRNA were obtained from 20 frog bladders. The mRNAs were precipitated by 2.5 volumes of cold ethanol after addition of 0.1 volume of 2.5 M potassium acetate (pH 5.0) and stored at -70 °C until further use.

For microinjection, mRNA was washed with cold 70 % (v/v) ethanol and finally dissolved in sterile water at 1 µg/µl.

Control poly(A^+)-RNA was prepared from rat brain using magnetic oligo(dT) microspheres [17].

Size fractionation of poly(A^+)-RNA was performed under denaturing conditions (1.2% agarose gel containing 2.2 mM formaldehyde). Parts of the gel were cut and electro-eluted at 200 volts for at least 90 minutes in an elutrap apparatus (Schleicher and Schuell, Dassel, Germany). The assay of "in vitro" translation was performed using rabbit reticulocyte lysate as described by Promega Biotec with [^{35}S]-methionine (1000 Ci/ mmol. Amersham, Les Ulis, France) as a marker for newly synthetized proteins.

Xenopus oocyte microinjection

Xenopus lævis females were anesthetized for 25 minutes in ice and the ovarian lobes dissected and placed in a Barth solution (88 mM NaCl, 1 mM KCl, 0.8 mM $MgSO_4$, 0.4 mM $Ca(NO_3)_2$, 7.5 mM Tris, 2.4 mM $NaHCO_3$, pH 7.6) containing 100 U/ml penicillin and 0.01 mg/ml streptomycin. Oocytes were defolliculated by a 2 hour incubation with collagenase (type IIa, Sigma, St Louis, MO, USA) and those in the Dumont stage V-VI of development [18] (1 mm diameter) were microinjected with 50 nl of sterile water or mRNA (50 ng of total mRNA or 5 ng of fractionated mRNA in 50 nl). They were then kept at 18°C until flux measurements, in a Barth solution with antibiotics changed daily. Control of the maturation state of the oocytes was carried out by incubating five oocytes with [^{35}S]-methionine for six hours and counting the labeled proteins according to the method of Gurdon and Wickens [19].

Urea transport measurements

The urea uptake assay was initiated by resuspending individual oocytes in 0.2 ml Barth solution containing 1 mM urea and 2 µCi [^{14}C]urea (C.E.A., Saclay, France) and 10 µCi (2 nM) [^{3}H]raffinose (NEN, Dreiech, Germany) as a control of oocyte integrity. 10^{-4} M nitrophenylthiourea or NPTU (Aldrich, Strasbourg, France) were or not present in the incubation medium. The uptake was generally stopped after 10 minutes' incubation by addition of 2 ml ice cold Barth solution, followed by rapid filtration through glass fiber filters which were washed twice with 5 ml of the same solution. Oocyte and filter were incubated in counting vials with 0.1 ml formic acid for 5 minutes, then mixed with 5 ml scintillation liquid (Ultima Gold, Packard, Downers Grove, IL, USA) to determine the radioactivity associated with the oocyte. When the raffinose permeability was higher than 10^{-7} cm.s^{-1}, the oocyte was discarded.

Urease activity was measured using the technique described by Morel and Bankir [20].

RESULTS

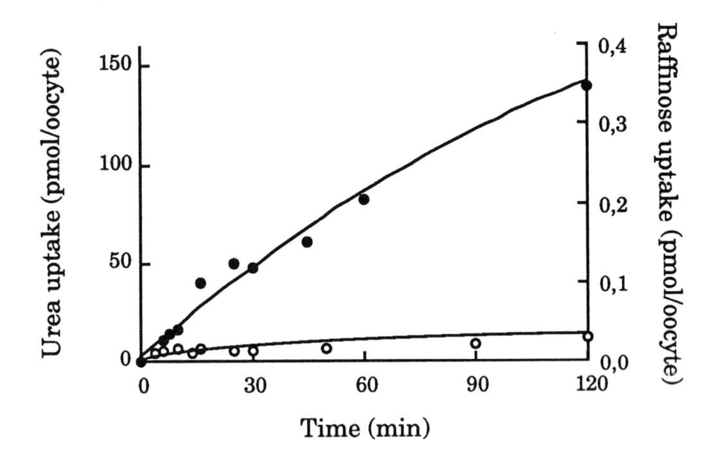

Fig.1 : Time courses of urea and raffinose uptakes in *Xenopus lævis* oocytes.
Urea uptake by Xenopus oocytes was measured as a function of time by a radiotracer technique. The incubation medium contained [^{14}C]urea and [^{3}H]raffinose. Black circles : urea uptake ; white circles : raffinose uptake.

Preliminary experiments were made to verify the absence of an intrinsic urea carrier in native, collagenase-treated oocyte. The time course of urea uptake into uninjected oocytes revealed that urea permeability was very low (P_{urea} ranging between 0.7 and 2 x 10^{-6} cm.s^{-1}, a range also found by Zhang and Verkman, [12]). Urea uptake was linear during the first 10 minutes (Fig.1) and saturation reached after about 180 minutes' incubation. The urea apparent volume corresponded to the oocyte volume as estimated from its diameter assuming the oocyte to be a smooth sphere. No urease activity was found during the experiments (data not shown).

Urea permeability was not altered by the presence of 0.5 mM 8-Br-cAMP added to the incubation medium 15 minutes before the beginning of the assay suggesting that there is no cAMP regulation of an endogenous urea carrier (Fig.2). Oocyte microinjection with water did not modify the urea uptake across the plasma membrane (P_{urea} was (1.10 ± 0.08) x 10^{-6} cm. s^{-1}, n = 28). On the other hand, when oocytes were artificially permeabilized by a 0.5 mg/ml amphotericin B pretreatment, the urea flux was increased four or five-fold (Fig.2), showing that urea permeability variations were satisfactorily detected by our experimental procedure.

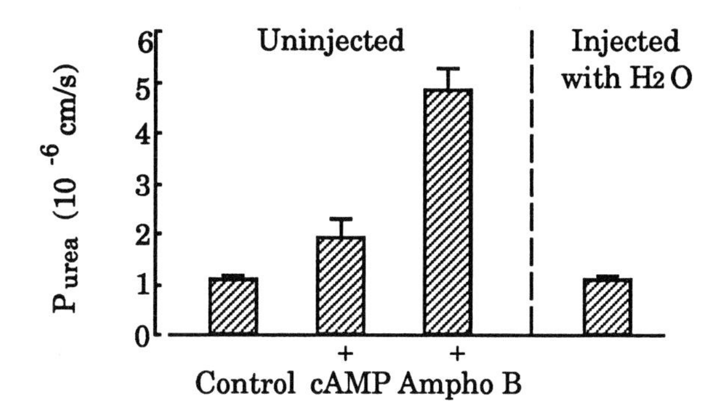

Fig.2 : Effect of 8-Br-cAMP, amphotericin B and water injection on oocyte permeability.
Urea uptake from the medium described in Fig.1 was measured during 10 min at 18°C. For 8-Br-cAMP and amphotericin B (ampho B) experiments, the oocytes were preincubated for 15 min in a medium containing either 0.5 mM 8-Br-cAMP or 0.5 mg/ml amphotericin B. The incubation medium also contained 8-Br-cAMP or amphotericin B.

The quality of the frog urinary bladder mRNA preparation and also the state of oocyte maturation were controled by measurement of the [^{35}S]-methionine incorporated in newly synthetized proteins in the reticulocyte lysate system and in freshly sampled oocytes respectively. The oocytes were then mRNA-microinjected and the maximal increase in urea uptake (Fig.3) was observed three or four days after injection (P_{urea} was $4.05 \pm 0.38 \times 10^{-6}$ cm.s^{-1}, n = 34).

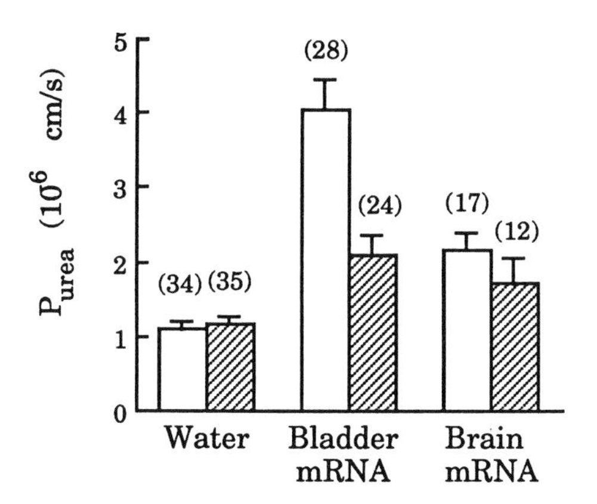

Fig.3 : Urea permeability of Xenopus oocytes injected by either water, frog urinary bladder mRNA or rat brain mRNA.
Four days after injection, the oocytes were incubated for 10 min, at 18°C, in a medium containing [^{14}C]urea and [^{3}H]raffinose, in the presence (hatched bars) or absence (white bars) of 0.1 mM NPTU. Each bar represents the mean ± S.E. of n determinations (n being given within brackets).

This increase was inhibited by the incubation of the oocyte in the presence of 10^{-4} M NPTU. The microinjection of mRNA from brain tissue which is thought not to contain urea transporters, did not increase urea permeability beyond that measured in not-injected or water microinjected oocytes. In these conditions, there was no observed effect of NPTU on the permeability.

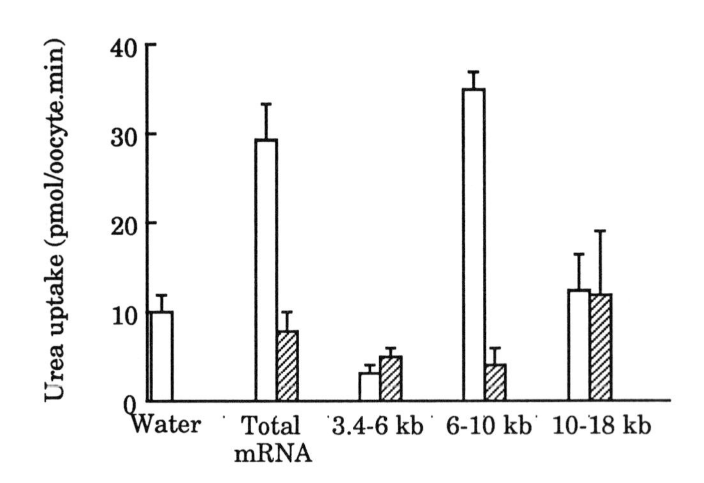

Fig.4 : Urea fluxes in oocytes microinjected with either water, total mRNA or different fractions of mRNA separated by electrophoresis.
4 days after injection, urea uptake was measured at 18°C with 1 mM urea, in the presence (hatched bars) or absence (white bars) of 0.1 mM NPTU. Transport values are the means of four determinations, with their respective S.E. values.

The NPTU-sensitive urea uptake of microinjected oocytes was also followed using size-fractionated mRNA. Figure 4 demonstrates that urea uptake was exclusively expressed by oocytes microinjected with 5 ng of fraction 6-10 kb. Fractions below 3.4 kb (data not shown) and fractions 3.4-6 kb and 10-18 kb did not increase the urea permeability of the oocytes which was not altered in the presence of NPTU. In contrast, NPTU drastically reduced the urea uptake expressed with the 6-10 kb fraction to a value close to the basal urea uptake. Because size fractionation was carried out from gels by electroelution, we were unable to determine from these experiments a more narrow molecular mass range of mRNA. Neither is it possible to exclude the involvement of multiple mRNAs in the functional expression.

DISCUSSION

The purpose of this study was to develop a functional expression assay to characterize the urea transporter. The experiments we describe here demonstrate that it is possible to observe the expression of the urea transporter in Xenopus oocytes.

The permeabilities were calculated assuming the oocyte to be a smooth sphere. Our values are therefore overestimated since the surface area amplification factor, due to the presence of numerous microvilli, is probably greater than 5 [21].

P_{urea} of native oocytes was in the range of values reported for P_{urea} in lipid bilayers [22]. Moreover, this permeability is not affected by urea analogues such as NPTU, a potent inhibitor of urea transport in red blood cells described by Mayrand and Levitt [23] ($K_i = 10^{-5}$ M). NPTU was chosen instead of phloretin which seems to be a poorly specific inhibitor due to possible actions of this molecule on lipidic structures [24]. Some experiments showing inhibition by phloretin have however been described in oocytes [12,13]. The artificial perforation of the plasma membrane with amphotericin B, also used by Micelli et al. [25] to increase urea permeability in toad urinary bladder, has proved to be efficient in the oocyte plasma membrane suggesting the possibility of measuring urea permeability variations without the complication of unstirred layers or the binding of urea to material contained in intervillar spaces or previtelline membrane.

Oocytes do not contain the means of expressing a cAMP-dependent endogenous carrier. Indeed, the incubation of oocytes in the presence of 8-Br-cAMP does not alter their urea permeability. In the same way, microinjection of mRNA isolated from cells in which there is no urea carrier, in this case brain cells, provides evidence that microinjection itself does not provoke the expression of an endogenous carrier. We can also eliminate the possibility that it is a development of urease activity that secondarily accelerates the urea uptake. Together, these results indicate that the measurement of a urea transport in oocytes after mRNA microinjection is probably due to the translation in a functional form of the carrier in the oocytes. However, the possibility that the mRNA encodes for a protein which catalyses the synthesis of a specific component necessary for the expression of the transport and not the transporter itself, cannot be excluded.

In our mRNA size-fractionation experiments, we found the expression of an NPTU-sensitive urea uptake in oocytes microinjected with mRNA in the 6-10 kb molecular size range. This high range seems surprising. Indeed, according to Gargus and Mitas [26], the screening of a cDNA library prepared from human reticulocytes for the gene coding the Jk antigen, the putative urea transporter in red blood cells, shows the presence of this gene in a fraction of about 2.5 kb. However, irradiation experiments [27] in red blood cells show that the molecular weight of the urea transporter could be around 459 kDa. Thus, we cannot conclude that our fractionation is not sufficiently precise to confirm that the urea carrier is a large protein or a subunit of a large complex. Further experiments will be necessary to confirm this point, in particular, the screening of a cDNA library derived from our mRNA preparation.

We conclude that the functional expression of mRNA coding for the frog urinary bladder urea transporter in the Xenopus oocyte is in favour of the existence of a urea carrier of proteic nature and this protein is suitable for expression cloning. Our study should be helpful for isolating and characterizing the cDNA encoding this urea carrier.

REFERENCES

1. Hediger M.A., M.J. Coady, T.S. Ikeda and E.M. Wright; Expression cloning and cDNA sequencing of the Na^+/glucose co-transporter. Nature (1987), 330: 379-381.

2. Sha'Afi R.J., G.T. Rich, D.C. Mikulecky and A.K. Solomon; Determination of urea permeabilty in red cells by minimum method.; J. Gen. Physiol. (1970), 55: 427-450.

3. Bindslev N. and E.M. Wright; Effect of temperature on nonelectrolyte permeation across the toad urinary bladder. J. Memb. Biol. (1976), 29: 265-288.

4. Cohen B.E.; The permeability of liposomes to nonelectrolytes. I. Activation energies for permeation. J. Memb. Biol.(1975), 20: 205-234.

5. Knepper M.A. and R.A. Star; The vasopressin-regulated urea transporter in renal inner medullary collecting duct. J. Physiol. (1990), 259: F393-F401.

6. Sha'Afi R.I., C.M. Garybobo and A.K. Solomon; Permeability of red cell membranes to small hydrophilic and lipophilic solutes. J. Gen. Physiol. (1971), 58: 238-258.

7. Macey R.I. and R.E.L. Farmer; Inhibition of water and solute permeability in human red cells. Biochim. Biophys. Acta (1970), 211: 104-106.

8. Brahm J.; Urea permeability of human red cells. J. Gen. Physiol. (1983), 82: 1-23.

9. Levine S., N. Franki and R.M. Hays; A saturable, vasopressin-sensitive carrier for urea and acetamide in the toad bladder epithelial cell. J. Clin. Invest. (1973), 52: 2083-2086.

10. Levine S., N. Franki and R. Hays; Effect of phloretin on water and solute movement in the toad bladder. J. Clin. Invest. (1973), 52: 1435-1442.

11. Noma Y., P. Sideras, T. Naito, Bergstedt, S. Lindquist, C. Azuma, E. Severinson, T. Tanabe, T. Kinashi, F. Matsuda, Y. Yaoita and T. Honjo; Cloning of cDNA encoding the murine IgG1 induction factor by a novel strategy using SP6 promoter. Nature (1986), 319: 640-646.

12. Zhang R. and A.S. Verkman; Water and urea permeability properties of Xenopus oocytes: expression of mRNA from toad urinary bladder. Am. J. Physiol. (1991), 260: C26-C34.

13. Martial S., P. Ripoche and C. Ibarra; Functional expression of urea channels in amphibian oocytes injected with frog urinary bladder mRNA. Biochim. Biophys. Acta (1991), 1090: 86-90.

14. Pisam M. and P. Ripoche; Redistribution of surface macromolecules in dissociated epithelial cells. J. Cell Biol. (1976), 71: 907-920.

15. Sambrook J., E.F. Fritsh and T. Maniatis; Molecular cloning. A laboratory manual; 2nd-Ed. (Vol. 1) ; Cold Spring Harbor Laboratory Press (1989).

16. Brown D.; Cell biology of the cycling of the vasopressin-sensitive water channel; In: Vasopressin (Vol.208), Jard, S. and Jamison, R.J. eds., (1991); 73-83.

17. Jakobsen K., E. Breivold and E. Hornes; Purification of mRNA directly from crude plant tissues 15 minutes using magnetic oligo dT microspheres. Nucl. Acids Res.(1990), 18: 3669.

18. Dumont J.N.; Oogenesis in *Xenopus lævis* (Daudin). I. Stages of oocyte development in laboratory maintained animals. J. Morph. (1972), 136: 153-180.

19. Gurdon J.B. and M.P. Wickens; The use of xenopus oocytes for the expression of cloned genes. Methods In Enzymol. (1983), 101: 370-386.

20. Levillain, O., A. Hus-Citharel, F. Morel and L. Bankir; Production of urea from arginine in pars recta and collecting duct of the rat kidney. Renal Physiol. Biochem. (1989), 12: 302-312.

21. Dick E.G., D.A.T. Dick and S. Bradbury; The effect of surface microvilli on the water permeability of single toad oocytes. J. Cell Sci. (1970), 6: 451-476.

22. Gallucci E.S., S. Micelli and C. Lippe; Non-electrolyte permeability across thin lipid membranes. Arch. Int. Physiol. Biochim. (1971), 79: 881-887.

23. Mayrand R.R. and D.G. Levitt; Urea and ethylene glycol-facilitated transport systems in the human red cell membrane. J. Gen. Physiol. (1983), 81: 221-237.

24. Lefèvre P.G. and J.K. Marshall; The attachment of phloretin and analogues to human erythrocytes in connection with inhibition of sugar transport.. J. Biol. Chem. (1959), 234: 3022-3026.

25. Micelli S., C. Massagli and E. Gallucci; Serosal and mucosal facilitated transport of urea in urinary bladder of Bufo Bufo: Evidence for an alleged water uptake. Comp. Biochem. Physiol. (1983), 75A: 157-162.

26. Gargus J.J. and M. Mitas; Physiological processes revealed through an analysis of inborn errors. Am. J. Physiol. (1988), 255: F1047-F1058.

27. Dix J.A., D.A. Ausiello, C.Y. Jung and A.S. Verkman; Target analysis studies of red cell water and urea transport. Biochim. Biophys. Acta (1985), 821: 243-252.

© 1992 Elsevier Science Publishers B.V. All rights reserved.
Molecular mechanisms of transport. E. Quagliariello, F. Palmieri, eds.

Toxin-Induced Membrane Channels: Characterization and Regulation by Cations

C. Lindsay Bashford

Division of Biochemistry, Department of Cellular and Molecular Sciences, St. George's Hospital Medical School, Cranmer Terrace, London SW17 0RE UK

INTRODUCTION

A number of haemolysins of animal, microbial or synthetic origin damage the plasma membrane of target cells in a characteristic manner [1-6]. Initially the toxins depolarise the cell. Subsequently net changes in intracellular cation content (Na^+ entry, K^+ loss) occur, followed by leakage of small, usually membrane-impermeant intermediates of metabolism such as sugar phosphates and nucleotides. Extensively damaged cells leak cytosolic enzymes such as lactate dehydrogenase. Such a pattern of damage is consistent with the induction by the toxins of non-specific pores in the membrane whose effective diameter increases both with time and toxin concentration [1-6]. Cells do not lyse as judged both by their appearance using optical microscopy and by their ability to pellet through oil of density 1.02. The membrane-damaging agents which we have studied are identified in Table 1. Although the pores are generated by apparently unrelated agents they share sufficient common features to allow different toxins to act synergistically [5] - concentrations of one toxin too low to elicit damage become damaging in the presence of subthreshold levels of another, membrane-damaging toxin.

Membrane-damaging toxins share other features in addition to the induction of membrane pores. For example, both divalent

Table 1
Agents that form divalent cation sensitive pores in membranes

	Agent	References
Viruses	Sendai	1-7
	Newcastle Disease	8,9
	Influenza	1,10
Bacterial Toxins	*Staphylococcus aureus* alpha	1,2,4-7,11,12
	Staphylococcus aureus delta	7
	Streptolysin O	13
	Perfringolysin (O-toxin)	11,13
	Eschericia coli haemolysin	11
Animal Toxins	Melittin (bee venom)	2,5,6,14
	Cytolysin (sea anemone)	12
Immune Proteins	Activated complement	1,5,15
	Cytolysin/perforin (T cell)	12,16
Synthetic Compounds	Polylysine	2,5,7
	Triton X-100	2,5,6,14

cations, such as Zn^{2+} or Ca^{2+}, and protons ameliorate toxin-induced damage by antagonising both the formation of pores [1-6] and the effective permeability of pre-formed pores [11]. Protons are the most effective antagonists (at about 0.01mM) followed by Zn^{2+} (0.1mM), Ca^{2+} (1mM) and Mg^{2+} (>10mM).

To follow the molecular events underlying pore formation and pore antagonism by cations we have used model membrane systems, such as liposomes and planar phospholipid bilayers, to study membrane-damaging toxins [5,11,14,17]. In planar bilayers pore-formation can be identified as the induction of ion-conducting channels. Toxin-induced channels share some common features with the ion channels of natural membranes in that they exhibit ion selectivity, single channel behaviour (unitary fluctuations) and specific current-voltage characteristics. In this paper we will consider the properties of two bacterial toxins of clinical significance, namely the alpha toxin of *Staphylococcus aureus* (implicated in the pathogenesis of septicaemia) and the thiol-dependent toxin, pneumolysin, of *Streptococcus pneumoniae* (a virulence factor of bacterial pneumonia). The toxins damage cells and lipid membranes in a cation sensitive manner and the molecular determinants of these properties are hinted at by the modulation of toxin properties introduced either by selective chemical modification of alpha-toxin lysine residues or by site-directed mutagenesis of pneumolysin.

RESULTS

Intact Cells

Lettre cells, an ascites tumour cell line passaged in mice, are convenient cells for monitoring membrane damage. They take up [3H]-choline and convert it to phosphoryl[3H]choline, a membrane-impermeable metabolite which leaks only from damaged cells [18]. Alpha-toxin and alpha-toxin with about 4 lysine residues modified by diethylpyrocarbonate (DEPC-modified alpha-toxin [19,20]) both induce leakage of phosphoryl[3H]choline from Lettre cells (Table 2). In this series of experiments the DEPC-modified toxin is less potent than unmodified toxin, probably because more lysine residues were modified than in previous studies [19]. The optimum pH for damage was 6.5

Table 2
Effect of H^+, Zn^{2+} and Ca^{2+} on alpha-toxin induced damage of Lettre cells

	concentration required for 50% P-choline leakage	optimum [H+](M)	50% inhibition by (M) H+	Zn2+	Ca2+
native toxin	3ug/ml	3.10-7	6.10-6	6.10-5	1.10-2
DEPC-modified toxin	30ug/ml	3.10-6	6.10-5	3.10-4	2.10-3

for native toxin and 5.5 for DEPC-treated toxin. At pH values approximately one unit lower than the optimum damage in both cases was reduced by 50%. The sensitivity of alpha-toxin induced damage to Zn^{2+} was reduced by the DEPC treatment, whereas the sensitivity to Ca^{2+} was enhanced. This suggests that lysine residues have important role(s) in conferring divalent cation sensitivity and that these roles are different for Zn^{2+} and Ca^{2+}.

Wild type pneumolysin (C428) damages Lettre cells at very low toxin concentrations (Table 3) and mutations introduced into the cysteine containing region common to all the thiol-dependent toxins [21] significantly modify both pneumolysin potency and sensitivity to divalent cations. The variant with tryptophan 433 replaced by phenylalanine (F433) showed the lowest sensitivity to Zn^{2+} and was selected for further, more detailed studies.

Cation antagonism of toxin-induced damage may occur at any or all of the steps leading to pore formation, namely binding of toxin to the cell surface, insertion of toxin into the cell membrane or directly on the pore itself. The first possibility (prevention of binding) seems not to be a general mechanism; cells incubated with toxin in the presence of inhibitory concentrations of cation start to leak when toxin and cation are removed by washing [11,15,16]. Indeed, under such circumstances leakage is often greater than in the complete absence of inhibitory cation, indicating that it is common for

Table 3
Effect of Zn^{2+}, Ca^{2+} and Mg^{2+} on pneumolysin-induced damage of Lettre cells

	concentration required for 50% P-choline leakage	50% inhibition by (M) Zn^{2+}	Ca^{2+}	Mg^{2+}
wild type (C428)	0.05ug/ml	3.10-5	1.10-3	>10-2
variant (F433)	0.2ug/ml	6.10-4	>10-2	>10-2

Table 4
Effect of Zn^{2+} on preformed pneumolysin pores in Lettre cells

	P-choline leakage (%)	Erythrosin positive cells (% +/- Zn^{2+}) 0	0.1	0.3	1	3mM
wild type (C428)	74 ± 6	74 ± 6	43 ± 7	3 ± 6	21 ± 7	13 ± 7
variant (F433)	70 ± 9	67 ± 10	48 ± 11	50 ± 10	47 ± 9	58 ± 7

P-choline leakage was measured after a 20 minute incubation with pneumolysin. The cells were then washed and erythrosin uptake $\pm Zn^{2+}$ assessed.

divalent cations and protons to promote (rather than inhibit) binding to toxins to cells. The last possibility can be assessed in the following manner: Labelled cells are treated with toxin and release of phosphoryl[^{3}H]choline monitored to confirm that membrane permeabilisation has occurred. After washing away excess toxin, cells are incubated in isotonic medium in the absence or presence of antagonistic cations. The proportion of cells that remain permeabilised is then scored by adding dyes, such as erythrosin or trypan blue, which only enter damaged cells. In the case of pneumolysins (Table 4) the sensitivity of pre-formed pores to Zn^{2+} is only marginally less than the overall sensitivity of pneumolysin-induced damage to Zn^{2+} (cf Tables 3 and 4). Interestingly, pre-formed pneumolysin pores are much less sensitive to Ca^{2+} than is the overall process, whereas with alpha-toxin the converse is true: pre-formed alpha-toxin pores retain their sensitivity to Ca^{2+} but are rather insensitive to Zn^{2+}. Nevertheless, with both toxins it is possible to demonstrate closure of pre-formed pores by divalent cations, which represents one facet of the protective action of such cations.

The size of the molecules which leak from toxin-treated cells provides a rough estimate of the effective size of toxin-induced pores. On this basis the effective diameter of alpha-toxin induced pores is about 2nm [22,23]. Unfortunately, there are few convenient "markers" between the impermeant dyes (MW around 800 for trypan blue and erythrosin) and cytosolic enzymes such as lactate dehydrogenase (MW 34000 for the monomer and 130000 for the tetramer). Pore size in natural membranes can also be estimated by studying the protective effects of extracellular solutes on toxin-induced haemolysis. Permeabilisation of the red cell membrane to small molecules leads to lysis in isotonic saline by a colloid osmotic mechanism. Such lysis can be prevented by including solutes in the saline that are too large to permeate the pores. The effective concentration turns out to be approximately equal to that of haemoglobin within the red cell namely 5mM. For alpha-toxin and pneumolysins protection is conferred by solutes of differing size (Table 5). Some protection against alpha-toxin damage is afforded by 6000 MW dextran, whereas pneumolysin C428 requires 15000 or 40000MW dextran and pneumolysin F433 requires >40000MW dextran. This result suggests that alpha-toxin pores are smaller than C428 pores which in turn are smaller than F433 pores.

Table 5
Effect of solutes on toxin-induced haemolysis

	Haemolysis (%) ± 10mM solutes as follows:					
	0	stachyose	dextran 1500	dextran 6000	dextran 15000	dextran 40000
alpha-toxin	100	100	100	80	40	20
pneumolysin						
C428	100	100	100	100	65	55
F433	100	100	100	100	100	100

In summary, studies with intact cells indicate that both alpha-toxin and pneumolysin induce pores in Lettre cell and erythrocyte membranes in a H^+, Zn^{2+} and Ca^{2+} sensitive fashion. Minor perturbation of toxin structure, by chemical modification or site-directed mutagenesis, significantly affects the ability of the toxins to form pores and, in addition, modulates the properties of the pores such as their effective size or their sensitivity to antagonistic cations.

Model Membranes

To explore the molecular basis of toxin-induced pore formation and its antagonism by cations we have studied the behaviour of alpha-toxin (native and DEPC-modified) and pneumolysin (C428 and F433) in planar phospholipid bilayers. Bilayers were prepared either by the Montal technique [24] with subsequent addition of toxin or by the Schindler technique [25] in the presence of toxin. In each case pore formation is indicated by the appearance of ion-conducting channels in the bilayer. Table 6 summarises the properties of toxin-induced pores in planar bilayers.

Alpha-toxin (native or DEPC-modified) induced channels are remarkably homogeneous in size, most channels have a conductance between 80 and 160pS in 0.1M KCl at pH 7, and remain open at most applied voltages [5,17,19]. Membranes containing many channels exhibit rectification (greater current at positive than at negative potentials) as do membranes containing a single channel [17]. The channels from unmodified toxin are anion-selective [17,19] and in the presence of divalent cations such as Ca^{2+} or Zn^{2+} [5,17] or protons [2] they exhibit voltage-dependent closing. Antagonistic cations do not significantly decrease the single channel conductance (on the contrary, they tend to increase it); rather they stabilise the closed configuration(s) of the channel. The DEPC-modified toxin differs from native toxin in two regards. Firstly it is much less anion selective [19]; secondly the voltage-dependent closing in the presence of protons is substantially diminished at positive (but not negative) applied potentials. This indicates that at least one of the lysine residues modified by DEPC is associated with the voltage-sensitive mechanism operating at low pH and positive potential. The identity of the

Table 6
Properties of toxin-induced channels in planar bilayers

	conductance (pS, pH 7 0.1I)	selectivity	rectification ratio	closure by H^+, Ca^{2+}, Zn^{2+}
alpha-toxin	80-160	anion	>1	strong
DEPC-alpha-toxin	80-120	weak anion	<1	moderate
pneumolysin	<50	cation	>1	strong
(C428 or	50-1000	weak cation	1	strong
F433)	>1000	none	1	none

relevant residue(s) remains to be established.

Pneumolysin, like the related tetanolysin [26] affects planar bilayers in a more complex fashion than alpha-toxin. Well-defined channels, which remain open, do appear when the toxin is applied to (or incorporated in) bilayers containing 50% w/w ergosterol. However, the size of the channels varies over a wide range, and from membrane to membrane, in a rather unpredictable manner. The channels can be broadly categorised according to their effective conductance. Small channels have low conductance (<50pS in 0.1M KCl) and are highly cation selective (t+ 0.9 - 1), intermediate channels (50pS - 1nS) are less selective (t+ 0.85) and large channels (>1nS) are unselective (t+ 0.5). Divalent cations induce voltage-dependent closing of small and intermediate but not large channels. At zero potential channels reopen, albeit somewhat more slowly than alpha-toxin channels under similar conditions. If divalent cations are applied to one side of the membrane only (the cis side, to which toxin is added) the channels close only at voltages (positive) which tend to attract the cation into the channel; voltages of the opposite polarity accelerate channel reopening. Under these conditions toxin-treated membranes exhibit rectification (greater current at negative than at positive potentials) probably due to closure of small and intermediate sized channels rather (as in the case of alpha-toxin) than to the particular properties of the ion conducting pathway. The channels induced by the F433 pneumolysin variant strongly resemble those induced by the wild type (C428). Indeed F433 is at least as potent as C428 in the planar bilayer system, which is in marked contrast to its activity on cells. As with C428, F433 induces small and medium pores, which close in a voltage-dependent fashion in the presence of Ca2+ or Zn2+, and large pores which are insensitive to cations. However, the probability that F433 will induce large pores is significantly greater than the probability that C428 will induce large pores. The greater tendency of F433 to create large pores is compatible with the observations in cells that F433 pores are larger and less sensitive to Zn2+ and Ca2+ than C428 pores.

DISCUSSION

Our results indicate that bacterial toxins can induce pores in cells and model membranes of characteristic size and ionic selectivity. For alpha-toxin neutralising positively charged lysine residues by reacting them with diethylpyrocarbonate reduces the anion selectivity of the induced channels without significantly changing channel size. At this stage it is not possible to determine the precise location of the modified residues in the lumen or the mouth of the pore. In either case the neutralizing of positive charges would be expected to reduce anion selectivity. In the case of pneumolysin substitution of a neutral (tryptophan) residue for another neutral (phenylalanine) residue has little effect on channel selectivity. It may be relevant that the differing selectivity of alpha-toxin and pneumolysin channels mirrors the differing overall charge of the toxins: alpha-toxin has a net positive charge (pI 8.5) and pneumolysin a net negative charge (pI 4.5) at neutral pH.

Can the altered toxins throw any light on the gating process? Neutralisation of lysine residues in alpha-toxin strongly reduces the voltage-dependent gating at low pH and positive potential (but not at negative potential). This suggests that at least one lysine is involved with "sensing" of positive potentials, perhaps in a manner analogous to that found in endogenous ion-channels [27]. The precise role of lysine residues in the sensing of Zn2+ and Ca2+ remains unclear, although changes in sensitivity to such cations is apparent, particularly in intact cells. The situation with pneumolysin seems to be the following. In model membranes the gating properties of wild type (C428) and variant (F433) toxins are broadly similar, in marked contrast to the results obtained in intact cells. In cells, however, channels induced by C428 are clearly smaller than those induced by F433 and in model membranes it is only the small and intermediate channels which exhibit divalent cation sensitivity. Thus the decreased sensitivity of F433 to inhibitory cations in cells probably reflects the geometry of the channels induced; ie it has a structural rather than a mechanistic explanation. Further studies of both pneumolysin and alpha-toxin are required to identify the parts of the toxins important for "gating" and voltage/inhibitory cation "sensing".

The relation between residue charge and ionic selectivity may be quite general, although it must be noted that the neutral pore-former, triton X-100, induces cation-selective channels [Dr. T. Rostovtseva, personal communication]. Furthermore, it may not be necessary for the critical residues to be located in or even near the pore because it has been shown in studies of enzyme mechanism that changing surface residues remote from the active site nevertheless significantly modifies enzyme activity [28]. It is quite possible that ions flow over the surface of the channel former and membrane prior to their flow through the lumen of the channel. The flow of ions over charged surfaces is complex and it may be that some of the general properties of such surface conductance [29] underly some aspects of ion-channel behaviour.

ACKNOWLEDGEMENTS

I am grateful to Glenn Alder, Cecilia Pederzolli and Yuri Korchev for assistance with these experiments, to Graham Boulnois for providing pneumolysins, to Charles Pasternak and Adolf Lev for encouragement and support and to the Cell Surface Research Fund for financial sponsorship.

REFERENCES

1 Bashford CL,Alder G,Patel K,Pasternak CA.Biosci Rep 1984;4:797-805.
2 Bashford CL,Alder GM,Graham JM,Menestrina G,Pasternak CA. J Membrane Biol 1988;103:79-94.
3 Bashford CL,Micklem KJ,Pasternak CA.Biochim Biophys Acta 1985;814:247-255.
4 Bashford CL,Alder GM,Gray MA,Micklem KJ,Taylor CC,Turek PJ, Pasternak CA.J Cell Physiol 1985;123:326-336.
5 Bashford CL,Alder GM,Menestrina G,Micklem KJ, Murphy JJ, Pasternak CA. J Biol Chem 1986;261:9300-9308.
6 Bashford CL,Rodrigues L,Pasternak CA.Biochim Biophys Acta 1989;983:56-64.
7 Mahadevan D,Ndirika A,Vincent J,Bashford L,Chambers T, Pasternak C.Amer J Path 1990;136:513-520.
8 Burnet FM.Nature 1949;164:1008.
9 Poste G,Pasternak CA.In:Poste G,Nicolson GL,eds.Membrane Fusion.Amsterdam:Elsevier/North Holland Biomedical Press,Cell Surface Reviews 1978;53:305-367.
10 Patel K,Pasternak CA.Biosci Rep 1983;3:749-745.
11 Menestrina G,Bashford CL,Pasternak CA.Toxicon 1990;28:477-491.
12 Pasternak CA,Alder GM,Bashford CL,Menestrina G.In:Karbara JJ, ed.The Pharmacological Effects of Lipids III.Champaign,IL: The American Chemists' Society,1989;64-68.
13 Thelestam M,Mollby R.Infect Immun 1980;29:863-872.
14 Alder GM,Arnold WM,Bashford CL,Drake AF,Pasternak CA. Zimmermann U. Biochim Biophys Acta 1991;1061:111-120.
15 Micklem KJ,Alder GM,Buckley CD,Murphy JJ,Pasternak CA. Complement 1988;5:141-152.
16 Bashford CL,Menestrina G,Henkart PA,Pasternak CA. J Immunol 1988;141:3965-3974.
17 Menestrina G.J Membrane Biol 1986;90:177-190.
18 Impraim CC,Foster KA,Micklem KJ,Pasternak CA.Biochem J 1980;186:847-860.
19 Cescatti L,Pederzolli C,Menestrina G.J Membrane Biol 1990; 119:53-64.
20 Pederzolli C,Cescatti L,Menestrina G.J Membrane Biol 1990; 119: 41-52.
21 Saunders FK,Mitchell TJ,Walker JA,Andrew PW,Boulnois GJ. Infect Immun 1989;57:2547-2552.
22 Bhakdi S,Muhly M,Fussle R.Infect Immun 1984;46:318-323.
23 Fussle R,Bhakdi S,Sziegoleit A,Tranum-Jensen J, Kranz T, Wellensiek HJ.J Cell Biol 1981;91:83-94.
24 Montal M,Mueller P.Proc Natl Acad Sci USA 1972;69:3561-3566.
25 Schindler H.FEBS Lett 1980;122:77-79.
26 Blumenthal R,Habig WH.J Bact 1984;157:321-323.
27 Stuhmer W,Conti F,Suzuki H,Wang X,Noda M,Yahagi N, Kubo H, Numa S.Nature 1989;339:597-60.
28 Russell AJ,Fersht AR.Nature 1987;328:496-500
29 Lev AA.In:Kuczera J,Przestalski S,eds.Proc Xth School on Biophysics of Membrane Transport.Wroclaw,Poland:Agricultural University,1990;231-250.

INDEX OF AUTHORS